# Forecasting Travel in Urban Americ

**Engineering Studies**

Edited by Gary Downey and Matthew Wisnioski

Matthew Wisnioski, *Engineers for Change: Competing Visions of Technology in 1960s America*

Amy Sue Bix, *Girls Coming to Tech! A History of American Engineering Education for Women*

Jessica Smith, *Extracting Accountability: Engineers and Corporate Social Responsibility*

Elizabeth Reddy, *¡Alerta! Engineering on Shaky Ground*

Konstantinos Chatzis, *Forecasting Travel in Urban America: The Socio-Technical Life of an Engineering Modeling World*

# Forecasting Travel in Urban America

## The Socio-Technical Life of an Engineering Modeling World

**Konstantinos Chatzis**

**The MIT Press**
**Cambridge, Massachusetts**
**London, England**

The MIT Press would like to thank the anonymous peer reviewers who provided comments on drafts of this book. The generous work of academic experts is essential for establishing the authority and quality of our publications. We acknowledge with gratitude the contributions of these otherwise uncredited readers.

This book was set in ITC Stone Serif Std and ITC Stone Sans Std by New Best-set Typesetters Ltd. Printed and bound in the United States of America.

Library of Congress Cataloging-in-Publication Data

Names: Chatzis, Konstantinos, author.
Title: Forecasting travel in urban America : the socio-technical life of an engineering modeling world / Konstantinos Chatzis.
Other titles: Engineering studies series.
Description: Cambridge, Massachusetts : The MIT Press, [2023] | Series: Engineering studies | Includes bibliographical references and index.
Identifiers: LCCN 2022033313 (print) | LCCN 2022033314 (ebook) | ISBN 9780262048101 (paperback) | ISBN 9780262374514 (epub) | ISBN 9780262374521 (pdf)
Subjects: LCSH: Urban transportation—United States—Mathematical models—History. | Traffic estimation—United States—Mathematical models—History. | Urban transportation—Forecasting—Social aspects—United States.
Classification: LCC HE308 .C53 2023 (print) | LCC HE308 (ebook) | DDC 388.40973—dc23/eng/20230221
LC record available at https://lccn.loc.gov/2022033313
LC ebook record available at https://lccn.loc.gov/2022033314

10 9 8 7 6 5 4 3 2 1

To Eirini Despina and Giorgos,
thanks to whom this book is not as good as it could have been

# Contents

## Series Foreword

We live in highly engineered worlds. Engineers play crucial roles in the normative direction of localized knowledge and social orders. The Engineering Studies series highlights the growing need to understand the situated commitments and practices of engineers and engineering. It asks: what is engineering for? What are engineers for?

Drawing from a diverse arena of research, teaching, and outreach, engineering studies raises awareness of how engineers imagine themselves in service to humanity, and how their service ideals impact the defining and solving of problems with multiple ends and variable consequences. It does so by examining relationships among technical and nontechnical dimensions, and how these relationships change over time and from place to place. Its researchers often are critical participants in the practices they study.

The Engineering Studies series publishes research in historical, social, cultural, political, philosophical, rhetorical, and organizational studies of engineers and engineering, paying particular attention to normative directionality in engineering epistemologies, practices, identities, and outcomes. Areas of concern include engineering formation, engineering work, engineering design, equity in engineering (gender, racial, ethnic, class, geopolitical), and engineering service to society.

The Engineering Studies series thus pursues three related missions: (1) advance understanding of engineers, engineering, and outcomes of engineering work; (2) help build and serve communities of researchers and learners in engineering studies; and (3) link scholarly work in engineering studies to broader discussions and debates about engineering education, research, practice, policy, and representation.

Gary Downey, Editor

# Acknowledgments

Since this book has been long in the making, I have the sincere pleasure to thank many people for various sorts of help along the way. The bulk of the field work took place in the United States in 2010–2012, a period during which I received, in addition to my regular salary as a tenured researcher at IFSTTAR (now *Université Gustave Eiffel*), a grant from the (then) French Ministry of Transportation: Both institutions are to be thanked for making the research leading to the book materially possible. John Krige and Ted Porter were my contact points at Georgia Tech and UCLA during the academic years 2010–2011 and 2011–2012 respectively, while I also had the good fortune to be invited by Ted for short-term research stays at UCLA during several summers after my return to Paris. I cannot imagine better hosts for a visiting scholar groping his way around in a new territory both geographically and intellectually. As the book is about a nationwide object whose trajectory spans many decades, I had to access a great deal of diverse documents held by several libraries throughout the United States. Besides the staffers of the Georgia Tech and UCLA libraries, I am also especially truly beholden to the librarians working at Cornell University, George Mason University, the Massachusetts Institute of Technology (MIT), Northwestern University, and the University of Montreal. They are too many to be thanked individually, yet even for those whose names I omit, I will never forget what my work owes to their diligence and expertise.

Although the research presented here is essentially based on written documents, I also conducted, especially during my American stay, a number of interviews with people heavily involved in urban traffic forecasting, the specific modeling domain whose history constitutes the book's subject matter. John Bowman, Daniel Brand, Robert Chapleau, Robert Dial, Michael

Florian, Michael Mahut, Lance Neumann, Tom Rossi, Howard Slavin, and Heinz Spiess shared with me their intimate knowledge of the field they had cultivated for decades and some of them still serve. Besides providing me with a series of hard facts impossible to find elsewhere, they also, and above all, help me in seizing "atmospheres," developing insights and advancing hypotheses. In addition to their role as informers, Bowman, Brand, Dial, Rossi, and Slavin as well two other characters of the story told here, Hani Mahmassani and Peter Stopher, agreed to read parts of, even the entirety of, successive versions of the manuscript, and kindly communicated their comments to me. It goes without saying that while they are to be praised for what is good in this work, I remain solely responsible for the book's imperfections.

Gary Downey and Matthew Wisnioski, the editors of the MIT Engineering Studies series that hosts *Forecasting Travel in Urban America*, have supported this editorial project from its incipient stage. They have been patient with me and never expressed the slightest doubt about my capacity for carrying the project successfully, despite the delays I have been responsible for. Their detailed comments and insightful suggestions on the first full version of the manuscript, around 60 percent more voluminous than the text I was supposed to deliver, proved instrumental in the shaping of the final document (and again, attributions of merit go to them, and demerit to me). Last but not least, I consider myself very lucky to have worked with Katie Helke, who also had her patience heavily taxed, as well as Laura Keeler, Kathleen A. Caruso, Julia Collins, and other MIT Press staff involved in the crafting and the release of the book.

As has been seen, in the course of working on this research and editorial project I have accrued numerous debts to many people. Neither my spouse nor our two beloved children can legitimately figure among them. Unlike the wives and offspring of so many colleagues, none of them patiently supported my efforts. During my field work and throughout the writing process, Laetitia brilliantly pursued her own career and made me proud, but she hardly read a single paragraph of the manuscript (besides this one). As for Eirini Despina (now nineteen) and Giorgos (already twenty-two), they constantly behaved as normal children do, hence forcing me to spend more time attending to their needs and desires than going about my own business. Given their rather negative impact on this editorial undertaking, my dedicating this book to them requires some explanation. Though it is true that Eirini and Giorgos robbed me of my time and consumed much of my energy as well, they also clearly taught their father that family matters much more than any intellectual endeavor, even professional success. This is why dedicating this work to them is the least I can do. So yes, thanks to you, the loves of my life, this book is definitively not as good as it could have been. And that is for the best.

# Abbreviations

| | |
|---|---|
| **AASHO** | American Association of State Highway Officials |
| **ABM** | activity-based modeling |
| **AMOS** | Activity-Mobility Simulator |
| **AMV** | Alan M. Voorhees & Associates |
| **ARF** | Armour Research Foundation |
| **ATIS** | advanced traveler (or traffic) information system |
| **ATMS** | advanced traffic (or transportation) management system |
| **ATR** | Alliance for Transportation Research |
| **AV** | autonomous vehicle |
| **BART** | Bay Area Rapid Transit |
| **BPR** | Bureau of Public Roads |
| **BRUE** | boundedly rational user equilibrium |
| **BSTR** | Bureau for Street Traffic Research |
| **CA** | cellular automaton |
| **CAAA** | Clean Air Act Amendments |
| **CATS** | Chicago Area Transportation Study |
| **CBD** | central business district |
| **CDC** | Control Data Corporation |
| **CEMDAP** | Comprehensive Econometric Micro-simulator for Daily Activity-travel Patterns |
| **CRT** | Center for Research on Transportation |
| **CT-RAMP** | Coordinated Travel–Regional Activity Modeling Platform |
| **CTS** | Center for Transportation Studies |
| **CV** | connected vehicle |

| | |
|---|---|
| **DASH** | Dynamic Activity Simulator of Households |
| **DMATS** | Detroit Metropolitan Area Traffic Study |
| **DODOTRANS** | Decision-Oriented Data Organizer-TRansportation ANalysis System |
| **DOT** | U.S. Department of Transportation |
| **DTA** | dynamic traffic assignment |
| **Dynameq** | DYNAMic EQuilbrium |
| **DynaMIT** | DYnamic Network Assignment for the Management of Information to Travelers |
| **DYNASMART** | DYnamic Network Assignment-Simulation Model for Advanced Road Telematics |
| **DynusT** | DYNamic Urban Systems for Transportation |
| **EMME (Emme)** | Équilibre Multimodal/Multimodal Equilibrium |
| **FAMOS** | Florida Activity MObility Simulator |
| **FAST-TrIPs** | Flexible Assignment and Simulation Tool for Transit and Intermodal Passengers |
| **FSM** | four-step model(ing) |
| **FHWA** | Federal Highway Administration |
| **GIS** | geographic information system |
| **GPS** | global positioning system |
| **HATS** | Household Activity-Travel Simulator |
| **HI** | Home Interview |
| **HOV** | high-occupancy vehicle |
| **HRB** | Highway Research Board |
| **HUD** | U.S. Department of Housing and Urban Development |
| **IATBR** | International Association for Travel Behaviour Research |
| **IIA** | independence of irrelevant alternatives |
| **ISTEA** | Intermodal Surface Transportation Efficiency Act |
| **ITE** | Institute of Traffic Engineers (now the Institute of Transportation Engineers) |
| **ITS** | intelligent transportation systems |
| **ITTE** | Institute of Transportation and Traffic Engineering |
| **IVHS** | intelligent vehicle-highway systems |
| **JHRP** | Joint Highway Research Project |
| **LANL** | Los Alamos National Laboratory |
| **MBE** | minority business enterprise |
| **McTrans Center** | Center for Microcomputers in Transportation |

| | |
|---|---|
| **MITSIM** | MIcroscopic Traffic SIMulator |
| **MITSIMLab** | MIcroscopic Traffic SIMulation Laboratory |
| **MNL** | multinomial logit |
| **MNP** | multinomial probit |
| **MPO** | metropolitan planning organization |
| **NCHRP** | National Cooperative Highway Research Program |
| **NEXTA** | Network EXplorer for Traffic Analysis |
| **NISS** | National Institute of Statistical Sciences |
| **NMSHTD** | New Mexico State Highway and Transportation Department |
| **O&D** | origin and destination |
| **ORNL** | Oak Ridge National Laboratory |
| **PB** | Parsons Brinckerhoff |
| **PBQD** | Parsons, Brinckerhoff, Quade and Douglas |
| **PwC** | PricewaterhouseCoopers |
| **PML** | Peat, Marwick, Livingston & Co. |
| **PMM** | Peat, Marwick, Mitchell & Co. |
| **POLARIS** | Planning and Operations Language for Agent-based Regional Integrated Simulation |
| **PRC** | Planning Research Corporation |
| **PTV** | Planung Transport Verkehr |
| **QRS II** | Quick Response System II |
| **QUAIL** | QUAlitative, Intermittent, and Limited dependent variable statistical program |
| **RP** | revealed preference |
| **RSG** | Resource Systems Group |
| **SAFETEA-LU** | Safe, Accountable, Flexible, Efficient Transportation Equity Act: A Legacy for Users |
| **SAMS** | Sequenced Activity-Mobility Simulator |
| **SE** | system equilibrium |
| **SFCTA** | San Francisco County Transportation Authority |
| **SMART** | Simulation Model for Activities, Resources and Travel |
| **SP** | stated preference |
| **STARCHILD** | Simulation of Travel/Activity Responses to Complex Household Interactive Logistic Decisions |
| **SUE** | stochastic user equilibrium |
| **TEA-21** | Transportation Equity Act for the Twenty-First Century |

| | |
|---|---|
| **TMIP** | Travel Model Improvement Program |
| **TMS** | Traffic Management Simulator |
| **T-PEG** | Transportation Planning Computer Program Exchange Group |
| **TRACC** | Transportation Research and Analysis Computing Center |
| **TRANPLAN** | TRANsporation PLANning |
| **TRANSIMS** | TRansportation ANalysis and SIMulation System |
| **TRB** | Transportation Research Board |
| **TRC** | Traffic Research Corporation |
| **TRIPS** | TRansportation Improvements Programming System |
| **TSD** | Transportation Systems Division |
| **TSU** | Transport Studies Unit |
| **TTI** | Texas Transportation Institute |
| **TWG** | TRANSIMS Working Group |
| **UE** | user equilibrium |
| **UMTA** | Urban Mass Transportation Administration |
| **USL** | Urban Systems Laboratory |
| **UTC** | University Transportation Centers |
| **UTDM** | urban travel demand model(ing) |
| **UTP** | urban transportation planning |
| **UTPS** | Urban Transportation Planning System, and (less frequently) UMTA Transportation Planning System |
| **VISTA** | Visual Interactive System for Transportation Algorithms |
| **VPA** | value perception analysis |
| **WSA** | Wilbur Smith and Associates |

# Introduction

## Being an Urban Travel Demand Modeler in Postwar America

> Great house, great office, great boss, great computer, great projects, great crew, great wife and two great kids: It was 1966 and, man, I was blessed.
> —Robert Barkley Dial, 2007

It was "the greatest moment of collective inebriation in American history," when "the clock of history reset and a whole people's aims [were] limited no longer by the past."[1] As often, the novelist got it right. The twenty-five years or so following the end of World War II were a period of high spirits for many Americans. Robert Barkley Dial was one of them (figure 0.1).

A Seattle native, Dial was born on September 26, 1934.[2] Benefiting from the massive growth in university enrollment in the 1950s and 1960s,[3] he entered the University of Washington to study mathematics, and on graduation he worked from 1961–1963 for the Puget Sound Regional Transportation Study, which was conducting one of the dozen comprehensive transportation studies completed or in progress in the larger metropolitan areas of the United States by then.[4] But since "good employment in transportation planning in those days required a degree in Civil Engineering, not mathematics," Dial recollected nearly half a century later,[5] he decided to put his young professional career on hold and enrolled again in the University of Washington, where he acquainted himself with civil engineering matters while enhancing his programming skills.[6] Dial had hardly received his MSE in the mid-1960s when he saw the professional dream he harbored coming true, as he was hired by Alan M. Voorhees & Associates (AMV), a transportation consultancy then at the vanguard of the "scientific"

**Figure 0.1**
A portrait of Robert Dial (born in 1934). *Source:* William F. Brown, Robert B. Dial, David S. Gendell, and Edward Weiner, eds., *Emerging Transportation Planning Methods* (Washington, DC: U.S. Department of Transportation, 1978), 46.

transportation planning of the day through the pioneering use of computerized mathematical models.[7] While Dial started working as a systems analyst, a $110,000 contract granted by the U.S. Department of Housing and Urban Development (HUD)—to build a means of forecasting ridership for a new mass transit system—gave him the opportunity to extend his range of tasks within AMV to include travel modeling. Thus, catapulted into a de facto manager position for the HUD project, Dial designed all the relevant algorithms, wrote most of the code, and prepared papers and documentation for the company's HUD client.[8] Only a few months after the transit software was delivered to the HUD client, AMV initiated a new project with the Ontario Department of Highways. Three programmers from Canada paid a call to the AMV premises and, under Dial's direction, they began designing a highway planning software system. The result was the TRansportation Improvements Programming System (TRIPS), probably the first integrated multimodal transportation planning computer package in the world. Launched in 1968, TRIPS immediately embarked on a tremendously successful global career that spanned several decades.[9]

In 1968, Dial was a highly promising member of a fledgling community of proficient transportation planners sharing the same passion for sophisticated mathematical tools and computer programming techniques. Small but vibrant, the group had already succeeded in infusing its members with a sense of belonging and collective identity through regular meetings and other social transactions and exchanges, in which participants could display their professional virtues, be they cognitive or moral, while vying for each other's esteem.[10] For example, they took part in informal contests for the most efficient algorithms and the fastest codes in the group's area of expertise. One of Dial's rivals in these intellectual competitions was Vergil G. Stover, the prime architect of BIGSYS, a powerful traffic software package sponsored by the Texas State Highway Department in cooperation with the federal Bureau of Public Roads (BPR).[11] Sometime in early 1968, when Stover dropped in on the AMV premises, Dial handed his visitor a deck of cards containing the code written for TRIPS, and asked his competitor to run it once back in Texas on his home institution mainframe, a powerful IBM 360/65 machine. A couple of days later, a phone call came from College Station, and a voice announced to Dial that his program had run in just one second, making it three times faster than BIGSYS![12]

The year 1968 was undoubtedly a good one for Dial the engineer and modeler. It also had a darker side, as the AMV employee had to depart for Seattle to spend more time with his illness-stricken mother. Dial decided to take advantage of his forced return home and enrolled in his alma mater for a third time, with a view to preparing a Ph.D. dissertation. The thesis was easily wrapped up within two years and was approved on June 9, 1970.[13] Some days afterward, *Transportation Research* received a paper of Dial's that drew heavily upon his dissertation. Published in June 1971, the article turned out to be an instant classic.[14]

Despite his new academic credentials, Dial turned his back on the idea of a university career. After a brief stint back at AMV, he left the private sector in 1971 to take a "great job"—his words—at the Urban Mass Transportation Administration (UMTA), a federal agency within the U.S. Department of Transportation (DOT). The 1970s were busy years for UMTA as the decade proved rather an exceptional period for advocates of mass transportation in a country so keen on cars. The 1956 Federal-Aid Highway Act had committed the nation to the construction of 41,000 miles of interstate highways, including a 5,500-mile subnetwork through urban areas. A quarter of a century later the original projected network was mostly completed, but the initial enthusiasm, for what came to be the largest building program in U.S. history, had already significantly dampened among the public at large, transportation planners, and policymakers as well.[15] Because freeways proved exceedingly difficult to build in urban settings without cutting through old neighborhoods and parks, the country witnessed a series of "freeway revolts" in the late 1960s and the early 1970s. Under pressure from rising grassroots protest movements, the nation and its leaders were on the point of (re)discovering the multiple benefits of mass transit, which had constantly been losing ground to the private car since the 1940s.[16]

It was therefore in a marked pro-mass transportation atmosphere that Dial channeled the bulk of his efforts into the development of the Urban Transportation Planning System (UTPS), a set of various planning techniques implemented in generalized computer software. At the core of the project lay a number of models collectively geared toward forecasting traffic for a proposed transportation system, be it a highway network, a mass transit one, or a multimodal system.[17] In 1976, UTPS was already a thirteen-module computer package, and probably the most powerful modeling toolkit for transportation planning ever conceived, since it incorporated the knowledge and expertise possessed by fourteen consultancies and other think tanks of the day along with the contributions of one university and two other public agencies.[18] Moreover, the entire suite was totally free, while a series of training courses and self-instructional texts strengthened the popularity of this federally sponsored computer package with transportation practitioners even more.[19]

After a period of ten years that proved "terrific" and during which he and "great people did fun things," Dial left the DOT for a full professorship at the University of Texas at Austin. From the early 1980s, DOT stopped furthering UTPS software improvement or development, and for several years, the software development in traffic forecasting would be dominated by a handful of proprietary software companies.[20] As for Dial, he was now delighted at the prospect of designing "real codes" again. In the mid-2000s, a retiree for good now, he was still eager to continue carrying out research. But, in an ironic twist of fate, the man who had fed plenty of federal money to all major American transportation consulting firms in the 1970s was now an "artist" in need of funding, with Howard Slavin, president of Caliper Corporation, a cutting-edge transportation planning-software company established in 1983, in the role of the "patron."[21]

This role reversal was not the only change Dial experienced in his long career. During his early years as a traffic modeler, he was evolving in a rather small community of practitioners and academics, who often called each other by their first names, even their nicknames.[22] As time went by, the professional world Dial faithfully served for over forty years grew much more populous,[23] while undergoing significant qualitative changes as well. The young Dial entered a still fledgling field and his first forays into it were those of a self-taught modeler who was learning a good deal of his "stuff" with the help of more seasoned practitioners and through hands-on apprenticeship. However, from the 1970s on the number of Americans earning a living in urban traffic forecasting swelled as steady flows of academically trained practitioners poured into transportation modeling.[24] Exclusively male in its beginnings, the field has also progressively opened up to women and, while populated essentially by Americans till the 1970s, it has become increasingly transnational, drawing members from all over the world, thus mirroring the evolution of other communities of academically trained people.[25]

## What Urban Travel Demand Modeling Is About

But, so much for individuals, for the moment at least. This is not a biography of Robert Dial. He was singled out for serving as a vehicle for exploring something much bigger than him: an entire academic and professional field usually now called urban travel demand modeling (UTDM), the true protagonist of the story this book seeks to tell.

Several years ago, the specialist of the American Revolution Gordon S. Wood quipped that historians are often teased about being interested in seemingly insignificant subjects as long as they cover a significant period of time, and sufficient information about them is available. Urban travel demand modeling—also labeled *urban traffic forecasting* in this book—has, beyond a shadow of a doubt, a long history to display and has unarguably produced an impressive array of texts to be delved into. Though inaccessible to the layperson, insofar as it incorporates large doses of scientific and engineering knowledge and expertise, urban traffic forecasting is by no means an insignificant subject for it has had and likely will continue to have a palpable importance for the daily life of millions of people, even if they have never heard of it. In fact, from the 1920s on most Americans have been living in cities and urban travel demand modeling, for better and for worse, has definitely been significant in shaping the urban landscape they inherited and now inhabit.[26] City dwellers, especially after the development of residential neighborhoods formed along class lines as homes and workplaces separated increasingly in the nineteenth century,[27] tend to be very mobile. Therefore, they need infrastructures to enable that, whether mass transit to travel in, expressways to drive on, or inviting roads to walk and to bike along. As transportation infrastructures are very costly entities that are to endure over a long period of time, it is in the best interest of everyone if they are properly sized, neither too large nor too small, having just the right dimensions for people using them to meet their various mobility needs. Urban travel demand modeling was born as a procedure enabling the planner and the engineer to properly *size* urban transportation systems. A specific practice-oriented modeling domain, it consists of an array of mathematical tools and practices[28] initially geared toward *predicting* traffic flows in urban transportation networks for a period up to fifteen to twenty years into the future, enabling therefore the engineer and the transportation planner to design the infrastructure in question accordingly. Since its inception as a specific field in the wake of World War II, UTDM, following successive refinements and substantial modifications in its mathematical structure and underlying hypotheses, has been utilized to help plan and design new transportation infrastructures, be they urban highways or subways. But over time, UTDM has been periodically remade, through innovation and (sometimes cutthroat) intellectual competition, to suit new uses and to face original challenges. Thus, new generations of traffic forecasting models, heavily drawing from multiple disciplinary bodies of knowledge, have progressively emerged to assist planners and transportation engineers to make optimal use of existing capacity instead of building a new transportation system, through a series of policy instruments, including the creation of priority facilities on freeways for high-occupancy vehicles (HOVs), and the use of intelligent transportation systems (ITS).[29] The main ambition of this book is to scrutinize this specific modeling domain in the United States and to investigate its trajectory in the long run, from the interwar period to the present.

## Modeling as a Production Process

There are many vantage points from which to observe and study models and modeling.[30] At the heart of the approach followed here is the assumption that as a problem-solving device used essentially for practical purposes, urban traffic forecasting, like other practical problem-oriented modeling fields, resembles very much "ideas" as they are seen from a pragmatist angle.[31] Like "ideas," models—a set of relationships between variables and other parameters, portrayed in mathematical terms and forming a large and elaborate set of equations to be fitted to large amounts of data—are neither entities just "out there" waiting to be passively recorded, nor more or less successful attempts to accurately represent some independent reality. They are instead tools that people, more precisely experts, devise in order to come to grips with the world they and their fellow citizens inhabit, explore, alter, and (re)shape. Like "ideas," models can be successful, may fail, and may even prove harmful by bringing forth adverse side effects as well as negative unanticipated consequences. As specific responses to particular circumstances, they must also adapt in order to continue operating efficiently. No existing modeling practices can, in fact, protect themselves fully and for long against all the impulses for change, whether they stem from external sources—a change in the environmental policies, the end of Cold War, or the advent of personal computers, for example—or from internal tensions. Last but not least, similar to every tool, models have to be conceived and produced before reaching their potential users and being enacted in local contexts. To put it another way, models are not viewed here as mere symbolic systems originating in people's brains, and representing independent states of affairs, but as entities involving humans as well as nonhumans, and collectively shaped through a specific *production process*.[32]

While the metaphor of *production process* can be intuitively understandable, it might need more analysis. To start with, adopting such a perspective means that the inquiry is not confined to the final *product* only, meaning, the model. Of course, the main formal characteristics of the model as well as the various ways in which broader social and cultural factors impact on and enter into its production will all be subject to close scrutiny. However, in addition to looking inside the "black box" of the model, the investigation extends to the *means of production* and the *raw materials* involved in its crafting and implementation. As a matter of fact, in order to calculate its ultimate outputs—the predicted values of future traffic, in our case—a model must be fed with large amounts of information (raw materials), while, for the model to perform such calculations, the production process must make use of many and diverse tools and techniques (means of production). This book therefore dedicates several pages to the specific raw materials and means of production involved in urban travel demand modeling. Thus, special attention will be paid to the various *travel survey methods* devised to obtain the critical information needed, such as the face-to-face interviews first designed in the 1940s and typically conducted with humble paper and pencil in respondents' homes, and the data-collecting systems built on remote-sensing technologies, including dedicated global positioning systems (GPSs) and ordinary devices such as smartphones. Regarding the *means of production*, the focus was put on the multifarious computing facilities that traffic forecasting practitioners have mobilized in their work—from the tabulating machines in the 1920s to the first mainframes dating back to the 1950s, to the personal computers (PCs) and finally, the supercomputers of today—as well as on the corresponding software packages. And there is more.

Any historical work worth its salt looks for human actors. A complex production process is very likely to be populated with and operated by plenty of people performing various tasks and functions. As the opening scenes featuring aspects of Dial's life and career clearly suggest, the number and diversity of the actors involved in a model-related production process can be impressively high. Indeed, the history of urban traffic forecasting has embodied the efforts of scores of individuals—running the gamut from a Nobel Prize laureate in economics to many anonymous modelers who have not written a single word about their activities and practices—participating, either simultaneously or serially, in a wide range of institutions and organizations: federal and state agencies, metropolitan transportation planning organizations, think tanks and foundations, university research centers and engineering departments, professional associations, consulting firms and proprietary software companies, and, especially for the most recent period, groups gravitating toward open-source software, social movements, and even judges as well. These individuals, displaying specialized knowledge and having specific skills, form a large network of cooperating people all of whose work, carried out according to an evolving division of labor, contribute to the production of traffic forecasting practices. Put together, they constitute a specific *social world* in sociologist's parlance,[33] in our case the "UTDM social world."[34]

## The Social World of Urban Travel Demand Modeling

By laying the emphasis on cooperation, I do not intend to play down the heterogeneity of the UTDM social world. Any social world is by virtue of definition divided into different, albeit imbricated, subgroups and segments, which may pursue specific objectives, have their own interests, and develop particular sets of practices that, though (ideally) converging on the same overarching goal—crafting good forecasting practices able to produce accurate predictions about future traffic, in our case—are not necessarily identical. Thus, scholars belonging to the academic segment of the UTDM social world are very likely to value innovation, and to be particularly concerned with cutting-edge modeling techniques. Yet, easiness of use, expediency, cost, and other practicalities may matter more for the purchaser of a commercial software package. However, the fact that the UTDM social world has not been a monolithic entity, the components of which share everything, intellectually or professionally with each other, does not mean that it was doomed to splinter into isolated and self-contained fragments:[35] as a matter of fact, the centripetal forces of cooperation proved strong enough in the long run to countervail the potentially disunifying forces rooted in the heterogeneity that characterizes UTDM as a social world.

Cooperating is a demanding endeavor, and its successful unfolding requires people engaged in it to undertake a wide range of actions. Forging and sustaining various kinds of links as well as coordinating and managing multifarious exchanges are some of them. Actors involved in urban travel forecasting very soon felt the necessity to create arenas where frequent and intense interactions could take place. They also developed a number of infrastructures whereby communication and circulation of locally produced knowledge and site-dependent practices are made possible, and thanks to which larger collective identities, transcending local ones, are built progressively. Thus, members of the UTDM social world working for different employment units set up informal networks while grouping themselves into various formal organizations, including professional and science-oriented societies. They also fashioned tools and forums of communication, formal and informal, ranging from face-to face interactions and extended personal contact to letters, and now emails, to conference proceedings and newsletters, to journals and manuals, to seminars, and now webinars, and workshops. Thanks to these forms of interaction and channels of communication, different traffic forecasting-related pieces of knowledge and various travel modeling practices could enter into dialogue, compete with each other, and spread across geographical distances nationwide.[36] It is also worth noting here that "boundary objects,"[37] such as standardized software packages common to scholars and more practice-oriented people, as well as "boundary institutions,"[38] even individual "brokers" and "go-betweens"[39]—Dial was such a broker—also emerged to ease the cooperative work within the urban traffic forecasting community and the circulation of its outcomes, serving therefore as interfaces between the multiple subgroups populating the UTDM social world.

However, the mere existence of such arenas and infrastructures cannot alone make a productive and lasting social world. Sustainable effective cooperation presupposes, in addition, the production and replenishment of an appropriate working force composed of individuals willing to work together, having the necessary skills to perform their part in the collective action, and sharing some common ground on which collaborative activities can be built and endure over significant periods of time. Thus, professors must transform the traffic forecasting practices of the day into codified digestible knowledge to be transmitted in undergraduate classes, and new recruits must be attracted and properly trained through school teaching or on the job. The would-be travel demand modelers also have to acquire the relevant skills, be taught to work with the available tools,[40] be prepared to develop new ones, and be instilled with habits enabling them to effectively respond to new challenges. They must, in addition, internalize and put into execution the specific "moral economy"[41] underlying the functioning of the UTDM social world, all these rules, customs, values, and norms defining appropriate behavior, regulating a variety of issues—ranging from access to funds and other key resources to the ownership of collective achievements to rewards for good work[42]—and helping handle "crisis" situations and resolve conflicts.[43] (Note that some of these conflicts may lie within the same person: Think of someone who can simultaneously have an interest in rendering public his findings qua a higher education academic and keep them quiet qua a member of a proprietary software company, thus being bound to undertake a calculated mix of openness and deliberate secrecy.[44]) As with any other social world, the UTDM one has also been confronted with the knowledge/power nexus,[45] because reorganizations of modeling knowledge and practice are likely to be intimately intertwined with new power arrangements. Indeed, innovating in urban traffic forecasting (and this applies to other modeling fields as well) does not only produce original procedures: with the advent of new lines of modeling some actors may rise to prominence while others may be pushed from the stage or relegated to the background.

## About the Book

This is a history book and hunting for changes in time is part and parcel of its aims and objectives. I attend therefore to the trajectory of American urban traffic forecasting and seek to narrate the main episodes punctuating the course of the many components constituting UTDM as a specific modeling field, be they intellectual and material, institutional and organizational, professional and cultural, or financial and social. Some changes take on the form of accumulations and small modifications occurring within an encompassing structure that remains more or less stable, while others are more akin to a (Kuhnian) "paradigm shift." Both kinds of changes are featured in this book. In hindsight, several innovations proved pivotal for the career of the American urban traffic forecasting. They deserve, and will indeed receive, a lot of space in these chapters. However, I sought to avoid writing a novelty-centered study focusing on more or less radical and successful breaks with the past.[46] Instead, I systematically tried to combine the themes of innovation and use, of novelty and routine, of success and failure. The result is an "intercalated" and "time-laminated" history,[47] in which the different components of urban traffic forecasting are certainly interconnected but without fusing and move forward together but at their own pace and according to their own dynamics, while changes in some regions of UTDM occur against the backdrop of (relative) stability and continuity in some others (although, if a part of UTDM undergoes dramatic changes, other parts of it are likely to follow suit). To use another metaphor, the history told here often resembles geology in that several layers of different ages can coexist at the same juncture, given that long after a new set of modeling techniques is implemented at some vanguard sites, old practices might continue surviving in other places for significant periods of time.

The fact that I am interested both in novelty and routine accounts for a series of choices and assumptions made about the topics and questions to which priority was given. To begin with, research findings as well as specific modeling work originating in local settings are explored only insofar as they were eventually made into common practices shared by a good number of practitioners, and largely put into use over space and time. This is why practice-oriented entities, such as consultancies, governmental transportation agencies, and proprietary software companies, to name the most material of them, are featured in the book on a par with university departments and other knowledge-centered organizations. For the same reason, highly inventive figures belonging to the elite fractions of the UTDM social world[48] are mixed with less-known practitioners and teachers, while original research studies published in highly ranked academic journals coexist with textbooks and other means of large-scale diffusion of knowledge and practices, such as turnkey software packages and their user guides, as well as with a number of rather mundane businesses, including counting cars and asking people elementary questions about their daily travel patterns. Because I do not favor originality over repetition, I have tried to identify and pin down mechanisms promoting and facilitating changes in the life of UTDM—such as competition and emulation among individuals and organizations, or the high mobility of scholars and practitioners coming from several academic backgrounds—*and* processes accounting for stability and giving existing configurations inertia, including the systematic pursuit of massive diffusion of "best" practices nationwide through teaching, training, and software packages.

Although the ambition of this author is to convey adequately the changing texture of American urban travel demand modeling in the last seventy years or so, through a series of historical episodes that combine the "big picture" and telling detail, this is by no means a comprehensive history pretending to completeness. Besides pure academic research leading to no practical outcomes, and very particular operational practices that remained confined to a few specific places, several other aspects of the story are left out of the book or were relegated to the background. First, given the general analytic framework put forward it comes as no surprise that the overall balance of the study is tilted toward the *production* side of the equation, while issues regarding *use* are rather scant. Thus, you will find no accounts of the various ordinary daily tasks performed by the modelers, such as calibrating a specific model and validating its outputs or adapting a standardized software package to fit local needs and situations. Second, the various receptions of UTDM by policymakers, the interactions they develop with traffic forecasters as well as the ways they tend to integrate modeling results into transportation policy and decision-making are also themes that are touched upon very little, if at all. Both the long-term perspective adopted and the decision to treat both novelty and routine also took their toll on the final shape of the book. Much to my chagrin, more than once, firsthand information material gathered during the field work period was omitted from the main body of the text or was relegated to the status of a mere reference in the notes. Another regret I have about the book is its monographic character. Having already worked on the history of urban traffic forecasting in France, I know only too well that UTDM very soon made itself into a vibrant transnational reality. Maybe, the highest ambition I can entertain in delivering the gathered material and its exploration to the reader is that other scholars will take an active interest in the subject matter of the book and will help expand the story told here.

The reader, especially if he or she happens to be involved in urban traffic forecasting in one way or another, should also be aware that some people have probably received less attention than they deserve given their personal record and reputation within the UTDM social world, and, conversely, some others received more lines than their due (most of the time, the main reason for these asymmetrical treatments was the lack of relevant information; but sometimes a less imposing figure in the eyes of somebody involved in the field can offer more productive material for the historian). However, regardless of the degree of recognition and fame reached by each member of the UTDM social world, I systematically sought to bring on stage as many individuals as possible; I even tried to let them speak for themselves, and not just to bring an abstract subject to life and enliven the story with human portraits. Though the book aims to unravel long-term impersonal developments and collective patterns transcending individual actors, the latter are not regarded as mere faceless captives of large deterministic processes. Figures such as Dial remind us that history is shaped by the outcomes of real choices that living people—assuredly evolving in circumstances they do not always choose (Marx)—actually made. By multiplying short biographical sketches, I wanted neither to lose sense of the causal power of particular and locally deployed actions nor to underestimate the impact of contingent events. Contingency—something that may or may not happen, such as Dial's return to Seattle in 1968 that led up to his doctoral dissertation, which gave birth to his much-quoted article in *Transportation Research*, which impacted greatly on the work of other members of the UTDM social world, and so on—is central to any historical process while vital to the success of our narrative strategies about both the past and the present.[49] And there is more. When a field is mature and serviced by a populous and well-trained group of competent people, specific individuals, no matter how gifted and talented, are not likely to matter greatly; however, when a set of practices is in its incipient stage the action of few persons can prove critical.[50] And last but not least, individuals matter also on a more symbolic level. As urban travel demand modeling grew into a mature world, it progressively developed a self-image, weaved a collective memory, and created its own genealogy. The iconization of a handful of modelers, going down in history as founding fathers and fine leaders of their group, has been part and parcel of the previously mentioned processes, often operating through eulogies, tributes paid to leading practitioners and inspiring teachers, and the establishment of awards named after them.[51]

As with any book, this one has its own history. It has been quite a long time in the making as it distills several years of research started in France and culminating in North America. I began to truly familiarize myself with its general subject matter in the mid-2000s, when I undertook a study of the history of French urban traffic forecasting from the 1950s on.[52] Uninterrupted field work over two years in the United States and (secondarily) in Canada from 2010 to 2012 allowed me to refine the analytical grid I first built for the French case, and to extend the empirical findings about the transnational history of urban travel demand modeling on both sides of the Atlantic. I defended a first draft of the book, including the French case, as the mainstay piece of my Habilitation thesis in Paris in December 2013.[53] Based on this first version, I completely rewrote, and grudgingly shortened in the process, the American part of the study for an English-speaking readership, with a view to producing a text able to appeal both to the layperson who has never thus far heard of urban travel demand forecasting and to people involved in this modeling field.

Thus, general educated readers will discover the biography of a technoscientific object that humanities and social sciences have treated as an aside only. They will acquaint themselves with a reality that, interesting in its own right, can also serve as a vehicle for the exploration of a number of themes in American history: The modern American city, through the transportation planning techniques that contributed to shaping the country's urban landscape in the twentieth century and the first decades of the following one; the emergence and consolidation of original engineering knowledge and practices, developed and put into use by new academic and professional communities; the state-building process, especially with regard to the role played by science and technology; the rise of new kinds of companies filled with "knowledge workers";[54] the question of the growing commercialization of academic knowledge; and the increasing transnationalization of science and engineering in post–World War II America, to name some of them. As for those readers involved in urban traffic forecasting, they will learn much about the past of the world they now inhabit and may even enhance their perception of the ordinary by starting to see their turf in the unfamiliar light provided by an outsider.

To the best of my knowledge, the present book gives the first detailed account of the history of urban traffic forecasting in the United States in the long run from the perspective of human and social sciences.[55] However, although terra incognita for the historian and the sociologist so far, the history of UTDM has not been so for those involved in this modeling field. Following in the footsteps of many working scientists in the nineteenth and twentieth centuries,[56] several travel demand modelers, most of them academics, have periodically embarked on producing historical surveys of their specific area of expertise.[57] And from the 2000s on, urban traffic forecasting literature has also been enriched with a significant number of works written by seasoned professionals or by people under their aegis, and expressing a reflective stance on the functioning and the trajectory of UTDM as a whole over the last sixty years or so. In this respect, *Forecasting*

*Urban Travel: Past, Present and Future*, published in 2015, assuredly constitutes a milestone in the annals of historical surveys on urban traffic forecasting conducted by people serving the field.[58] The authors of the book, the American David Boyce (Ph.D. awarded in 1965) and the Briton Huw Williams (Ph.D. awarded in 1971)[59] have obviously succeeded in acquiring an intimate knowledge of the history of the field, and have authored an erudite piece of work.[60] Fortunately, their book and mine are quite dissimilar, and, despite a series of inescapable overlaps, differ from each other in many and significant ways both in content and form.[61] Suffice it to say that I intended to write a historical and sociological study that is easy to read for people who are totally unfamiliar with its subject matter and have a limited, if any, command of mathematics and engineering sciences. In contrast, by crafting a book geared principally, if not totally, to people involved in this field of modeling, Boyce and Williams wanted to help them invent a better future for their common domain of inquiry through an active dialogue with the past. No wonder they advise the reader to "read [their] book in conjunction with a technical textbook on travel forecasting and a formal course in order to augment the discussion and fill in the gaps."[62]

## Overview

A few words are in order on the structure of this book. A historical work, its dominant direction is to move forward from the past to the present. As a result, the different parts progress from the interwar period to the present while within each chapter the chronological order is also observed. Yet, from time to time the rules of strict chronology are broken to the advantage of a more thematic approach. Thus, two successive chapters or smaller units within them may be concerned with the same time frame while focusing on different themes. Yet, different and autonomous as the various chapters of the book might be, together they form nevertheless a coherent whole. The encounter of the accumulating historical material with the general analytical framework[63] developed here—modeling as a production process operated by a specific "social world"—gave birth to a narrative organized around a single long cycle, which is, in all likelihood, about to reach its end. As a matter of fact, each chapter expounds in significant detail specific episodes of the "rise and fall" of the most popular comprehensive outcome of the action of the American UTDM social world so far, the four-step model (FSM) (box 0.1). The latter occupies, both metaphorically and literally, center stage in this book. Thus, the first two chapters (part I) trace the emergence, development, and rise to prominence of the four-step model on the American transportation stage, from the interwar period to the late 1960s. However, although having already acquired paradigmatic status, at the turn of the 1970s the FSM would face a number of problems, some of which emanated from external sources while others stemmed from internal tensions. These challenges prompted the UTDM social world of the day to introduce a series of novelties in the general FSM framework, which proved plastic enough to significantly change without this modeling approach losing its soul. Chapters 3 and 4 (part II) offer a detailed account of the major shifts American urban traffic forecasting knowledge and practices as well as the social world that crafted them underwent during the 1970s and 1980s and that, taken together, gave the four-step model a new lease on life. The last three chapters of the book (part III) cover the post-1990s period and are dedicated to a series of attempts to supersede the rejuvenated FSM with newer and alternative modeling lines. Pressured by new legislation, especially regarding air pollution, and faced with the increasing inability of traditional procedures to cater to a series of social justice issues, the UTDM social world would first experiment locally and later deploy on a large scale a multitude of original modeling practices; in hindsight, the latter brought about a profound rupture with the past, thus making it possible to talk about a "paradigm shift."

Here is a more detailed overview of the book, chapter by chapter.

Though urban travel demand modeling—as it has come to be seen through the lens of today—can be traced back to the first decade following World War II, I decided nevertheless to start the story earlier. Chapter 1 is concerned with the interwar years, during which several preconditions, be they professional, organizational, material, or ideological, for the postwar blossoming of urban traffic forecasting came to be met, and a number of elements of the would-be UTDM, albeit mostly in an emerging state, appeared for the first time. Confronted with soaring numbers of traffic accidents and rising levels of congestion in the major urban centers of the nation, a constellation of actors, some of them already old while others newer on the transportation stage, sought to contrive an original science, called *traffic engineering*. In their bid to fight against these two scourges caused by successive tidal waves of motorization, they collectively devised,

**Box 0.1**
The Four-Step Model

The four-step model seeks to forecast the traffic flows on an urban transportation network located in a given urban area. As its name indicates, this modeling procedure is based on a combination of four stages or steps: trip generation, trip distribution, mode choice, and traffic assignment. Each step addresses a specific question and the outputs of each step become inputs for the following step.

1. How many trips will be made? The aim of *trip generation* is to predict the number of trips to and from the different zones making up the urban area under investigation.
2. Where will the trips go? Once the total number of trips originating and ending in each zone is known, the objective of *trip distribution* (or destination choice, or zonal interchange analysis) is to predict the pattern of movement between the different zones. This stage makes it possible to determine the trips corresponding to various possible origin-destination pairings within the entire urban area.
3. What mode—private car, mass transit, bicycle, walking, and so on—will be used for the trip? *Mode choice* (or modal split or mode split) computes the relative proportions of these movements by alternative modes.
4. And, finally, what specific route will be taken? *Traffic assignment* (or route or trip assignment) breaks down movements over the transport network studied.

Take note: In light of the foregoing, it is obvious that the four-step model is not a model strictly speaking, but instead a general modeling approach. As will be seen throughout the book, several different models have been produced within each of the four steps.

The reader should also keep in mind that in order for the four-step model to be internally consistent (coherent) feedback loops must be added to the system. To understand the reason why such loops are necessary, let us rapidly treat the cases of the trip distribution and traffic assignment steps (steps 2 and 4). As a matter of fact, the modeling procedures developed for step 2 generally involve the time required to move from one urban zone to another. However, these times will only be accurately calculated during traffic assignment, step 4. Hence the need to return (feedback loop) to the trip distribution step with the new times and run the distribution models again. Unfortunately, this iterative process is difficult to carry out in operational situations (owing to financial constraints, delays, difficulty in achieving the desired convergence). As a result, for several decades the four-step model was used without such feedback loops.

among other things, the first traffic surveys aiming to identify and quantify the American household's travel patterns, and they even made some quick forays into the realm of travel prediction.

Spanning more than a quarter century, from the early 1940s to the turn of the 1960s, chapter 2 occupies a pivotal place within the general organization of the book. It traces a long process leading from a series of sporadic and somewhat uncoordinated efforts to forecast urban traffic in specific local settings to the creation of a nationwide, full-fledged modeling domain, which at the turn of the 1960s had succeeded in morphing itself into both a specific professional activity and a particular field of academic training and research. More precisely, chapter 2 relates the emergence, growth and rise to prominence of a specific line of travel demand modeling, the now legendary four-step model. Seen through the lens of the "social world" perspective, the FSM as well as the travel data collection techniques providing this general modeling approach with the necessary information for it to operate, are analyzed as the collective outcome of the many cooperative yet often competitive activities of a host of actors, including individuals and more collective entities, knowledge-centered and more practice-oriented organizations, and for-profit and not-for-profit institutions. In addition to laying out the historical setting in which the FSM emerged and expounding its core features at the different stages of its development, the different sections of the chapter deal with the structure, the workings, and the various resources enlisted by the UTDM social world involved in crafting, developing, sustaining, standardizing—through manuals, training courses, and, especially, the distribution of free computer packages—and using the FSM in the United States over the period under consideration.

Chapters 3 and 4 pick up the trajectory of the four-step model where chapter 2 left off. Though both cover on the whole the same time span, the 1970s and 1980s, each chapter focuses on a different stage within the overall structure of the FSM. Chapter 3 deals with the mode choice step. Against a backdrop of multiple "freeway revolts," increasing social concerns, and the first oil crisis, U.S. society was on the point of (re)discovering the multiple benefits of mass transit. Consequently, while it had initially focused on forecasting automobile traffic, the UTDM social world became increasingly concerned with the question of *mode choice*. Strongly *aggregate*, calculating *total* flows between the various zones making up the metropolitan area under investigation, the FSM has considerable difficulty in effectively accounting for the heterogeneity of *individual* mobility decisions and travel patterns. A new sort of modeling, the *disaggregate approach* that regards trip-makers as separate individuals who seek to maximize the welfare derived from their travels, emerged, developed, and increasingly had pride of place within the mode choice step. Chapter 3 explores in detail how a holy alliance between the university segment of the UTDM social world and a new sort of consultancy cultivating strong bonds with academia would progressively transform, with the help of the federal administration, what started in the early 1960s as an academic curiosity, mostly cultivated within economics departments, into a powerful and long-standing operational tool in the hands of the transportation professional. Though the chapter is mostly dedicated to disaggregate modeling, an entire section is devoted to original travel data collection and treatment methods associated with the new modeling approach. First originating in marketing studies, these new methods helped enrich the fabric of the UTDM social world through a number of new actors.

Chapter 4 treats a series of novelties taking place within the fourth and last step of the FSM: traffic assignment. Like disaggregate modeling, these new undertakings, among them the so-called user equilibrium (UE) assignment, owed very much to research studies performed by academics, most of whom were now coming from operations research backgrounds. As occurred with disaggregate modeling, the dissemination of these new assignment procedures among practitioners was widely associated with the emergence of new actors, especially a series of proprietary software firms cultivating strong ties with knowledge-related institutions and rushing to fill the void left after the Federal Highway Administration (FHWA) decided, in the early 1980s, to no longer maintain, let alone further enrich, the library of computer programs progressively built in the previous decades.

While chapters 2, 3, and 4 each deal in a specific way with the four-step model, together they depict, in Kuhnian parlance, the creation and workings of a "normal (engineering) science."[64] Thus, chapter 2 traces the route to the FSM and its rise to prominence within the community of travel forecasters, and chapters 3 and 4 are devoted to "puzzle-solutions" the UTDM social world had to develop in order to respond to a series of "anomalies" that the FSM encountered over time. And, though both the disaggregate approach and the UE assignment techniques introduced significant and lasting changes into the FSM general framework, the latter did preserve its encompassing initial four-step structure. The three following chapters 5, 6, and 7 form another large organic unit also relating to the four-step model, albeit in a more indirect way. Resorting to the standard Kuhnian picture of the science again, chapters 5, 6, and 7 tell the story of a "paradigm shift," since each analyzes original lines of travel modeling attempting to collectively supersede the FSM.

Nothing better illustrates the huge swathes of time and money spent as well as the vast amounts of efforts deployed by the UTDM social world to replace the FSM with a new set of practices than the TRANSIMS (TRansportation ANalysis and SIMulation System), a large federal government–sponsored research project. At the same time, nothing better embodies the difficulties encountered in building a global modeling approach that radically departs from long-standing practices than the same TRANSIMS, the core subject matter of chapter 5. With the end of Cold War, a great deal of existing resources, both material and human, had to shift from military projects to civilian ones; as a result, in the early 1990s a large team of scientists from Los Alamos National Laboratory (LALN) began to familiarize themselves with travel demand modeling, and set to work on building the TRANSIMS project. A quarter-century later, and thanks to multimillion-dollar federal funding, TRANSIMS was probably the earliest successful example of open-source software development in urban travel forecasting. However, it eventually failed to become the preferred modeling tool for the transportation planning agencies across the nation. Yet, TRANSIMS can be described as a sort of "productive failure," since it helped the UTDM social world to renew itself, and to envision new operational tools that have already been put to extensive use, especially via other original lines of traffic forecasting, the most popular of which is addressed in chapter 6.

"Zones do not travel, people travel," the motto goes. But who are these people? For the proponents of the disaggregate approach (chapter 3), they are separate individuals acting like the proverbial *homo economicus* portrayed by neoclassical microeconomic theory. For the adepts of activity-based modeling (ABM), also yearning to replace the aggregate FSM with new modeling practices more focused on the individual, flesh-and-blood travelers are not autonomous, self-centered maximizing machinery (not all the time at least), but rather are creatures embedded in manifold social relations. Erected on the general idea that travel is not a good desired for itself, ABM is about scrutinizing and modeling *homo sociologicus's* travel behavior. Originating in the university segment of the UTDM social world, ABM took much time to take off from an operational point of view, because its much greater behavioral realism compared to other modeling lines has long translated into serious computational complexity. Chapter 6 relates the ins and outs of the activity-based approach. It especially reconstitutes the rather torturous path leading from the first theoretical insights to the latest practical outcomes, obtained with the help of *new* data collection procedures and treatment—including GPS-based techniques for the most recent periods—and, in an ironic twist, thanks to techniques first emerged within the "old" utility-based disaggregate approach, so sharply criticized by early proponents of activity-based modeling.

The activity approach aspires to supersede the *first three stages* of the traditional four-step model. The seventh and last chapter of the book relates a series of developments essentially occurring with the fourth step of the FSM: traffic assignment. Despite path-breaking modeling work carried out within academia in the 1970s, progressively incorporated into operational procedures in the following decades (chapter 4), today's most popular operational assignment techniques are still subject to severe limitations. Being static, and therefore only concerned with traffic flows remaining *stable* over time, they are unable to successfully address time-varied phenomena, such as congestion formation and day-to-day traffic variations. Faced with a series of new problems, including the impact on congestion—and hence on air pollution—of changes in departure time, as well as the effectiveness of policies aiming to manage in real time specific traffic events, such as emergency evacuation in case of natural disasters, the UTDM social world began to experiment with dynamic traffic assignment (DTA) techniques, able to model *time variations* in traffic flows and conditions. DTA, the origins of which can be traced back to the 1970s, has emerged as a vigorous area of research within academia since the advent of ITS technology in the 1990s, while practitioners can now make use of DTA-related operational computer suites designed either by proprietary software companies or by new collective actors cutting across the public/private divide and revolving around open-source DTA platforms. Although till now the existing operational activity-based models have been usually coupled with *aggregate* static assignment models, while disaggregate DTA techniques have been mostly used in tandem with *aggregate* travel demand models originating in the FSM era, recent developments in modeling techniques have made the perspective of ABM-DTA *integration* one of the most promising avenues in urban traffic forecasting. Are we about to witness the demise of the renowned four-step model?

While the various chapters of the book are concerned with specific episodes in the history of American urban traffic forecasting, the conclusion takes a more global look at this particular modeling field. In doing so, it pursues a threefold goal. By undertaking a high-altitude flight over the UTDM territory, it intends, first, to identify and summarize some general core elements and patterns of UTDM development in the long run, and second, to offer some explanations of these same elements and patterns. A worthy subject in its own right, UTDM can also speak to larger questions and contribute to their study. The third aim pursued in the conclusion is to place American travel forecasting and its history within broader contexts in order to use the subject matter of the book as a lens through which to further observe and analyze a number of more general phenomena currently under investigation by historians and other social scientists, ranging from modeling complex systems to the history of the American State, and from the increasing commercialization of academic knowledge to the development of knowledge-intensive firms, to name a few.

# I The Emergence, Development, and Rise to Prominence of the Four-Step Model, Interwar Years–1960s

# 1 Counting and Forecasting Traffic in the Interwar Years

## The American City and the Traffic Predicament

In 1900, there were only eight thousand or so registered motorized vehicles on American roads. At that time, they were probably viewed with amused tolerance, and treated both as an object of curiosity and a plaything for the rich. Very soon this was a thing of the past and two decades later the United States could pride itself on displaying one motorized vehicle for every 7.8 Americans aged fifteen years and over, a figure rising to a spectacular 3.2 ten years later. With such numbers on the rise,[1] it was obvious to everyone that the automobile was no longer the preserve of the wealthy, and it was already a cause for serious concern. As the numbers of motorists, essentially white men but also a small minority of women and black people,[2] steadily grew, stiff new demands were placed on existing urban streets. Having been designed for a carless landscape, the American city proved, more often than not, unable to respond adequately. As a result, from the early 1910s on the specter of congestion started haunting urban America. In 1913, New Yorkers suffered twice-daily traffic jams, and the situation had gotten so bad a decade later that, despite an increase in vehicles, the number of trips per auto was actually down.[3] "The traffic problem of New York . . . is the problem of all our cities," argued an observer in 1916. Furthermore: "Los Angeles feels it; it is an acute issue in Chicago; both are simply representative of dozens of other cities in which congestion, narrow streets, and automobiles have created a situation that requires a drastic remedy."[4]

Yet, congestion was not the only footprint of the automobile on the interwar urban landscape. Motorized vehicles very soon sadly distinguished themselves as an efficient deadly weapon as well. By 1915, automobiles slew 659 people in New York, comparing therefore extremely "favorably" with murder (260 people),[5] while some fifteen years later the death toll from motor vehicle accidents had risen to 32,929, a number that included 5,081 children.[6] Judged by any standards, these figures were definitively unbearable even for a "tolerant people" ready to express "willingness to suffer for the sake of progress."[7]

And there was more. As downtown traffic jams became more pronounced, and parking problems kept worsening,[8] numbers of businessmen in the central business district (CBD)[9] followed in the footsteps of many fellow downtown-dwellers by making themselves into suburbanites,[10] and experimented with new locations outside the city center.[11] This automobile-related exodus from downtown would in turn upset people who had financial interests in the central city as they saw the value of their properties dropping. And since most American cities of the time critically depended on their CBDs to feed the municipal budget with taxes, local politicians and urban managers also considered with much anxiety an eventual economic devitalization of the city center.[12] The auto industry also feared congestion, especially when sales began to fall by the end of 1923 for the first time.[13] In the early 1920s, the federal government also began to envision congestion and accidents as an economic disease to be treated imperatively. And as the car proved to be a material object, massively causing injuries and deaths, people with the professional habit of reading human lives through the cold language of numbers, namely insurers, were also seriously worried about the compensations they had to pay.[14]

In the 1920s, a wide range of "relevant social groups"[15] were variously confronted with the new car-induced urban landscape. Undoubtedly, the interests at stake were diverse, and the precise nature and intensity of concerns and worries expressed greatly differed from one group to another. Yet, a common sense of urgency about the need to take energetic action against the adverse effects of motorized vehicles upon the American city would take hold of them all.[16]

## The New Science of Traffic Engineering

Belief in the capacity of science to solve all kinds of problems, and to produce "the greatest good for the greatest number"[17] was a lasting legacy of the period known as the Progressive Era (ca. 1895–ca. 1920).[18] Unsurprisingly, it was therefore to science that a nation, gravely concerned with the problems caused by the massive introduction of motorized vehicles in its major cities, would turn for help. Designated *traffic engineering*, the new science, immediately charged with the task of fulfilling the "two-fold purpose of facilitating movement and avoiding accidents,"[19] had no difficulties in finding people to serve it.

On the "supply side," both American universities and colleges were producing increasing numbers of graduates specializing in the various branches of engineering.[20] On the "demand side," there was a host of potential employers ready to listen to the claims and willing to enlist the services of professionals able to address traffic issues. As a matter of fact, starting from the late 1910s on and throughout the interwar period, many chambers of commerce, automobile clubs, car dealers, and public transport companies throughout the country regularly funded urban traffic surveys and other actions aimed at curbing congestion and accidents.[21] In addition to these local actors attached to a particular city or region, several nationwide stakeholders, including the National Automobile Chamber of Commerce, established in 1913, the Automobile Manufacturers Association (1911), the Automotive Safety Foundation (1937), and the insurance industry[22] would also systematically employ traffic engineers.

Besides working on a more or less permanent basis for large private actors, traffic engineers could also, following in the footsteps of urban planners,[23] establish themselves as private consultants, or sell their labor power to the first traffic engineering firms. Harland Bartholomew and Associates, set up in Saint-Louis in 1919,[24] and De Leuw, Cather & Company, founded in Chicago,[25] were among the first American consultancies specializing in traffic engineering. They would play, especially the second one, a significant role in the blossoming of urban traffic forecasting in the post–World War II years.

Not being content with playing second fiddle to the private sector, very soon all levels of government would also distinguish themselves as a home to traffic engineers. Hired by the city council of Pittsburg at the end of 1924,[26] Burton W. Marsh (1898–1989), a civil engineer who graduated from Worcester Polytechnic Institute and with a supplementary year of study at Yale University, is believed to have been the first full-time city traffic engineer in the United States.[27] In 1930, obviously enchanted by the prospect of seeing his salary shifting from $5,000 to $7,500, Marsh decided to change employer, and headed for Philadelphia.[28] There he worked for the city till 1933, before embarking on a thirty-one-year career at the American Automobile Association in Washington, DC. Marsh's case was not unique. By 1928, a dozen American cities had already resorted to the services of traffic engineers,[29] while some twenty years later there were more than eighty metropolitan areas and about a dozen counties employing this new sort of technical expert or having set up traffic divisions.[30]

It was through the Bureau of Public Roads, established in 1893 as the Office of Road Inquiry, that traffic engineers became members of the federal administration family.[31] In the middle of World War I, reflecting the emphasis that the Progressive Era reform movement placed on technical expertise, Congress passed the 1916 Federal-Aid Highway Act, which committed federal money to building rural roads for the first time.[32] Under the terms of the act, the states were charged with designing, building, and maintaining roads adapted to the new automobile era, whereas the BPR was vested with the authority to set technical standards while inspecting and approving plans, specifications, and construction. In order to perform its new missions, the federal agency began to expand, and, in April 1930, the permanent staff of the organization amounted to 976 persons, including 461 engineers.[33] States were also obliged to have a competent highway department in order to qualify for participation in the federal aid program.[34] By 1922, all state highway departments along with the BPR had joined the American Association of State Highway Officials (AASHO),[35] which was set up in 1914 with the purpose of providing mutual cooperation and assistance to the state highway departments and the federal government.[36] In 1921, a new Federal-Aid Highway Act shifted the emphasis from rural mail delivery to a national network of primary and secondary roads between cities while consolidating the partnership between the federal government and the states on the basis of a mutually accepted division of labor. Thus, state highway departments would continue being responsible for providing facilities for the use of vehicles, and the BPR, which in the early 1920s set to work on making itself into a research powerhouse in highway engineering,[37] would apportion the federal aid money among the state highway departments, approve their plans, supervise their projects, and provide state engineers with intellectual guidance based on "the best of the results of its researches, studies and efforts periodically."[38] Following the two acts, the BPR and, especially, the state highway departments hired a significant number of traffic engineers.[39] As a result, at the moment when the United States was about to engage in World War II, nearly 40 percent of the 580 or so traffic experts identified by Maxwell Nicoll Halsey (born in 1902), a graduate of the University of California at Los Angeles and Harvard University and a leading figure in the field of traffic engineering in the interwar period, were working for state highway departments.[40]

It would have been astonishing if this growing demand for traffic engineers in both the private and public sectors had not prompted the national academic system to react.[41] Traffic engineering seems to enter American higher education in 1920, when the University of Michigan offered a graduate course called Highway Transport Legislation and Traffic Regulations. At the turn of the 1930s seven universities at least, including MIT, were offering semester courses in traffic engineering, some of them on a graduate basis, while many others were providing some limited instruction in traffic engineering in connection with courses in highway and transportation engineering or city planning. Two-week intensive courses and conferences lasting for several days were also held at various places and at different points in time.[42]

In the midst of all these courses and conferences a specific organization stood out. Four years younger than the Eno Foundation for Highway Traffic Control established by William Phelps Eno (1858–1945) in 1921,[43] the Bureau for Street Traffic Research (BSTR) would very soon transform itself into a nationwide beacon of traffic engineering education and research.[44] The BSTR was originally set up at the University of California at Los Angeles in the summer of 1925 by Miller McClintock (1894–1960), the first Ph.D. holder in the field of traffic science.[45] Thanks to a grant of $10,000 received from Studebaker, the then well-known car company, McClintock and his research center moved to Harvard University in the same year. In 1938, the BSTR relocated again, and it was now housed by Yale University. While from 1926–1936 its one-year training program was on a graduate research basis only, starting from the academic year 1936–1937 formal traffic engineering courses were also given—there were five on offer in 1939–1940. By June 1939, the BSTR had already trained seventy-five men through its one-year graduate program, and approximately one hundred other people already in office were trained through its short courses established in the mid-1930s.[46] Bankrolled by various nongovernmental actors, the bureau was transferred in 1968 to Pennsylvania State University. Before it was eventually dismantled a dozen years later for lack of funding, it would produce some of the first manuals in the field,[47] and turn out 823 traffic engineers and transportation planners,[48] among them several famous urban travel forecasters in the post–World War II period (chapter 2).

Undoubtedly the major academic institution in the field of traffic engineering till the early 1950s, the Bureau for Street Traffic Research, which had been renamed the Bureau of Highway Traffic when it was housed by Yale University, would quickly become part of a wider network of actors aiming to promote research and efficient practices in highway and traffic engineering. The appointment of Thomas Harris MacDonald (1881–1957), a graduate of Iowa State College in civil engineering in 1904,[49] as head of the BPR in May 1919 would play a decisive role in the constitution and action of this network. In addition to cultivating the research potential of the organization he faithfully served for thirty-four years, MacDonald would actively participate in the creation of the Highway Research Board (HRB) in 1920 while figuring among its key backers over many years.[50] Bringing together all actors, both public and private, that mattered in transportation engineering at that time, HRB was eventually established as a unit of the Division of Engineering of the National Research Council under the charter of the National Academy of Sciences. Largely financially supported by the federal administration, HRB lacked the necessary human resources to design and conduct in-house research and development projects.[51] However, through its many committees—including the Committee on Highway Traffic Analysis created in 1922[52]—and its various meetings and multiple publications, starting with the annual *Highway Research Board Proceedings*, HRB proved instrumental in the development of highway, and, more generally, transportation research as well as in the communication and dissemination of new knowledge and original practices to both academics and practitioners.[53]

In light of the foregoing, it is obvious that as far as traffic engineers were concerned, the 1920s witnessed the emergence and the progressive consolidation of a series of elements and factors generally participating in the creation of a new profession.[54] To begin with, a vast array of anxious, even infuriated actors—think of the many grieving parents whose children had been victims of car accidents—exposed to increasing congestion and growing numbers of accidents, were actively looking for experts able to address the new traffic-related urban problems, while, among them, a number of potential employers were willing to hire traffic specialists and use their competences. The university system and other research-oriented institutions and organizations had also proved responsive to this growing social demand for traffic expertise. From the end of the 1910s, a series of periodical publications—be they specializing in highway engineering,[55] dedicated to planning and city management topics,[56] or aiming at a general public audience[57]—would also regularly welcome in their columns the prose of traffic experts.

Another attribute, and probably among the most material ones, characterizing a fully constituted profession, would see the light of day at the very end of the 1920s. For years, traffic engineers through the country had discussed the creation of a professional organization for sharing information and advancing their interests. In 1930 some of them decided it was time for them to take action. The Institute of Traffic Engineers (ITE)—now the Institute of Transportation Engineers—was set up in October 1930 as the professional association of American traffic engineers, and greatly helped make a collection of more or less separate practitioners into fellows of a single organization.[58] ITE was born as a tiny organization. Only nineteen persons participated in its foundation while the Great Depression made the first years of the institute into a moment of struggle for survival and recognition. Yet, ITE did survive as it saw several leading figures in the field, including McClintock, Marsh, and MacDonald, among its founders or joining the organization during the early period of its existence. As time went by membership numbers rose. In October 1944, the ITE numbered 294 traffic engineers,[59] while three decades after its creation it had more than 1,200 members, most of them graduates of a civil engineering department (67 percent) and employed by some level of government (71 percent).[60]

Following in the footsteps of the many engineering associations that had preceded it, the ITE immediately equipped itself with its own periodical press.[61] And there is more. Any transportation practitioner today is very likely to come across the *Traffic Engineering Handbook* (7th edition, 2016), the first two editions of which date back to 1941 and 1950 respectively.[62] Providing traffic engineers with "basic traffic engineering data as a guide to best practice,"[63] the *Handbook* probably also served to remind other professions in competition with traffic experts, especially city planners, that traffic engineering had definitively come of age, and traffic engineers were now a force to be reckoned with.[64] The original *Traffic Engineering Handbook* had fifteen chapters and two appendices, while Halsey's textbook, also published in 1941,[65] incorporated twenty-four thematic units. Within less than two decades (1920–1940), American traffic engineering proved able to produce enough knowledge and practices to fill hundreds of pages with words and mathematical symbols as well as diagrams and charts. Among the many products traffic engineers collectively crafted during the interwar period, some would pave the way for the birth and growth of urban travel forecasting after World War II. Deserving of special mention are *traffic surveys*, especially those designed to identify auto trip origins and destinations, and a series of early, albeit rather timid, inroads into *travel forecasting*.

### Traffic Surveys in the Interwar Years

Judging from the opinions expressed by leading traffic experts in the early 1920s, such as the New Yorker Nelson P. Lewis (1856–1924) and McClintock himself, a mood of optimism was prevalent about the capacity to erect a "traffic control system making possible a safer and more convenient use of *existing facilities*."[66] Along with several traffic-control techniques introduced in the 1910s, including stop signs, one-way streets, left-turn bans, or painted lines to delineate traffic lanes,[67] the 1920s witnessed the rapid development and implementation on a large scale of new technical and legal undertakings aiming to make the most of the existing urban road system, ranging from various systems of automatic traffic lights along major urban thoroughfares to the design of new traffic codes accompanied by educational efforts targeted at the various street users.[68]

Yet, as increasing flows of cars kept flooding the streets of American cities, traffic engineers started looking for solutions going beyond the existing facilities. Considered one such solution in the 1920s was the "Major Traffic Street Plan" that typically called for infrastructure improvements such as opening new roads, and organized streets into a hierarchical system on the basis of their traffic burden.[69] Along with traffic control techniques, Major Traffic Street Plans did provide some temporary congestion relief, but as more and more American households were adopting the automobile as their favorite means of transportation, the Plans eventually proved unable to eradicate traffic jams and other car-related scourges. Traffic engineers did not lay down their arms, however, and by the early 1930s a radical solution was already looming on the horizon in the shape of an original "techno-political"[70] object: The *urban freeway*, depicted by its proponents as the ideal artefact to help people move fast yet safely.[71] The term seems to have been coined in 1930 by planner Edward M. Bassett (1863–1948) to denote freedom of movement, and referred to a specific kind of highway to which there was no vehicular access from abutting properties, and for which at widely separated points the authority in charge would locate entrance and exit points. Although before the outbreak of World War II only a few places had constructed any significant portions of this new kind of urban street,[72] the urban freeway, labelled also by contemporaries as expressway, express highway, throughway, motorway, limited way, major traffic highway or superhighway, was here to stay.[73]

Yet, in order to properly *locate* and *size* new urban freeway networks one had better know a lot about the transportation patterns and mobility needs of its potential users in the *present*, and, even better, in the *future*. Traffic surveys, hopefully providing such information, were among the iconic achievements of American traffic engineering in the interwar period. As with any practice-oriented science worth its salt, the then fledgling science of traffic was in urgent need of data gathering in order to "substitute accurate facts of an engineering character for guesswork in the formulation of plans for the reduction of street friction and conflicts."[74] As a matter of fact, most of the actions undertaken and the tools devised in the 1920s with a view to mitigating the effects of traffic-related scourges largely relied on an "avalanche of printed numbers"[75] collected through a great variety of traffic surveys. Among them, the "cordon counts," and, especially, the origin and destination (O&D) traffic studies, following a series of changes and refinements, would become, after World War II, part and parcel of urban travel demand modeling.

Since urban mass transit had long antedated motorized vehicles in the American city,[76] it is no wonder that the first significant traffic surveys were conducted by mass transportation companies.[77] According to contemporaries, it was the city of Philadelphia that carried out, with the help of the consulting office Ford, Bacon & Davis, the "most comprehensive traffic count ever undertaken" in 1912.[78] The data collected were later transferred to Hollerith punched cards, and were processed by the Statistical Service Company with the help of the electric tabulators of the day.[79] The resulting origin and destination matrices, in modern parlance—where each cell represents total traffic between two particular sections of the city—were then used to draw a number of conclusions as to the probable ridership on the proposed extension system of the city's subway and elevated lines. Philadelphia's example was soon emulated by a number of other large metropolitan areas. Detroit and Boston, both in 1915, then Chicago (1916), and Pittsburgh (1917) undertook in their turn origin and destination surveys for their mass transit systems, often with the help of local university resources.[80]

Some of these undertakings, in Pittsburgh and Chicago for example, included a car component as well. But as time went by and car numbers on the urban streets soared, the automobile began to generate its own traffic investigations. It was Chicago that pioneered a cordon study to count all vehicles entering and leaving the downtown district in June 1916,[81] while other large cities followed suit, including Pittsburg in 1920[82] and New York in 1924.[83] By 1930, cordon studies had already become a routine device in the field of traffic engineering.[84] Useful, even indispensable, as they might be, however, cordon counts can report "*only the number*, and perhaps the type, of vehicles passing certain points within specified periods of time. Such counts will show where there is necessity for providing additional highway facilities, but they are not enough to *indicate the best route that such highways should follow*. For this purpose, it is necessary to know both the *origin and destination* of all the vehicles concerned."[85] These O&D surveys of automobile traffic first emerged in the first half of the 1920s, before witnessing a dramatic development and extension over the next decade. More complicated and advanced in both design and execution than the cordon counts, since they determine not only the number of vehicular trips but also the locations where they begin and end, and, often, the purpose for which travel is undertaken as well as the route taken and the intermediate stops eventually made, O&D surveys would complement cordon counts to soon become the principal method of obtaining data for planning major urban traffic improvements.[86] The various methods used for conducting O&D studies in the interwar period can be subsumed into two large families. The first groups together survey techniques that, like cordon counts, allow information to be collected on the origin and destination of motorists' trips in an *indirect* way, and without *soliciting* any interaction with the driver.[87] For example, in a large traffic survey carried out in the city of Detroit in 1936, the license numbers of cars parked in the city's CBD—that is, the destination of the trips—were recorded, and the place where the parked car was usually garaged, as indicated by the registry list, was assumed to be the origin of the trip to the central city.[88] If the approach illustrated by the 1936 Detroit traffic study isolates the observer from the observed, the second large family of O&D surveys requires, on the contrary, the active collaboration of the driver as the main source of information. Though not the only actor involved,[89] the Bureau of Public Roads would distinguish itself as a past master at designing and carrying out—often with the help of the states and other local governments—O&D surveys based on *interviews* with drivers.

The measurement and analysis of highway traffic were among the challenges that the federal agency attacked as a matter of priority, starting with a complete investigation of the California highway system in 1920.[90] Among the first surveys of the period, the one for Connecticut that was conducted between September 1922 and September 1923 deserves a special mention. Immediately hailed as the "most comprehensive program ever attempted,"[91] it departed from the previous studies carried out under the aegis of BPR in that it was much more research-oriented and the first to employ tabulation machines and make use of an interview-based O&D survey. Given its research orientation, BPR engaged John Gordon McKay, the author of a recent Ph.D. dissertation in economics in 1922.[92] Installed as chief of the BPR's Highway Economics Division, McKay was to work for the federal agency for several years, while his name featured many times in the member list of the HRB's Committee on Highway Traffic Analysis in the 1920s and 1930s. McKay's engagement would not be an isolated act. Throughout the interwar period, BPR hired several persons with academic credentials who participated in the many transportation surveys[93] done under its direction before World War II.[94] The first transportation surveys conducted by the BPR were statewide in scope and had little, if any, urban components. Although the first comprehensive traffic survey in an urban region involving the bureau was for Chicago in 1924, it was the Cleveland Regional Area Traffic Survey of 1927 that proved a milestone in the slow transformation of BPR from a mostly rural actor into a significant urban one. A very extensive survey that counted traffic at 973 points on the roads of the regional area, the Cleveland study was not just another routine interview-based origin and destination survey because along with the origin and destination of their trip, drivers were also asked a series of additional questions, ranging from the exact route followed to their reasons for choosing it.[95]

Ironically, the Great Depression proved overwhelmingly favorable to American roads. With a quarter of the civilian labor force in the throes of unemployment, roads proved an excellent place where large numbers of people could be "put to work,"[96] and work relief money be usefully spent. Transportation surveys, regarded as a weapon in their own right in the fight against the unemployment of "white-collar personnel,"[97] greatly benefited from these Department of Public Works-friendly circumstances. They especially received a strong stimulus from the Hayden-Cartwright Act passed by Congress in 1934, which allowed up to 1.5 percent of federal aid highway funds appropriated for any year by any state to be used for studies into the nation's highway system. One of the prime objectives of these actions was to assemble factual data that would help the federal and state administrations to fulfill a twofold purpose: to determine "where highway-user revenues[98] should be spent to *benefit the greatest number of motorists,*"[99] and to devise a "rational method for apportioning road-user taxes assuring adequate maintenance and needed development of state-administered highways and benefits to urban and rural taxpayers in *strict proportion to the tax payment of each group.*"[100] A new kind of survey, named the "Road-use Survey," was therefore devised with a view to achieving these objectives. While traditional traffic surveys involved "the actual observance of the vehicles at some point of their travel on a given trip," the new Road-use Surveys depended upon "interviews with the owners of motor vehicles[101] during which a complete enumeration is made of all or a large part of the travel performed in the vehicles of the interviewed owners throughout a specified period of time, usually twelve months."[102] It was during the same Road-use Surveys that the Bureau of Public Roads specialists worked more systematically than they did in the past to classify the various automobile trips of American households according to the *purpose* for which they were undertaken.[103] Both the interview approach and the travel classification schemes based on the trip purpose would be greatly developed within post–World War II urban traffic forecasting.

Suggested by the BPR, which assigned a resident engineer to each state to this end,[104] and conducted by the state highway departments in the second half of the 1930s, Road-use Surveys were based on an impressive number of interviews. During a first wave covering seventeen states, no fewer than 198,809 drivers of private cars and another group of 71,941 truck drivers spoke at length about their mobility practices.[105] These surveys produced an unprecedented amount of information about the travel patterns of American households of the day. In an ironic twist of fate, while these state-wide planning surveys were largely undertaken to evaluate the needs of the *rural* (intercity) highway system, they eventually revealed and emphasized the importance of *cities* in the production of traffic on the nation's roads. The gathered data showed, in fact, that most trips were short and traffic was primarily urban in nature.[106]

What were the BPR staff members to do with all the numbers produced? How would they translate pieces of information into a course of action? Here is the answer given by the intellectual father of post-1935 highway planning surveys, Herbert Sinclair Fairbank (1888–1962), a graduate of Cornell University in 1910, the editor-in-chief of *Public Roads* for many years, and a truly "educated man" who "read a great deal," and "tried to bring economists" into the Bureau of Public Roads as well as economics into the "thinking" of its staff.[107] Fairbank argued that since data clearly show that the city is the "place where the most people *want* to travel short distances,"[108] and given that traffic engineers' ruling concern "is to *read aright* the manifestation of *human desires* in respect to highway movement,"[109] public servants have a duty to provide their fellow citizens with what the latter obviously are yearning for: urban freeways enabling—and also constraining, it should be added[110]—urbanites to move fast and safely across their metropolitan areas.[111] Vested with the authority quantification provides,[112] considering itself the bearer of the "common good" in the domain of mobility, consistently developing a rhetoric of public interest, the BPR would throw all its weight, based on the power of numbers combined with that of federal money, behind the urban freeway, hammering home the message that it was the "one best way" of curbing urban traffic-related scourges while meeting at the same time the desires of American households. The BPR's insistence on urban freeways eventually paid off with the Federal-Highway Act of 1956, which translated what started as an interwar innovation into a massive urban reality in the 1960s (read more on this in chapter 2).

## From Current Traffic Counts to Travel Forecasts

Surveys tend to account for the present, but they can also help to take care of the future. Not content with gathering evidence for current traffic on American roads, traffic experts sought to forecast traffic on the network for the years to come, since a facility that "will not meet traffic demands during [its] expected life is a poor investment, resulting in traffic congestion and early reconstruction."[113] It was during the Maine transportation survey, the field work for which was done from July 1 to October 31, 1924, that a *statistical model*, based on the method of least squares, was first used by the Bureau of Public Roads to estimate future highway traffic.[114] The inputs of the model were the traffic counted at several stations for one week each year from 1916 to 1924 inclusive in combination with the evolution of the registered cars and the population of the state during the same period. Despite the smallness of the sample, BPR researchers were able to show that "traffic and registration are in nearly constant ratio from year to year."[115] Moreover, they also succeeded in establishing another close statistical relationship between the annual total population of the state and the total of motor vehicle registrations each year. Using the predicted population of some future year, obtained from the Bureau of the Census for example, BPR could first forecast future vehicle registration, and, from this, future traffic on the highway system.[116] BPR would apply the aforementioned forecasting approach in several other surveys conducted in the 1920s,[117] and obtained rather encouraging results.[118] At the beginning of the following decade, while keeping the same general statistical-centered methodology, BPR specialists would change the variables used for traffic forecasting, as the study of the accumulated data up to that time seemed to indicate "a much closer relationship between gasoline consumption and traffic volume than between registration and traffic volume."[119]

Although the BPR unarguably dominated the landscape of interwar traffic engineering, it was by no means the sole actor. As has amply been seen, American cities and traffic consultants working on their behalf were also significant parts of the traffic engineering community of the day and contributed to expanding the body of knowledge and practices in respect of traffic counting and even forecasting. Two major metropolitan areas, New York and Boston, would prove even more adventurous than the Bureau of Public Roads in their attempts to estimate future traffic. While the federal agency was generally content with assuming a five- to ten-year projection into the future, planning documents drafted on behalf of these two cities and dating to the 1920s contain traffic predictions for the year 1965 (figure 1.1)!

In order for such a bold move to be undertaken, a new kind of traffic forecasting approach was devised by consultants working for these two metropolitan areas. In 1926, BPR specialists were arguing that the "most scientific method of future traffic prediction is by projecting past traffic trends" since this "method has been found accurate in the prediction of population, business conditions, railway traffic, and other economic factors."[120] The leading traffic engineer and planning consultant, Ernest Payson Goodrich (1874–1955), a graduate of the University of Michigan and the first president of the Institute of Traffic Engineers,[121] was more skeptical about the scientific virtues of such a modeling approach, which relied on mere statistical correlations between variables without providing any explanation for the traffic patterns observed.[122] While offering his services to the Regional Plan of New York and Its Environs in the mid-1920s,[123] Goodrich would develop instead a new modeling line, named by its author as the "distribution method." The latter "assumed that the traffic radiating from any one district to all other districts in the area, a county being the unit considered, is *inversely proportional* to the distance between the districts and *directly proportional* to the automobile ownership in the districts."[124] At about the same time, another leading figure of the community of city planners of the day, Robert Harvey Whitten (1873–1936), then working for the city of Boston, would build on Goodrich's "distribution model" to propose an even more pronounced Newtonian version of it.[125] From an analysis of the data gathered through an extensive origin and destination study conducted in Boston in 1927,[126] Whitten stated that the traffic between any two pair of districts varies directly with regard to the motor vehicle registration of those districts and *inversely* with regard to the *square* of the distance between them.[127] Although the historian can find some (rare) traces of this modeling approach in the 1930s,[128] the traffic engineering community was to wait another thirty years or so before the ideas put forward by Goodrich and Whitten in the 1920s resurfaced in post–World War II America (see chapter 2).

By comparison with the tremendous development that urban traffic forecasting would undergo in America after the end of the World War II, the

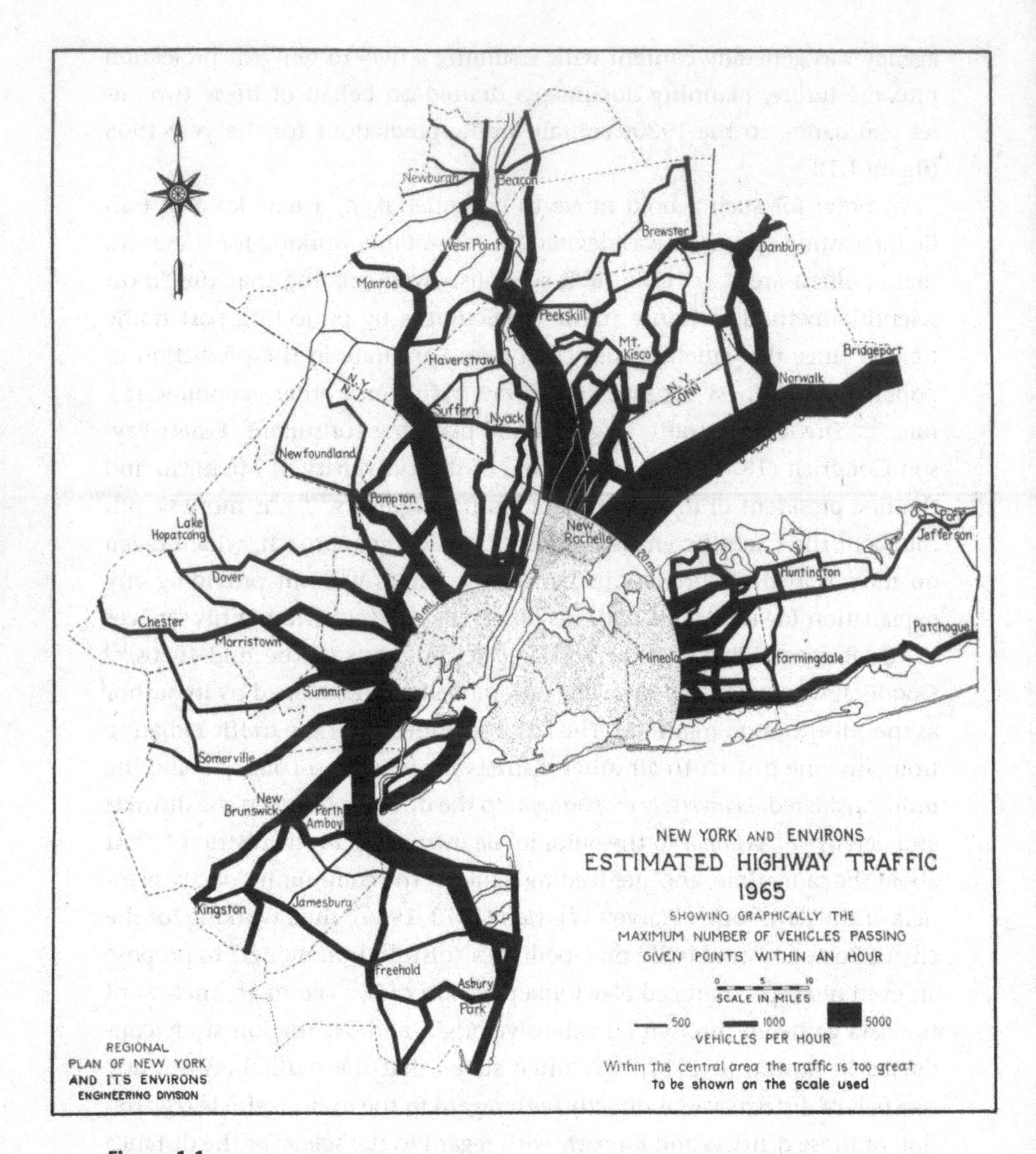


**Figure 1.1**

New York and Environs Estimated Highway Traffic for Year 1965. *Source:* Harold M. Lewis (in collaboration with Ernest P. Goodrich), *Highway Traffic in New York and Its Environs, Including a Program, by Nelson P. Lewis, for a Study of All Communication Facilities Within the Area* (New York: Regional Plan of New York and Its Environs, 1925), 107.

harvest of the interwar years seems rather poor. However, it was during the Roaring Twenties and the Great Depression that the seeds of the future were planted since a number of preconditions[129]—be they professional, organizational, material, or ideological—for the postwar blossoming of urban travel demand modeling came to be met, and several elements of the would-be UTDM social world, albeit mostly in an emerging state, appeared for the first time.

# 2 The Aggregate and Zone-Based Four-Step Model Takes Center Stage

## Toward the Interstate Highway System

Everywhere today the air is filled with post-war planning.

—Theodore M. Matson, "Uses of Traffic Surveys and Studies in Planning," in *1943 ITE Proceedings*, ed. Institute of Traffic Engineers (New York: Office of the Secretary, n.d.), 32

The end of World War II marked for the American automotive world both a return to the years leading up to the global conflagration and a new beginning. On the one hand, as the various rations that were introduced during the course of the war, including those on fuels and rubber, were abolished, and the major automobile manufacturers that had converted to the production of military equipment during the hostilities now came back to their previous civilian functions,[1] the evolution of the nation's fleet of vehicles caught up quickly with its past upward trends. Thus while in 1944 there were 30,479,300 registered motorized vehicles on American roads, in 1946, 1950, and 1970, the corresponding figures rose to 34,373,000, 49,161,600, and 108,407,300, respectively.[2] On the other hand, the war seemed to convince a large part of the political system and public opinion as well that the federal government should throw all its weight behind the nation's economy, much more than it had in the past. As a matter of fact, the three decades that followed the end of World War II witnessed the expansion of the prerogatives of the federal government, which was clearly reflected in the federal share in total government expenditures.[3] Cars and the various infrastructures they rely on were among the beneficiaries of the new responsibilities the federal government was vested with after World War II.[4]

*Interregional Highways*, the report of a committee appointed by President Franklin D. Roosevelt in April 1941 and chaired by MacDonald, the then chief of the Bureau of Public Roads, was published in January 1944. Building on the vast amounts of data amassed by BPR in the second half of the 1930s,[5] the committee recommended the construction of a national system of highways with a large urban component.[6] A year later, in December 1944, Congress passed the Federal-Aid Highway Act of 1944 designating a National System of Interstate Highways not exceeding 40,000 miles, approximately 5,200 miles of which were to be located in urban areas. Yet, the money required to materialize such a huge infrastructure project would be lacking until the enactment of the Federal-Aid Highway Act of 1956. This provided around twenty-five billion dollars—the equivalent of some $230 billion in 2018—over the period 1957 to 1969 with a 90 percent federal share for the construction of a system comprising 41,000 miles of highways and capable of accommodating the volumes of traffic forecast for the year 1975.[7] This was the famous Dwight D. Eisenhower National System of Interstate and Defense Highways, commonly known as the Interstate Highway System, the largest public works program ever adopted in the United States until then.[8]

## Home Interview: A New Kind of Traffic Survey

Though big federal money would not be available until 1956, the Bureau of Public Roads was able to start seriously envisioning the future Interstate Highway System thanks to the funds released by the Federal-Aid Highway Act of 1944. The BPR appropriated in each of the three postwar years $125 million in federal dollars to be matched by state and local funds for urban-aid highway construction.[9] Part of that money could be used for metropolitan traffic planning studies, and this was in fact accomplished. Between March and June 1944, nine American cities launched, with the help of the BPR, a new kind of large-scale O&D travel study designed to replace the interwar traffic survey practices, regarded now as less effective when operating within an urban setting (see figure 2.1).[10]

Quickly labeled the Home Interview (HI) traffic study, the new method aiming to capture both the *actual travel patterns* and the *mobility desires* of American households was the outcome of joint efforts by the BPR and the Bureau of the Census, already a national past master at conducting large

- 5 -

To aid in development of the method of sample selection the Public Roads Administration was fortunate in having the cooperation of the Bureau of the Census. That bureau, through the Division of Special Surveys, regularly conducts surveys to collect widely varying types of data, generally using some sampling technique. Their recommendation was to select a sample purely on a geographical basis, on the theory that in a sample so selected all other factors would be automatically included in proper proportion.

Thus, for a travel habit survey by this means, the first requisite is the selection of a sample inflexibly chosen as to geographical distribution, and adhered to in the interviewing without the slighest deviation. The natural tendency of an interviewer, on finding the occupants of a designated house absent, to call on a neighbor must be strictly avoided. It should be obvious, of course, that the travel habits of a person easily found at home must be quite different from those of a person seldom there, but this distinction is frequently overlooked unless its importance is stressed.

The size of the sample varies with the size of the city. In the smaller cities in which surveys have been conducted, those with populations up to about 150,000, a ten-percent sample has been used. As the size of the city increases, and as the volumes of travel with which we must deal also increase, a smaller sample is adequate. In cities around 500,000 population a five-percent sample has been found to be sufficient, and for larger cities in which studies are now contemplated, it is probable that the sample will consist only of one address in forty.

The manner of selecting the particular addresses to include in the sample varies with the city and with the material there available that is useful for the purpose. Generally the Sanborn maps have proved most helpful. Where coverage by these maps is complete and they are reasonably up to date, the street and number of each unit to be interviewed may be listed directly from the maps. These listings may be checked by a variety of means such as city directories, Census statistics, water or other utility company records, assessors' records, and other sources. No single method of sample selection is arbitrarily determined in advance. Instead the sources in each city are reviewed and the most complete and accurate used as a base, with other less detailed records used as a check. In newly developed or outlying areas it is sometimes necessary actually to list all addresses from a ground survey, and to select those for interviewing from the list.

Whatever method is used, a sample is selected generally by working entirely around each block and advancing block by block throughout each census tract. The census tract is used as a basic unit of area because it is usually of a suitable size to serve as a useful zone of origin or destination of travel for analysis purposes, and also because of the large amount of data on population, housing, and other trends that are available for all cities by census tracts. These data can obviously be of material value in estimating the trends of travel in the various sections of the city.

**Figure 2.1**

Extract from a paper authored by Thomas H. MacDonald, the then Head of the Bureau of Public Roads, in which he announced the creation of a new transportation design tool crafted with the help of the Bureau of the Census, the famous Home Interview. *Source:* Thomas H. MacDonald, "Analyzing Transportation Needs of Urban Areas—A New Technique" (A paper presented in the Convention in Print of the American Transit Association, published November 3, 1944); T. H. MacDonald Papers, volume 12, in Papers by Herbert S. Fairbank—Frank Turner—T. H. MacDonald, https://planeandtrainwrecks.com/Search.

surveys on the basis of representative samples of various target populations.[11] As a good deal of the meetings taking place between the two federal agencies were placed under the auspices of the Highway Research Board, the latter set up an Origin & Destination Survey Techniques Committee in late 1943, with John T. Lynch, a senior highway economist with the BPR, as its first secretary. Immediately setting itself to work, the committee issued, in 1944, a report laying out the core elements of the HI as well as the results of the first applications of the tool—twenty-four traffic studies in total for the year 1944, either completed or underway. The first technical manual describing in detail the procedures for conducting a Home Interview traffic study saw the light of day in the same year with the authorship of the BPR, while new, revised editions of the manual would appear in 1946, 1954 (reprinted in 1957), and 1973 (reprinted in 1975). Other significant actors, especially metropolitan transportation agencies involved in traffic planning studies, would also help spread HI principles nationwide through a variety of publications in the decades following the first edition of the BPR manual. No wonder the majority of the 130 or so O&D traffic surveys carried out in the United States by 1960 followed the Home Interview's particular method.

Following up on an idea first suggested by the Bureau of the Census, according to which unbiased representative samples of a target population can be obtained on a solely geographical basis and, hence, regardless of factors such as occupation, gender, size of family, or race, BPR experts promoting HI decided to use the household-dwelling unit as the basis of sampling. Constituted with the help of the available urban data sources at the time, among them the unrivaled Sanborn maps,[12] as well as the city directories, the size of the samples varied inversely with that of the area under study: While for cities with a population of under fifty thousand a large sample of 20 percent was deemed necessary, a 4 percent sample was regarded as sufficient for large metropolitan regions with more than one million inhabitants. Once the sample was selected, interviewers—often women, as they proved the most effective and also more easily available than men at that time—were given a list of the dwelling units in their specific territory. Equipped with a uniform and carefully designed interview forms, the interviewer was instructed to fill them in with standardized information. The latter was about the household's general characteristics as well as all the *trips*, defined as the *one-way* travel from one point (*origin*) to another (*destination*) for a particular *purpose*, undertaken by each member of the household aged five years or above on the day previous to the interview. For each individual concerned, the interviewer had to directly record on the appropriate questionnaire, explicitly designed so that most of the entries made by the interviewer were self-coding,[13] the origin and the destination, the start and arrival times, the type of transportation—whether as a car driver, a rider in a car, or as a passenger on public transport—as well as the purpose of the travel (there were nine options[14]). Unsurprisingly, the aforementioned conception of trips, as separate and independent one-way movements, the means of transportation envisioned, and the trip purposes considered all had a direct relationship with the social practices and the internal division of labor regarding mobility within the then typical American household. As a matter of fact, neither trips by bike nor travel on foot were to be recorded by the interviewer,[15] while for the designer of the Home Interview traffic study the husband was supposed to work, and the wife to do the shopping and chauffeur the children to school.[16]

To ensure reliable results, a variety of checking procedures were conceived and implemented, from the selection of the sample to the final tabulation of the data gathered and their automatic processing. Given the eventual reluctance of the interviewees to reveal, or their inability to recall all of the trips made the day preceding the interview, the issue of the completeness with which travel was reported was felt as being highly critical from the very start and prompted BPR experts to contrive special tests to cope with the problem.[17] Yet, however meticulously conceived and designed, the Home Interview traffic study still remained a complex human undertaking, the final success of which relied heavily on flesh-and-blood people. Thus, in order to elicit cooperation with the interviewee, HI organizers did not hesitate to resort to the various advertising practices that accompanied the flourishing of mass consumption in postwar America,[18] ranging from sending postcards to each sample dwelling unit beforehand to placards advertising the upcoming Home Interview traffic study in stores and on mass transit vehicles. However, it is obvious that the resident's willingness to cooperate cannot guarantee by itself the success of the operation. Judging from the multiple changes and revisions introduced in the successive editions of the BPR's manual as well as by personal accounts of people involved in HI traffic studies, the interviewers were considered to be one of the most important, critical factors in the success of the whole undertaking. In equipping the interviewer with standardized sheets, HI designers sought therefore to reduce him/her to an "inscription device" that passively recorded the mobility pattern of each household. The interviewer was also placed under close surveillance thanks to a number of elaborate mechanisms that were crafted with the view to unearthing bad practices and to limiting chicanery. The obsession with control was such that, in the 1960s, BPR experts would even write a computer program, named QUALCON, to automatically assess the quality of the interviewer's work by detecting suspicious data that differed greatly from average ones.[19]

Benefiting from continuous improvements since its inception in the mid-1940s, and despite its high cost in terms both of time spent and money consumed, the Home Interview method rapidly gained ground on its competitors,[20] and by the early 1970s it reigned supreme within the family of O&D traffic surveys in the United States. Yet scarcely had it come into being when a rather unanticipated problem arose, that of finding a practical way of coping with the vast swathes of data gathered. Undoubtedly, sophisticated equipment could aid, but this was not enough since before treating information with the help of machines people had to represent it. New ways of representing households' mobility patterns therefore began to circulate within the UTDM social world, in order to help traffic engineers and planners so that they would not be engulfed by tidal waves of trip information. Hence the development of another traffic planning tool in the 1940s, known as the "desire-line chart," representing the theoretical routes *minimizing* distance, and thus travel time, between the various origins and destinations within an urban area.[21] By plotting, often with the help of mechanical and, later, electronic devices the totality of individual desire lines of travel on a map,[22] the transportation planner could therefore easily, and visually, directly *locate* the "best" (future) transportation network from the point of view of the sovereign consumer-citizen,[23] the one who was satisfying, as far as possible, his or her travel needs and wants (figure 2.2).

## Predicting Traffic Flows Attracted by a New Highway

Useful, even indispensable as it may be for *locating* the new infrastructure facilities that would be able to meet the "desire lines" of American households, the Home Interview approach nevertheless provides, by way of definition, little help to the transportation engineer and planner wishing to

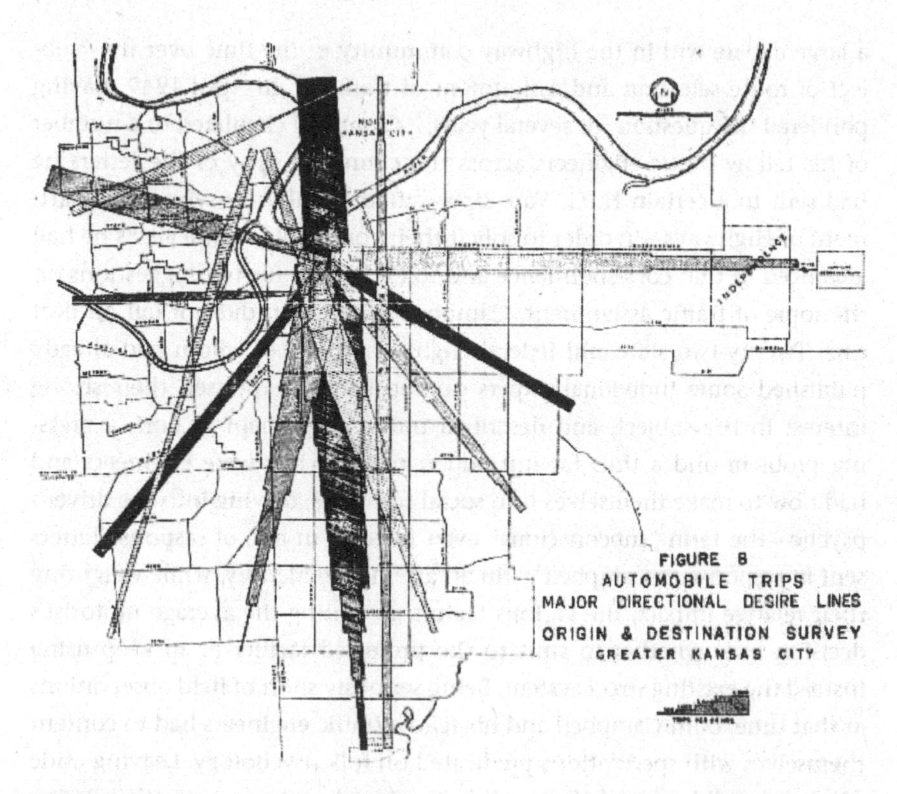


**Figure 2.2**
Example of early desire lines. *Source:* J. M. Picton, "Application of Origin and Destination Traffic Data in Planning Highway Facilities for Greater Kansas City," *Highway Research Board Proceedings* 25, HRB, National Research Council, Washington, DC, 1946, figure 8, 287. Copyright © National Academy of Sciences. Reproduced with permission of the Transportation Research Board.

properly *size* the planned facility. For that, they need to know, in addition to the major spatial distribution of trips revealed by the Home Interview O&D survey, the *volume* of traffic that will use the *new facility* as well. At the end of the 1940s, the problem of traffic attracted by a projected facility[24] was still in search of a solution but already had a name. In the words of traffic engineer Moses Earl Campbell, "THE ESTIMATED allocation of traffic to a proposed highway facility is commonly termed 'traffic assignment.'"[25]

A 1946 graduate of the Yale Bureau of Highway Traffic,[26] Campbell was hired by the Highway Research Board in March 1947, where he enjoyed a thirty-year career.[27] Shortly after he became a HRB staffer, he orchestrated a large debate within the highway community of the time over the "subject of route selection and assignment of traffic."[28] In April 1949, having pondered the question for several years,[29] Campbell circulated to a number of his fellow traffic engineers across the country a copy of the letters he had sent to a certain H. G. Van Riper—then at the Pennsylvania Department of Highways—in order to solicit their comments on the views he had advanced in that correspondence and to elicit their further suggestions on the topic of traffic assignment. Campbell's invitation did not fall on deaf ears. Twenty-two state and federal engineers, a few of whom had already published some individual papers on the issue,[30] expressed their strong interest in the subject, and described traffic assignment as both a pressing problem and a thus far untrodden path.[31] They were engineers and had now to make themselves into social scientists, delving into the driver's psyche—the term "subconscious" even features in one of response letters sent in response to Campbell[32]—in an attempt to identify, while weighting their relative impact, the various factors governing the average motorist's decision over whether to shift to the projected facility or to keep using instead the existing street system. Being seriously short of field observations at that time, both Campbell and his fellow traffic engineers had to content themselves with speculations predicated on folk psychology. Leaving aside "such intangibles as comfort, safety, tension, beauty, investigative desire, habit, and many other factors of human psychology,"[33] the HRB staffer was nevertheless able to develop a series of *theoretical* S-shaped curves depicting the probable attraction values of the two principal factors that were thought to drive an automobile operator's behavior in respect of route selection, namely *time saving*, and *distance saving*,[34] two measurable items that engineers could reasonably hope to succeed in putting into numbers.

Among the many recipients of Campbell's theoretical considerations was MIT graduate Edward (Ted) Henry Holmes, then chief of Division of Highway Transport Research at the BPR, which he had joined in 1928. Though he found no reason to criticize Campbell's S-curves and their theoretical underpinnings, Holmes felt that the "greatest need in this field of traffic assignment is an actual case or cases in which we can test various proposed methods."[35] Under the intellectual guidance of Darel L. Trueblood, a highway research engineer at the BPR, a study seeking to quantify the effect of travel time and travel distance on the driver's choice was conducted in the summer of 1950, with the urban portion of Henry G. Shirley Memorial

Highway in Arlington and Fairfax Counties, Virginia, serving as an outdoor experimental site.[36] During the study, an average of 23,249 passenger cars passed the six stations, and interviews were obtained from about 67 percent of the drivers. The information obtained was combined with the O&D data collected in the Washington metropolitan area in 1948, and supplemented with travel time and distance measurements. With the help of the curves, this made it possible to relate, albeit imperfectly, the percentage of traffic using the freeway with the ratio of travel time or distance by the freeway to the travel time or distance by an alternate route.

Obviously, more research was needed, and both the Bureau of Public Roads and the Subcommittee on Factual Surveys of the American Association of State Highway Officials strongly encouraged state highway engineers to initiate new empirical studies on urban highways already in operation.[37] Upon completion, the results of such studies were forwarded to the BPR in Washington, DC, for correlation and summarization.[38] Not later than 1952, the BPR and AASHO provided the traffic engineering community with *standardized* curves, aptly named *diversion curves*, both of them favoring travel time saving as the central factor affecting the driver's choice. However, though very popular with transportation planners, very soon these first assignment constructions had to compete with new and more sophisticated diversion curves that took into account factors other than the sole time, including comparative distance and speed for the projected freeway and an alternate route as well as time and distance differentials (figure 2.3).[39]

Different as they may be, all these curves that piled up on highway engineers' drawing boards nevertheless shared a common feature: their operational use was time consuming. To save time, the staff who conducted the now famous Detroit study from 1953 to 1955 (discussion of this follows) therefore developed a highly mechanized procedure—including coding, keypunching, data processing, and tabulation of assigned volumes with the help of an IBM 604 mainframe[40]—that made a complete assignment possible within three weeks only, a remarkable feat.[41] Detroit was by no means an isolated case, and transportation planning agencies frequently enlisted the services of mechanical and electronical devices to load the traffic on the proposed freeway and the alternate routes.[42] In order to encourage widespread use of diversion curves, the BPR even commissioned the consulting firm Edwards and Kelcey, established in 1946, to write a program for an IBM

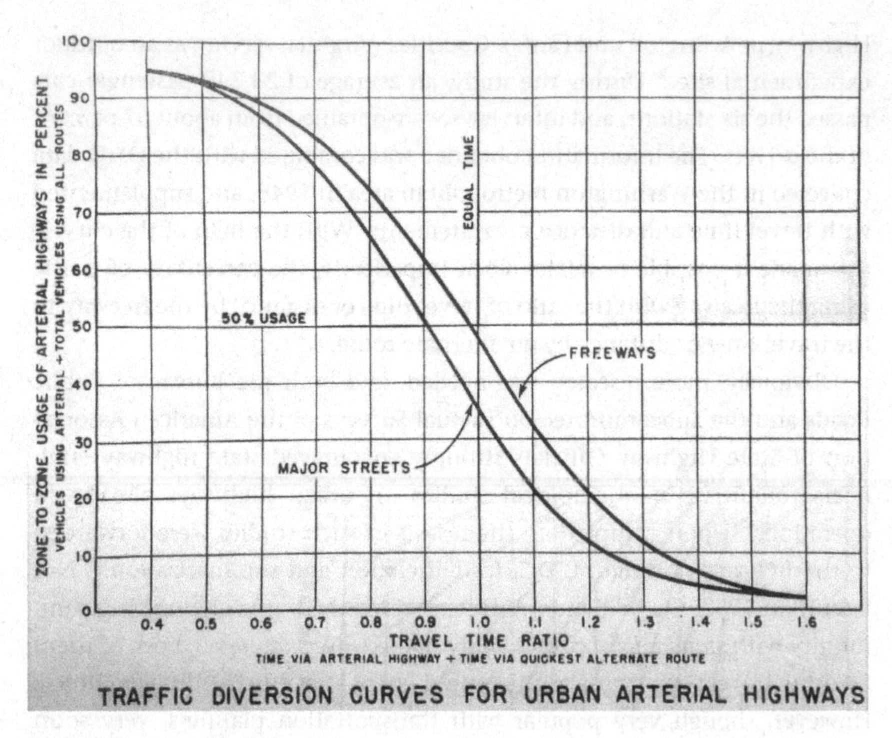


**Figure 2.3**
An early popular "Traffic Diversion Curve," built upon the basis of travel time (ca. 1957). *Source:* Bureau of Public Roads, *Electronic Computer Program for Assignment of Traffic to Street and Freeway Systems* (BPR Program No. T-6) (Washington, DC: n.p., n.d.), 3 (George Mason University Libraries. Special Collections Research Center, William L. Mertz Papers: C0050, Box 18, Folder 10).

650 machine using the AASHO traffic diversion curve but easily adaptable to any time-ratio assignment curve.[43]

## Forecasting Interzonal Traffic Flows: The Gravity Model

Assignment curves unarguably constituted a powerful tool in the hands of the traffic engineers and transportation planners of the time, as they allowed them to estimate the volume of traffic that was likely to be diverted from the existing streets to the projected facility. However, as the latter would continue serving for several years after it first started operating, it should also be able to accommodate not only present but *future* travel patterns too. Even as various diversion curves were being heaped up, a few traffic engineers set to work to devise methods for forecasting urban traffic several years into the future, thus assigning the estimated future traffic, with the help of diversion curves, to the planned facilities.

One of these traffic engineers was Thomas Fratar (1913–2001), who in 1936 graduated with a civil engineering degree from the Rensselaer Polytechnic Institute and was an active member of the Institute of Traffic Engineers.[44] According to his own recollections, Fratar began to investigate traffic issues, especially the question of the distribution of traffic on competitive routes in a road network, in other words, the assignment problem, as early as in 1941. At the suggestion of Yale professor Hardy Cross (1885–1959), the author of a general method for analyzing complex engineering structures, including frames and pipe networks, by a process of successive approximations,[45] Fratar sought to apply Yale academic's method to his own traffic problem.[46] Some ten years later, the traffic engineer, then an associate of the New York engineering firm Knappen-Tippetts-Abbett-McCarthy, was grappling with a significantly different issue as he sought to forecast for the year 1975 the distribution of trips between the various zones into which Cuyahoga County, Ohio, was divided up. Bearing in mind Cross's insights, while benefiting from discussions with Glenn E. Brokke (1912–1992), who was supervising the Cleveland Metropolitan Area Traffic Study on behalf of the BPR, Fratar was able to provide his fellow traffic modelers with a significantly improved variant of the then existing forecasting techniques known under the label of "growth factor methods."[47] According to these, the present travel patterns, revealed by O&D traffic surveys, could be projected into the future on the basis of traffic growth rates[48] imparted to each zone forming the area under study.

Immediately after Fratar devised his own model, it started to be used in New York City as well.[49] Moreover, in the mid-1950s the Bureau of Public Roads showed great interest in the new approach and included Fratar's contribution in a vast research project it undertook to test four different members of the growth factor family then in use.[50] The test called for roughly twenty-five million computations, and if ordinary desk calculators were to be used the project would require, "on a very optimistic basis," some thirty man-years for completion. Clearly the BPR was badly in need of employees knowledgeable about computers. William L. Mertz (1920–1993) was such a staffer. After completing the three-year BPR home training program,[51] he first participated in some road-test projects, when realized that his true calling was urban transportation planning (UTP). He therefore began a series of assignments and embarked on several research studies[52] that would eventually bring him to national prominence in this field, still in its infancy at that time. In the mid-1950s, Mertz moved to Washington, DC, where, from 1955 to 1959, he increasingly grew into "something of a computer nut."[53] It was during his internship at the National Bureau of Standards (NBS)—then equipped with a home-made machine called SEAC[54] and highly coveted by federal agencies lacking in computer resources of their own at that time[55]—that Metz would develop the library of programs necessary for this testing. However, given the amount of data to be processed and edited, he felt the necessity of shifting from the old SEAC to "the big one—the IBM 705." As it happens, DuPont had recently installed a brand new IBM 705 at its Delaware headquarters, and, for a while, the BPR engineer took the eight o'clock train to Wilmington every morning. The efforts deployed by Mertz and his collaborators paid off, since the test was a success. Though all members of the growth factor family, with the sole exception of the "uniform factor" method, gave results of equal accuracy, the Fratar variant required fewer iterations, hence consuming less computer time, still a scarce (and expensive) resource then. Having made himself the "computer-man" within the Bureau of Public Roads, Mertz started therefore traveling around the country running the Fratar program for several highway projects under way,[56] following the 1956 Federal-Aid Highway Act.

Though sponsored by a powerful federal agency, very soon the Fratar method nevertheless found itself struggling with an alternative modeling approach. As early as 1954, a young traffic engineer, Alan Manners Voorhees (1922–2005), drew the attention of his colleagues to a key assumption of the Fratar method, which he found highly problematic.[57] Remaining faithful to a central tenet of the general growth factor approach, Fratar had posited, in fact, that "trips between zones will change in accordance with variations in a 'growth factor' predicted for each zone *regardless* of modifications that may occur when the percentage of various types of trip purposes change between zones. However, such an assumption seems to be contrary to recent traffic studies,"[58] argued Voorhees, a civil engineer from the Rensselaer Polytechnic Institute in 1947, and a master's degree holder in city planning from MIT two years later.[59] After serving as Colorado Springs' first city planning engineer, he enrolled in the Yale Bureau of Highway Traffic, from which he graduated in 1952, having defended a research project dealing with work trips patterns.[60] In 1954, while criticizing the underlying assumptions of the growth factor approach, Voorhees was already exploring an alternative modeling path, more able, in his opinion, to account for the observed travel patterns as determined by origin and destination traffic studies.[61] His "General Theory of Traffic Movement" appeared in the proceedings of the Institute of Traffic Engineers as the "1955 Past Presidents' Award Paper,"[62] and provided the traffic engineering community of the time with the first articulate presentation of what would go down in history as the "gravity model," a forecasting procedure that, after refinements and further developments, is still in use today by transportation modelers in many parts of the world. While ambitious in terms of the objectives pursued in his paper—remember that the author wanted to establish nothing less than a *general* theory of traffic movement—Voorhees proved rather cautious about the ways to achieve them. As a matter of fact, "A General Theory of Traffic Movement" features neither heavy mathematics nor theory-ridden works in its footnotes;[63] it presents itself instead as a highly empirical undertaking the theoretical findings of which are predicated upon solid facts established by O&D traffic surveys. Most data used by Voorhees was concerned with shopping-related travel patterns.[64] The latter seemed to indicate that the "various shopping areas exert a 'pull' on shoppers," whose strength "seems[65] to follow the *gravitational principle*. In other words, the pull of a shopping area on a group of shoppers seems to be related to the *size of the attractor* (shopping area) and *inversely to the distance* the shoppers live from the area." Based on the available data, best results were obtained if distance was expressed as "auto driving time." Seeing that the gravitational principle seemed to work for shopping trips, an emboldened Voorhees sought to extend it to other kinds of travel. For work-related movements, the principle of gravitation seemed also to work satisfactorily, "but in this case the distance factor was a square root function rather than a squared function," while the "size of the 'pull' is best expressed by the number of workers employed, including shift workers." And when it comes to "social trips, the number of people in a residential area should be used to indicate the size of the 'attractor' and the distance factor should be raised to the third power."[66] Buoyed by the rather warm welcome from his colleagues, Voorhees continued arguing in favor of the theoretical soundness of his modeling approach throughout the 1950s, while underscoring its ability to produce accurate estimate trips during the *peak hour* as well.[67] Solicited as a consultant, he would also operationalize the gravity model, and by 1960 the latter had already been or was about to be used for forecasting traffic in several cities, including Frankfort (Kentucky), Boston, Baltimore, Hartford, seven Iowa cities, and Toronto (figure 2.4).[68]

In the second half of the 1950s, several central elements of the four-step model were about to be put in place. Thanks to a series of diversion curves, the transportation engineer and planner could easily *assign* traffic to the projected highway facilities (fourth step of the FSM), while the Fratar method and the gravity model enabled them to forecast *zone-to-zone movements* within an urban area (second step). Other elements of the would-be FSM, including those concerning the first and third steps, were also circulating within the traffic engineering and planning communities of the day, albeit in a rather inchoate form.[69] It was in Chicago that the central act of the FSM play would be put on and, as will be seen, the performance proved memorable in many ways.

## Chicago and the Birth of the Four-Step Model

On September 15, 1955, Joseph Douglas Carroll, Jr. (1917–1986) was named head of the Chicago Area Transportation Study (CATS), a planning agency set up in 1955.[70] Carroll was charged with organizing a traffic survey and developing a transportation plan for the Chicago metropolitan area, and would be reimbursed for his services with $1,500 monthly (around $14,000 in 2018).[71] This was a nice salary at the time, but Carroll was already a rising star in the field of transportation planning. A graduate of Dartmouth College in 1938, the young Carroll served in the U.S. Navy during World War II. As did many Americans of his age, after the global conflagration was over, he took advantage of the educational benefits provided by the federal government to veterans, and successfully applied under the G.I. Bill[72] for admission to the Department of Regional Planning of the Harvard

**Figure 2.4**
The gravity model in action in the mid-1960s. *Source:* Bureau of Public Roads/U.S. Department of Commerce, *Calibrating and Testing a Gravity Model for Any Size Urban Area* (Washington, DC: U.S. Government Printing Office, 1965), 4.

Where: $T_{ij}$ = trips produced in zone i and attracted to zone j

$P_i$ = trips produced by zone i

$A_j$ = trips attracted by zone j

$d_{ij}$ = spatial separation between zones i and j. This is generally expressed as total traveltime ($t_{ij}$) between zones i and j.

b = an empirically determined exponent which expresses the average areawide effect of spatial separation between zones on trip interchange.

C. Gravity Model Application

In applying a gravity model trip distribution formula to urban studies, it is necessary to develop the parameters in the gravity model formula for each urban area under study. Furthermore, these parameters are developed for each of several different categories of trips. These categories take into account the basic purpose for making trips and are generally referred to as trip purpose categories. Past experience has demonstrated that the exponent of traveltime is not constant for all intervals of time. Thus it is necessary to work with a gravity model formula which differs from that shown previously. This revised formula is expressed as follows:

$$T_{ij} = \frac{P_i \, A_j \, F_{ij} \, K_{ij}}{\sum_{j=1}^{n} A_j \, F_{ij} \, K_{ij}}$$

Where: $F_{ij}$ = empirically derived traveltime factor which expresses the average areawide effect of spatial separation on trip interchange between zones which are $t_{ij}$ apart. This factor approximates

$$\frac{1}{t^n}$$

where n would vary according to the value of t, and where t is the traveltime between zones.

$K_{ij}$ = a specific zone-to-zone adjustment factor to allow for the incorporation of the effect on travel patterns of defined social or economic linkages not otherwise accounted for in the gravity model formulation.

And where: $T_{ij}$, $P_i$, and $A_j$ are the same as previously described.

The use of a set of traveltime factors to express the effect of spatial separation on zonal trip interchange, rather than the traditional inverse exponential function of time, simplifies the computational requirements of the model. It also takes account of the fact that the effect of the spatial separation on trip making generally increases in a more complex manner than can be represented by the single exponent.

I-3

Graduate School of Design.[73] Carroll clearly appreciated the atmosphere of academic research at Harvard. Three years after he got his master's degree, he defended his Ph.D. dissertation in city and regional planning in 1950.[74] However, even before he had completed his academic work, he became engaged in a collaboration, between 1948 and 1953, with the University of Michigan as resident director of the Social Science Research Project. In this capacity he organized and participated in field research on urban and transportation issues likely to be of interest to the city of Flint.[75]

Detroit and Flint are only about sixty miles apart, and on August 2, 1953 the *Detroit Free Press* informed its readership that a traffic survey was about to start, under the direction of Carroll. From 1953 to 1955, he would head the Detroit Metropolitan Area Traffic Study (DMATS), the first of a series of ad hoc organizations created by the states with the strong encouragement of the Bureau of Public Roads in order to develop transportation plans for their major metropolitan areas.[76] As Carroll would recall several years later,[77] by deciding to shift from a university position to a more practice-oriented activity he found himself in uncharted waters. Thanks to the BPR's Home Interview manual, he and his collaborators in Detroit were able to conduct, from August through December 1953, the largest Home Interview O&D traffic survey up to that point with no significant difficulty.[78] In contrast, however, they had to demonstrate more imagination and creativity in putting the data gathered at the service of the overall project they were hired for, namely the design of a system of expressways capable of accommodating the year 1980 traffic. Fortunately, the team secured good ideas and assistance from several other actors present in highway planning at that time, including the California Division of Highways, the BPR, the personnel of the HRB, including Earl Campbell,[79] and Fratar himself, to name a few. Carroll and his collaborators obviously proved able to put the help they received to good use, since they eventually contrived the Detroit growth factor variant as well as a series of original assignment diversion curves.

These were incremental improvements in the common pool of the then available tools. However, Carroll and his DMATS team partners would also initiate some new ways of thinking about forecasting urban traffic, which would fully blossom shortly after they left Detroit for Chicago. It was in Detroit that Carroll seemed to recognize the need to conceive the various transportation facilities in a metropolitan area as many interacting and interdependent parts of a wider *system*, and to plan them accordingly.[80]

Another original idea the DMTAS can widely be credited with was the "land use-transportation method" in forecasting future traffic.[81] As has already been seen, in the early 1950s transportation planners wanting to estimate future travel were essentially basing this on existing traffic patterns and projecting them into the future; think of the growth factor method, or the even older approach developed by BPR in the interwar period, which consisted of relating traffic to trends in growth of other items, such as gasoline consumption, national income, or gross national product.[82] By marked contrast with this approach, Carroll and his collaborators, along with some other traffic planners of the day, thought of directly relating urban traffic to land use[83] as they had become increasingly cognizant of the fact that the type as well as the number of trips originating in and destined to each part of the metropolitan area depended heavily on the amount and kind of activity, in other words, the *land use*, located there. And, if these relationships proved regular and stable, they continued, then the quantity of various kinds of land use could be a good measure of both current and future urban travel.[84] A study sponsored by the Bureau of Public Roads and conducted in the early 1950s by Robert Buchanan Mitchell (1906–1993) and Chester Rapkin (1918–2001),[85] two academics who appear to have met with Carroll in Detroit,[86] provided additional theoretical justification for the land use-transportation method and proved instrumental in its favorable reception among transportation planning professionals.

When Carroll arrived in Chicago in 1955, he was therefore already a highly versatile professional, equally at ease in the realm of theory and the world of practice, capable of rubbing shoulders with planners and engineers, economists, computer programmers, geographers, and sociologists (more on this to follow). In common with many members of the so-called G.I. Generation (commonly known as the "Greatest Generation") born in the first quarter of the twentieth century and who successfully fought World War II, Carroll, like Mertz, Voorhees, and several other characters in this book cast in the same mold, would display "an exemplary willingness to tackle difficult and costly tasks," and "a relentless optimism," with the two combined to produce "a great capacity for teamwork."[87]

Carroll did not leave Detroit alone. Among the people who had worked under his direction at DMATS and who accompanied him to Chicago were Earl Wilson Campbell (1925–2007), a civil engineer and a 1951 graduate of the Yale Bureau of Highway Traffic; Howard W. Bevis, an economist from the University of Michigan; John Robert Hamburg (1928–2000), holder of a master's degree in urban sociology; R. E. Vanderford, a computer programmer; and Garred P. Jones, a geographer by training with experience as a cartographic editor for the U.S. Corps of Engineers.[88] Since Chicago was much bigger than Detroit, Carroll felt the need to increase the CATS crew with several new recruits, starting with his two assistant directors: Roger Lamont Creighton (1923–2017), holder of a 1950 master's degree in city planning from the Harvard Graduate School of Design; and Peter J. Caswell, an electrical engineer from the University of Detroit and former IBM employee who had even sold to the DMATS team their accounting machines. Dayton Jorgensen, appointed head of the projected O&D survey, brought to CATS the skills and the know-how of an expert who had previously worked for the Bureau of the Census. Three other people would play a decisive role in the Chicago study: Louis E. Keefer, a graduate of the Yale Bureau of Highway Traffic in 1955; Irving Hoch, born in 1926 and with a Ph.D. in economics from the University of Chicago in 1957;[89] and the highly gifted Morton Schneider (1928-1993[90]), a graduate of the University of Chicago with a brief research experience in physics as well as experience as a taxi driver, who was hired by Carroll as a systems analyst in 1956.[91]

Judged by the standards of its time, CATS was a rather unusual organization in the field of transportation planning. Uncommonly large, it was composed of many "knowledge workers"[92] from a wide variety of disciplinary backgrounds, and who had to organize and oversee the work of some 360 foot soldiers,[93] charged with tasks as various as interviewing, coding, and processing several million pieces of information, or producing voluminous reports in the shape of luxurious coffee table books.[94] Yet, its originality did not prevent CATS from distinguishing itself as a site of intensive original modeling production,[95] which, all in all, surpassed its fellows no less than its predecessors. A wealth of materials help depict the Chicago Area Transportation Study as a formal modern bureaucratic organization endowed with a detailed organization chart, having several hierarchical levels,[96] functioning on the basis of a meticulous division of labor and responsibility, fraught with mechanisms for controlling the quality of the work being done,[97] and producing huge swathes of paperwork in the shape of internal reports or more publicized documents (figure 2.5).

Carroll seemed also to think, along with the managers of some of the largest industrial firms of his time,[98] that even the R&D activities—and the

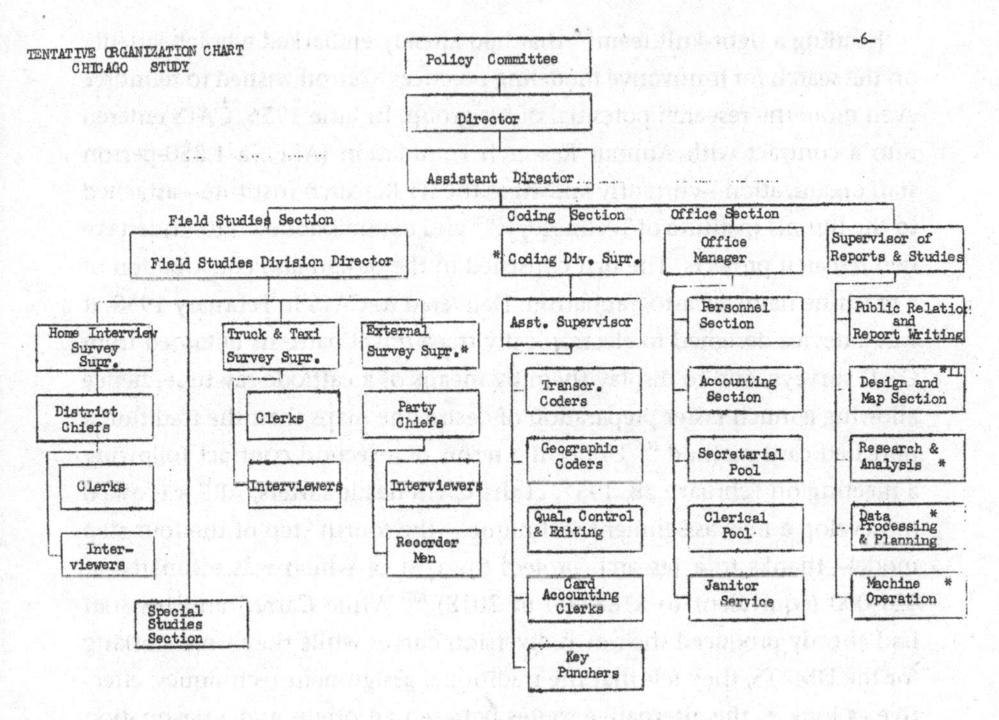


**Figure 2.5**
Tentative Organization Chart, Chicago Study (1955). *Source:* J. D. Carroll Jr., "Outline of Proposed Chicago Traffic Study," typescript, May 17, 1955, 6. Cornell University Library. Division of Rare and Manuscript Collections, J. Douglas Carroll, Jr. Papers: 4178, Box 1, Folder 10.

CATS staff would do a lot of research—could be, with some adjustments, scientifically planned, organized, and performed.[99] Be that as it may, as several of the Chicago Study's staffers recollected some years later, CATS never turned into a rigid bureaucratic structure. Carroll proved a charismatic and inspirational leader, able to articulate and communicate to both his collaborators and subordinates his vision of the future, cultivate a close relationship with his troops, galvanize them, and leave (at least some of) them plenty of room for maneuver.[100] The memories of some of Carroll's former collaborators equally convey the impression that CATS was "a very good place to work,"[101] where intense exchanges and interactions, often informal and not always work-related,[102] were able to create strong bonds based on mutual trust, the role of which cannot be overstated in knowledge making and the production of new scientific and engineering practices.[103]

Heading a tight-knit team[104] that had already embarked wholeheartedly on the search for innovative modeling practices, Carroll wished to reinforce even more the research potential of his group. In June 1956, CATS entered into a contract with Armour Research Foundation (ARF), a 1,250-person staff organization—currently known as the IIT Research Institute—attached to the Illinois Institute of Technology,[105] and commissioned it to undertake two research projects. The first consisted in the design and construction of a machine named Cartographatron. Delivered to CATS in February 1958, it was a device designed to electronically trace travel patterns obtained from O&D surveys, and to display them by means of a cathode ray tube, hence allowing a much faster preparation of desire line maps than the traditional punched card method.[106] Under the terms of a second contract following a meeting on February 28, 1957, at the CATS headquarters, ARF was asked to develop a new assignment technique— the fourth step of the four-step model—thanks to a research project the cost of which was estimated at $21,000 (equivalent to $188,000 in 2018).[107] While Carroll and his staff had already produced their own diversion curves while they were working for the DMATS, they felt that the traditional assignment techniques, effective as long as the alternative routes between an origin and a destination are very limited, were likely to perform poorly when applied to a metropolitan area as large as Greater Chicago, where for each origin-destination pair there was a great deal of possible routes. At the beginning of 1957, personnel from ARF's Operations Research Group of the Mathematical Services Section along with colleagues from the Computer and Data Processing Group set to work on assigning traffic to *an entire network* (and not just to a single new transportation facility).[108] Very shortly, they would come across two recent papers addressing the issue of devising an *algorithm* capable of finding the *shortest path*, either in terms of distance or of time, between any two nodes of a network.[109] Edward Forrest Moore (1925–2003), one of the founders of modern automata theory, had worked between 1951 and 1956 at the Bell Telephone Laboratories on telephone circuits. Following his stint there, he gave in April 1957 a lecture entitled "The Shortest Path through a Maze" at the International Symposium on the Theory of Switching, April 2–5, 1957.[110] At the same time, George B. Dantzig (1914–2005), the intellectual father of modern linear programming in mathematics and at the time a researcher at RAND Corporation,[111] published an article in which he elaborated on, among other issues, the "problem of the shortest route in a network."[112] In June 1957, the ARF group delivered to CATS their "minimum path algorithm," embodying the behavioral hypothesis that people generally, if not always, choose the route that minimizes the time needed to reach their destination. Programmed for an IBM 650 machine, and applied to a fictitious network of eighteen nodes, ARF's algorithm was mostly based on Moore's lecture even though the mathematical notations used were borrowed from Dantzig's article.[113]

Carroll and his staff were immediately seduced by ARF's mathematical product. However, as the latter was operationally unusable for a real transportation network, Morton Schneider embarked on developing an in-house "ingenious modification of this minimum path program to bring the traffic assignment within the range of computational feasibility" of the day. His intellectual labor resulted in a computer program able to assign trips to the entire Chicago metropolitan area transportation network after an eleven-hour run on an IBM 704 machine located hundreds of miles away from the CATS premises, in the General Electric Offices in Cincinnati, Ohio.[114] Schneider did not rest on his laurels, and very soon the former taxi driver would make another substantial improvement to the method initially devised by ARF, through succeeding in taking into account the fact that any route has a limited capacity, which means that travel duration, and therefore the shortest path, depends each time on the state of *congestion* in the network (but more on that in chapter 4).[115]

And the CATS "innovation machine" would keep on working. After dealing with the assignment step, the Chicago team would then come to grips with the issue of zone-to-zone movements (the second step of the FSM). Once again it was Schneider who sought to wander off the beaten track. The CATS systems analyst was not satisfied with the available traffic distribution models of the day. Thus, for him, the growth factor methods were just "more or less arbitrary algorithms of small theoretical interest." Though coming from a natural sciences background, he also held the gravity model in contempt, since there is "no real kinship between a gravitational field and a trip generating system. Newtonian gravity is an energy-force field characterizing the motions of particles, not their *intentions*."[116] While his CATS mates Bevis and even Carroll himself were both attracted to the gravitational principle,[117] Schneider would eventually devise a new technique for modeling zone-to zone trip movements that would truly take the traveler's intentions into account. Though he did not seem to refer explicitly to Samuel Andrew Stouffer (1900–1960), Schneider's own approach shares much with the conceptual framework the American sociologist proposed in the late 1930s in his attempt to address the problem of distance in human mobility. According to Stouffer's theory, predicated on the concept of intervening opportunities, "the number of persons going a given distance is directly proportional to the number of opportunities at that distance and inversely proportional to the number of intervening opportunities."[118] Labeled shortly after it was first used in Chicago in 1958 as the "trip opportunity model," subsequently mostly known as the "intervening opportunities model (or method)," the approach put forward by Schneider operationalized Stouffer's ideas on the basis of two assumptions: That "the probability of a trip finding a terminal in any element of a region is proportional to the number of terminal opportunities contained in the element; and that a trip prefers to be as short as possible, lengthening only as it fails to find a terminal" fulfilling the purpose for which it was undertaken.[119] Claiming that his own modeling technique was much more behaviorally sound than the gravity model—a statement Voorhees vividly contested in a panel discussion organized by the Bureau of Public Roads at the 1960 Highway Research Board annual meeting—Schneider also found that his formula turned out to be both "computationally convenient" and "most well-behaved mathematically" while faring better than its rival in terms of empirical accuracy, at least as far as zone-to-zone traffic data from the Chicago area was concerned.[120]

Yet, the most sustainable contribution of the Chicago study in the then still-fledgling field of urban traffic forecasting was probably the invention, and the first implementation, of the now classic four-step model (see the introduction to this book), which would prove strong enough to dominate urban travel demand modeling for decades. The discerning reader has already observed that two of the four stages within the general FSM framework, the trip distribution and traffic assignment steps, have already been extensively analyzed in the previous pages. Let us turn now to the other two steps of the FSM. The trip generation step that connects land use and travel has also been mentioned. Remember that in the early 1950s, the idea of linking trips originating in or destined to an urban zone with the characteristics of such a zone started to gain ground among transportation planners. Building on the work already done in Detroit, also drawing inspiration from an increasing number of studies addressing the issue of generation that saw the light of day later in the decade,[121] Carroll and his team at CATS would propose an original trip generation model based on the assumption that the number of trips taken by the residents of a zone was a function of auto ownership and population density. According to the model, the typical resident was expected to undertake 2.3 trips per weekday in 1980, compared to 2.0 trips in 1956.[122] These trips were then allocated to six land use types, and distributed to the various zones on the basis of a land use forecast.[123] Having estimated the amount of travel originating in and destined to each zone, Carroll and his collaborators proceeded to project the numbers of person trips to be taken by mass transit and by private vehicle. In introducing a *mode choice* module to their overall modeling framework, they delivered to their fellow traffic forecasters the first full implementation of the FSM, shortly after having laid out an explicit three-step model consisting of the stages of trip generation, trip distribution, and traffic assignment (figure 2.6).[124]

As in the case of the trip generation step, automobile ownership, which was expected to be on the rise, was an important parameter of the CATS mode choice module, while another factor put forward by Chicago study staffers was distance to the central business district.[125] It is worthy of note that the CATS mode choice module was designed to accept the outputs of the trip generation model while in what would become the standard version of the four-step model, the mode choice step usually comes after the trip distribution step (figure 2.7).[126]

## The Practice-Oriented Segment of the UTDM Social World

When the third and last volume of the monumental Chicago Area Transportation Study *Final Report* was eventually published in April 1962,[127] CATS was on the verge of substantial organizational changes since many of the Carroll team had already or were about to part ways and to fan out into the newly created urban travel demand modeling "market." Keefer had moved to Pennsylvania in 1959 to direct another major study of the day, the Pittsburgh Area Transportation Study (PATS), while, in the very early 1960s, both Creighton and Hamburg headed for the New York area. Here, Creighton was appointed director of the Niagara Frontier Transportation Study with Hamburg in the position of deputy director. Carroll himself left Chicago for the New York region in 1963 in order to hold top positions on

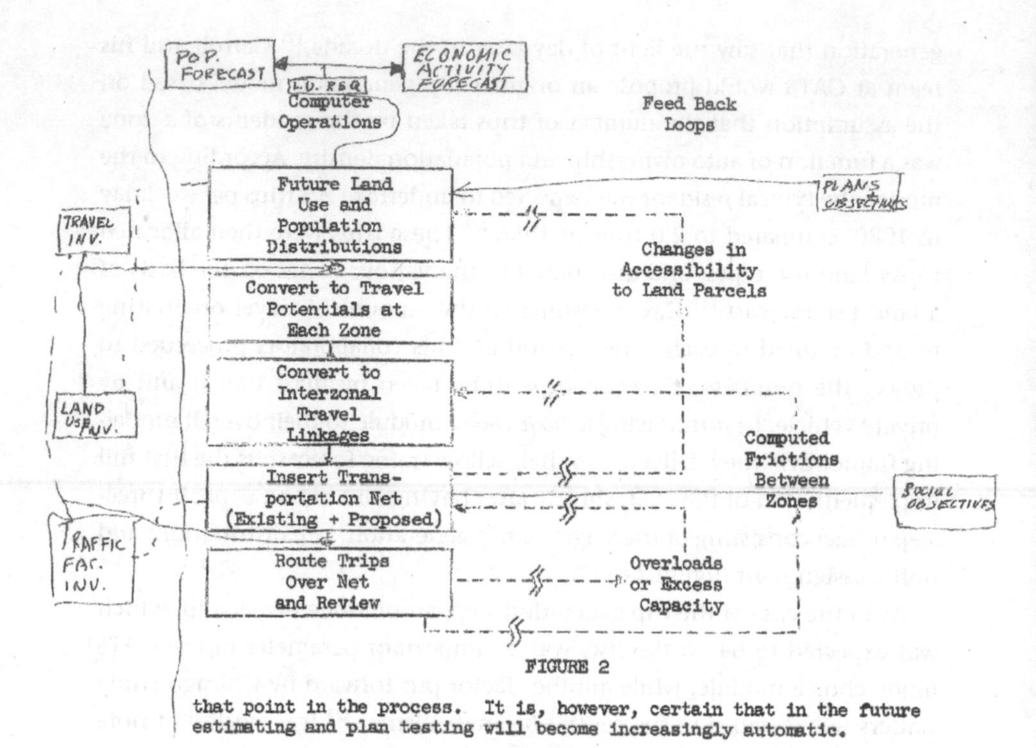


FIGURE 2

that point in the process. It is, however, certain that in the future estimating and plan testing will become increasingly automatic.

If any such projection is to be at all reliable its design and that of the traffic simulation must follow a most critical study of a city's functioning. Of itself, this preliminary study gives new insights uncovering lines of further exploration.

MEASURING PEOPLE'S TRAVEL DESIRES

People travel to satisfy their needs and desires. They travel to work, to shop, to movies, and for many other reasons. Against this basic need to travel, two factors operate to modify the individual's vehicle trip making potential:

1. Income which, to a point, measures the ability to pay for travel. As income and the corrolary of car ownership rise, more trips will be made.

2. Home address, if it be in a crowded area fewer vehicle trips will be made. If it be in a sparsely developed area where spatial imbalance is greater, more vehicle trips will be made.

Page 5

**Figure 2.6**

The CATS "Three Step Model" (trip generation, trip distribution, and traffic assignment). *Source:* J. Douglas Carroll Jr. and Howard W. Bevis, "Predicting Local Travel in Urban Regions," typescript, December 17, 1956, 5. Cornell University Library. Division of Rare and Manuscript Collections, J. Douglas Carroll, Jr. Papers: 4178, Box 1, Folder 16.

TRAFFIC RESEARCH CORPORATION
Neal A. Irwin reporting:

PAGE 9

1. MULTI-MODE TRAVEL FORECASTING MODEL

This model is available for estimating future travel flows of people, goods and vehicles via private and public transportation modes. Based on projected land use patterns and proposed transportation facilities, the model forecasts travel flows and travel speeds on all sections of the transportation network, for peak hour or 24 hour periods.

The model was originally programmed and tested on the IBM 650. In 1960 a much expanded version was programmed for the 7070, and this model has been applied to test runs and production runs in a number of urban regions with good results. As shown in Figure 1, it comprises a series of sub-programs which carry out respectively the calculations of zonal trip production and attraction rates, interzonal travel volumes, interzonal travel mode usage, usage of alternate routes, flows of passengers and vehicles on all road and rail links, and resulting travel speeds and times on all road sections.

Unique features are its ability to estimate mass transit and truck traffic as well as that by private automobile. Capacity restraints and flow properties of all roads are included so that the program is able to take into account the influence of traffic congestion on travellers' choice of destination, route, and travel mode.

(a) PROGRAM FEATURES

Travel Modes

It is possible with this program to deal with any one of the three basic travel modes: automobile, mass transit, trucks, or with any combination of the three. If only one mode is dealt with, calculating times are considerably reduced; however, the results obtained are approximate in that general assumptions are necessary concerning usage of the other modes and their inter-relationships with the mode being studied.

Capacity Restraints

If an unrestricted assignment is desired, capacity restraints can be eliminated. This also decreases machine calculating times considerably. However, the traffic patterns so obtained can in no way be considered "real"

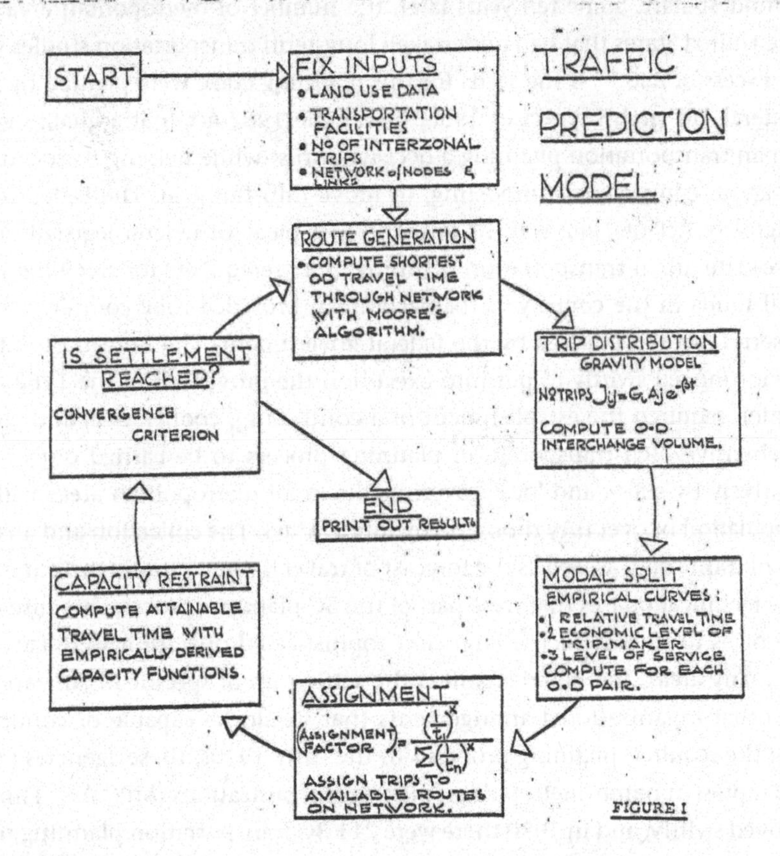


**Figure 2.7**

An early example of the four-step model framework with *feedback*, proposed by Traffic Research Corporation. The modeling suite was originally programmed and tested on an IBM 650 machine, while in 1960 a much-expanded version was programmed for the IBM 7070. *Source: Newsletter of the Transportation Planning Computer Program Exchange Group* 1, no. 1 (July 1963): 9.

the Tri-State Transportation Committee,[128] from which he would receive an annual wage of $30,000 in 1964, and where another CATS staffer, Schneider, would work as a research associate in the mid-1960s.[129]

This rather intense geographical and professional mobility, which helped make CATS a template for many other transportation agencies across the nation, can be easily accounted for once placed within the broader historical context of the time. By November 1962, of the sixteen U.S. metropolitan areas with over 1,300,000 inhabitants, twelve had begun and some already had completed a major transportation study.[130] And that was just a modest start. Some ten years later, the number of metropolitan areas in the United States that had undertaken long-term transportation studies was in excess of 250.[131] If the seeds for this planning craze were planted by the Federal-Aid Highway Act of 1956, it was the 1962 act that arguably gave urban transportation planning a decisive thrust while helping its scientific weapon, computerized modeling, to move into full gear. The Federal-Aid Highway Act of 1962 was, in fact, the first piece of federal legislation to mandate urban transportation planning as a prerequisite for receiving federal funds in the country's urbanized areas, provided they complied with a series of rules enacted by the federal government. The Bureau of Public Roads moved swiftly to put into execution the provisions of the 1962 act, which required the establishment of a continuing, cooperative, and comprehensive (3C) transportation planning process to be carried out cooperatively by states and local governments in all metropolitan areas with a population of over fifty thousand by July 1, 1965. The collection and analysis of traffic data as well as the forecast of travel figured prominently among the technical phases that were part of the 3C planning process. Because the agencies to mount such a large and sophisticated operation were lacking in many areas, the BPR demanded the setting up of specific organizations or other organizational arrangements that would be capable of conducting the required planning process—in the early 1970s, these agencies took their present name, metropolitan planning organizations (MPOs).[132] Things moved swiftly, and in 1970 there were 211 3C transportation planning programs throughout the nation.[133]

Though some of the established transportation agencies, following in the footsteps of the Detroit and Chicago studies, did acquire the necessary resources for meeting the 3C process demands, talent to implement the process was lacking in many places across the country. State highway departments and local governments therefore turned to consultant firms,[134] which started practicing urban traffic forecasting on behalf of the government, mostly, if not exclusively, within the FSM framework first developed and experienced in Chicago. Some of the consultants solicited were well-established transportation engineering companies dating to the interwar period. One was Parsons, Brinckerhoff, Quade and Douglas (PBQD), founded as early as 1885, which in the 1960s employed a talented modeler named Henry Dean Quinby (1925–1978).[135] De Leuw, Cather & Company also falls into this category. At the turn of the 1960s,[136] the company employed some 1,500 persons throughout the country and overseas, while a few years later, in 1977, it became part of Parsons, the gigantic planning, engineering and construction firm established in 1944.[137]

However, the majority of firms that were to distinguish themselves in the field of urban travel demand modeling in the 1960s were founded after the end of World War II. Barton-Aschman Associates was set up in 1958 as a Chicago area-based business by George W. Barton (1906–1987), a traffic engineer with a civil engineering degree, and Frederick T. Aschman, a lecturer in city and regional planning at Northwestern University.[138] In the mid-1960s the staff of the consultancy included some thirty-eight full-time professional engineers and planners, supported by technicians, draftsmen, and other clerical personnel.[139] According to Fred Schweiger, who holds degrees from the U.S. Military Academy and Northwestern University, and who entered the firm in June 1958 as a drafter and field survey crew member and left it in July 1994 as the consultancy's president, in 1973 there were about a hundred and fifty people working for the firm coming from a dozen disciplinary backgrounds. Following in the footsteps of De Leuw, Cather & Company, Barton-Aschman Associates was also acquired by the same Parsons corporation in 1985.[140] Creighton, the former Chicago Study staffer, founded its own consulting firm, Roger Creighton Associates, in the mid-1960s, while his CATS-mate Hamburg became the firm's partner for several years before establishing John Hamburg & Associates in the mid-1970s.[141] A few years later, Hamburg joined Barton-Aschman Associates, where he became vice president and principal associate from the early 1980s to the mid-1990s. It is worth noting that both Creighton's and Hamburg's firms could rely on the modeling competencies and skills of another CATS hero, Morton Schneider.

Wilbur Smith and Associates (WSA) came into corporate being in 1952.[142] While working for the South Carolina Department of Highways,

Wilbur Stevenson Smith (1911–1990) enrolled in the Bureau for Street Traffic Research at Harvard from 1936–1937. In 1943, he was even appointed the associate director of the bureau, then at Yale University, while accepting a few small engineering consulting contracts. Four years after WSA saw the light of day, the active members of the professional crew already numbered twenty-five or so people, including some theory-inclined practitioners, such as F. Houston (Hugh) Wynn (1916–2018), congratulated in 1955 by Carroll for having produced "one of the most-meaningful papers yet presented on basic urban-travel patterns and their predictability."[143] Holder of a B.S. degree from the University of Oregon in 1939, Wynn joined the Yale Bureau of Highway Traffic at the very end of the 1940s, and WSA in 1952. He went down in history as the author of the "interactance formula," an early interzonal distribution traffic modeling approach combining the gravity model principle with origin and destination survey information.[144] In 1956, Wynn became the project manager for one of the most significant transportation studies of the time, which took place in the Washington, DC, metropolitan area. Like other transportation consulting firms of the day, WSA largely benefited from the Federal-Aid Highway Act of 1962 and entered into dozens of municipal contracts across the nation. The consultancy started in the 1970s with twenty-one offices in the United States and nine in foreign counties, while the number of its employees soared to 635 in 1980. In fall 1961, the firm stepped front and center onto the worldwide transportation engineering stage by starting a traffic study for Greater London in association with the British company Freeman, Fox and Partners, a collaboration that was converted into a joint company, called Freeman, Fox, Wilbur Smith and Associates, in 1965.[145] The London study proved a crucible for several innovations in urban traffic forecasting,[146] which WSA practitioners would import into the United States very quickly, often in the form of computer programs.[147] While for the Washington study the firm had to enlist the services of Ernest E. Blanche & Associates, the statistical analysis and data processing firm founded by mathematician and Ph.D. holder Ernest E. Blanche (1912–1993) in 1955,[148] very soon WSA modelers could familiarize themselves with computers by running planning and transportation programs, first on various mainframes owned by other organizations[149] and, from January 1966, on machines installed in the company's premises, including a brand-new IBM 360/30 acquired in May 1967.

In 1940, a young man reached Canadian shores, only to be sent immediately to an internment camp. Two years later, Josef (sometimes Joe) Kates (1921–2018) was recognized by the Canadian government as a victim of the Nazis and was able to start a normal life on New World soil. In 1949, he obtained his master's degree in applied mathematics and his Ph.D. in physics followed in 1951.[150] One of the founding fathers of the operations research community in his adopted country, Kates set up KCS Data Control Limited in 1954, which began in Toronto by designing the world's first computer-controlled traffic signaling system in the very early 1960s.[151] Traffic Research Corporation (TRC) was born in 1960 with the objective to develop and apply the advanced methods for transportation planning and traffic control pioneered by KCS.[152] It immediately sought outlets for its innovative services beyond the small domestic market, and opened offices in Boston, New York, San Francisco, and Washington, DC. Like the Chicago Area Transportation Study, the TRC team, numbering around one hundred fifty employees in the mid-1960s, was largely multidisciplinary, featuring among its ninety or so professionals many engineers, mathematicians, economists, physicists, and statisticians, including Neal A. Irwin, holder of a bachelor of applied science degree who headed the New York branch from 1960 to 1965; Hans G. Von Cube (1919–2008), a German-born economist who came to Canada in 1952 and obtained a master's degree in economics in 1955; Donald M. Hill, a master's degree holder in statistics; and Daniel Brand, holder of a bachelor's degree and master's degree in civil engineering from MIT, who opened the Boston office in 1965. Well-endowed with computer resources,[153] TRC staffers would make significant contributions to urban travel demand modeling in the 1960s. They especially enriched the fourth (traffic assignment),[154] and the third (mode choice, known also as modal split) steps of the FSM with original, and often path-breaking forecasting procedures. In the first half of the 1960s, while working for Toronto and Washington, DC, the TRC team produced probably the most comprehensive modeling approach for the mode choice stage until the early 1970s.[155] Taking account of five basic parameters,[156] TRC's "modal split curves," particularly reminiscent of the older assignment diversion curves regarding the highway system,[157] represented the first major attempt to capture the decision process underlying mode choice, and to integrate this process into the whole FSM framework.[158] Though TRC proved a rather short-lived consultancy, its impact on

American urban travel demand modeling would be a lasting one, especially through the action and heritage of the many gifted modelers of the firm. In 1966–1967, the transportation staff of its U.S. offices integrated into Peat, Marwick, Livingston & Co. (PML), a company whose origins date back to 1960, when J. Sterling Livingston (1916–2010), then a professor at the Harvard Business School, founded Management Systems Corporation, which affiliated with the well-known international accounting corporation Peat, Marwick, Mitchell & Co. (PMM) five years later, in 1965.[159] TRC's heritage would also survive through James H. Kell (1930–1991), the firm's principal traffic engineer from 1964, and later a PML staffer. In 1971, Kell, who received a master's degree from Purdue University in 1952,[160] and worked as research engineer and lecturer at the University of California at Berkeley from 1954 to 1964, set up his own consultancy under the name JHK & Associates.[161] By the time of its founder's death, JHK & Associates had grown to a large firm employing 270 people and with offices in nineteen cities.

The author of a groundbreaking paper on the use of the gravitational principle in traffic forecasting, Voorhees was not a prolific author, and, according to a close collaborator, he was even almost incapable of writing.[162] But, the intellectual father of the gravity model had many strengths by which he could offset his problems with written expression, among them, a high intellect and the capacity to pull the right people together.[163] Indeed, much of the success of Alan M. Voorhees & Associates—the transportation consultancy that saw the light of day in the basement of Voorhees's house in 1961,[164] and which quickly became one of the largest firms of its kind in the country—lay in the quality of people with whom Voorhees surrounded himself. Robert Dial, the main character of this book's introduction, was one of them. Walter G. Hansen (1931–2008), Walt to his colleagues and friends and the consultancy's cofounder, proved another brilliant collaborator. A graduate of the South Dakota School of Mines and Technology, Hanson joined the BPR in the early 1950s. A few years later, his federal employer sent him to MIT, where Hansen received a master's degree in city planning in 1959.[165] His thesis made the future collaborator of Voorhees into a small celebrity within the then (small) UTDM social world, since for his research work the author developed the concept of *accessibility* (to an activity). Defined as directly proportional to the size of the activity in question, and inversely proportional to the distance to the latter's location, the notion of accessibility has been widely used since then to evaluate the services offered by a transportation system. Another early AMV modeler was Gordon Wilford Schultz (1937–2006), who received a master's degree from Northwestern University in 1962, a practitioner about whom a fellow traffic forecaster, Bill Davidson, said much later that "if you could think it, Gordon could model it."[166] Other AMV staffers in the 1960s and early 1970s were Bevis, the former member of the Carroll team in Chicago, Richard H. Pratt, who set up his own consultancy in 1969 before being absorbed by Barton-Aschman Associates in 1977, Gordon A. Shunk (who passed away in 2008), a Ph.D. recipient from Purdue University in 1967, and Thomas Blackburn Deen (1928–).[167] Deen was a graduate of the University of Kentucky and, like many modelers of his generation, of the Yale Bureau of Highway Traffic,[168] who served as director of planning for the National Capital Transportation Agency in Washington from 1960 and 1964. In that capacity, he closely collaborated with people from the BPR, including William L. Mertz, as well as from Traffic Research Corporation. When he joined the Voorhees office in 1964, he brought to AMV his skills and experience in mass transit modeling, thus complementing the expertise of the firm's founder in highway traffic forecasting techniques. Like TRC, in 1967 AMV was acquired by a bigger firm, Planning Research Corporation (PRC).[169] AMV remained however an independent subsidiary, with some 130 employees in 1977,[170] before merging into PRC Engineering in the late 1970s. PRC was acquired in turn by Ashland Technology Corporation in the mid-1980s, but a few years later, in 1990, the firm's approximately three thousand employees decided to form AECOM Technology Corporation,[171] the large firm today known as AECOM (for architecture, engineering, construction, operations, and management), which has been very present in urban traffic forecasting in the 2000s and the 2010s (see part III).

### The Academic Segment of the UTDM Social World

In August 1962 a Ph.D. dissertation titled "A Study of Highway Traffic Assignment" was submitted to the Graduate School of the Agricultural and Mechanical College of Texas by a certain Donald Edward Cleveland. Though most of the about 110 items listed in the bibliography had been published in professional engineering periodicals, one could also come across a number of master's theses and articles that had appeared in scientific journals.[172] The first issue of *Transportation Science* was released in February 1967. Just a few months later, in May, *Transportation Research* also saw the light of day.[173] Though older periodical publications, including those published by the Institute of Traffic Engineers and, especially, by the Highway Research Board—renamed Transportation Research Board in 1974[174]—would remain very popular within the UTDM social world,[175] the two newly-founded journals, the creation of which took place in the midst of a global publishing boom for transportation-related academic journals,[176] quickly became home to urban traffic forecasting literature. Just a couple of years after their first issues, both *Transportation Science* and *Transportation Research* published a review of *Traffic System Analysis for Engineers and Planners*,[177] probably the first academic transportation treatise in which the field of urban travel demand modeling was allotted significant space.[178]

These publications as well as data on the number of master's and Ph.D. theses granted by American (and British) universities from 1962 to 1981[179] clearly show that by the end of the 1960s urban traffic forecasting had not only already grown into a full-fledged professional practice, but also was about to make itself into a dynamic and vibrant academic reality. It is not by chance that the expression "travel demand," which is now widely used to refer to this modeling field, dates back to the 1960s.[180] With the rise to prominence of this expression, older terms started taking on new meanings. Thus "traffic engineering," which originated in the interwar period to denote the entire science of traffic (chapter 1), increasingly started to refer to *short-term* traffic improvement techniques and operations only, while "travel demand modeling" and other similar terms, including "travel demand forecasting," began to be associated with comprehensive urban transportation planning involving major increases in the capacity of all modes (including transit).

No visible hand intentionally designed and carefully planned academic research into urban travel demand modeling. Individual and more or less separate initiatives within academia[181] progressively created research sites for this modeling field within existing structures. Some of these individual actions coalesced into more permanent structures, embarking, in turn, on an autonomous institutional lifespan. Since the transportation of people and goods as a practical issue historically has been associated with the professional figure of the civil engineer, it is not by chance that several civil engineering departments across the nation became home to urban traffic forecasting in the 1960s, as they had done for the (more general) science of traffic engineering during the interwar period[182] and the first post–World War II decade.[183] In order to train travel demand modelers, and generate new practices in urban traffic forecasting, these departments had to import, or cultivate more intensely than they had in the past, bodies of knowledge and techniques originating in two broad academic disciplines that were not specific to civil engineering: (1) economics and (2) operations research (OR). Again, this should come as no surprise. On closer examination, travel can be considered a sort of *economic* problem: the two basic ingredients of modern economic theory—*supply*, represented here by the existing or the projected transportation facilities and services, and *demand* for travel, as well as their interaction through the price/cost mechanism—are indeed present in the decision-making of the traveler. As Marvin Manheim, an influential academic in the field of transportation studies (to be discussed further), would put it, with the help of an explicitly economic lexicon in 1968, the "general framework for the prediction of network flows is that of the *equilibrium* between *supply* and *demand*."[184] In the early 1950s, transportation practitioners were already well aware of the fact that when a new expressway was built, therefore lowering the cost of a travel trip in terms of time, there was indeed an increase in the number of trips.[185] The history of economic thought also informs us that professional economists had been interested in issues directly related to urban traffic forecasting as early as the 1920s (see chapter 4). Last but not least, the very structure implicit in the four-step modeling approach is fundamentally that of the supply/demand equilibrium framework, with the "demand functions" performed by the sequence of steps one, two, and three (trip generation, trip distribution, mode choice) and step four, traffic assignment, representing the attempt to find equilibrium in the network given the travel demands and the supply characteristics of the transportation system under study, especially the traffic capacities of the links forming the network.

It was in the early 1960s, amid the highway construction boom, that transportation in general and urban traffic forecasting in particular became increasingly fashionable subject matters for professional economists in the United States. In the summer of 1960 an ambitious research program, partly financed by the famous Ford Foundation,[186] began within RAND,[187] while its most famous end product, the book *The Urban Transportation Problem*, appeared in 1965, at the time when rapid suburban growth, central business district decline, and increasing highway congestion were about to become issues of national concern.[188] The book's cover listed John R. Meyer (1927–2009) as one of the authors, credited today with creating the field of transportation economics in the United States. Meyer was a professor in the Department of Economics (1955–1968) at Harvard and director of the Harvard Transport Research Program, in the 1960s.[189] While he and his collaborators were busy with addressing the urban transportation problem at RAND, another group of economists located on the East Coast was about to make deep inroads into traffic forecasting territories. Created in 1958 and heavily populated by academics from Princeton University, MATHEMATICA was a research-based firm that could boast among its employees a series of leading figures in both OR and economics.[190] A major federal project in the early 1960s, the Northeast Corridor Transportation Project, offered MATHEMATICA's economists the opportunity to become involved in the field of travel demand modeling.[191] MATHEMATICA, along with Traffic Research Corporation and a series of other consultancies and companies,[192] was, in fact, one of the organizations called upon to devise traffic forecasting models.[193] The main outcome of the efforts deployed by MATHEMATICA group, led by two Princeton economists, William J. Baumol (1922–2017) and Richard E. Quandt (born in 1930),[194] was the so-called abstract mode model, an approach integrating within a single framework the first three stages of the four-step model, and in which the different modes of transportation are represented with the help of general attributes, such as cost, speed, frequency, and convenience, without any reference to their specific physical properties.[195] In addition to RAND and MATHEMATICA, a number of other research centers, including the Stanford Research Institute in Menlo Park, California, set up in 1946,[196] and various university economics departments from Chicago to Berkeley, often solicited by actors well versed in operational urban travel demand forecasting, including the Bureau of Public Roads and several metropolitan planning agencies, also mobilized their economists on issues related to traffic forecasting during the 1960s (more in chapters 3 and 4).

Along with economics, the other major body of disciplinary knowledge through which urban travel demand modeling took root in academia was operations research, a field whose members included several professional economists.[197] Just as economists regard travel through their specific conceptual lens, meaning, as (individual) demand for travel that encounters a supply of transport services, so too OR specialists consider trips as a specific case of "Flows in Networks."[198] An offspring of World War II, OR rapidly expanded its potential application scope beyond the military.[199] Like many other civilian actors of the day, planners and transportation engineers were rapidly seduced by a discipline that promised to provide solutions to complex problems with the help of sophisticated science-based quantitative methods.[200] At the same time several OR specialists felt that traffic-related issues could be a good client for their services. In 1962, ten years after its foundation, the Operations Research Society of America created a Transportation Science Section, the members of which immediately used the society's journal to publish their work, and to promote travel modeling within the OR community.[201] From 1959 onward, OR specialists interested in transportation issues could also gather together on a regular basis at the International Symposium on Transportation & Traffic Theory. As has already been seen, operations research practitioners initially familiarized themselves with traffic forecasting-related issues through the notions of network and optimization—remember the pioneering work done by the Armour Research Foundation for CATS on the basis of the shortest-path algorithms devised by OR scholars Moore and Dantzig. But the issue of the shortest path[202] would not be the only point of intersection between operations research and urban travel demand modeling. OR specialists would also put their mathematical and computational skills at the service of a series of other operations, such as calibrating a model,[203] testing its validity,[204] or handling model-related data.[205]

Drawing upon economics and OR, urban traffic forecasting embarked on, and even made serious headway in, the process of its *academicization* in the 1960s. While by the mid-1950s the Yale Bureau of Highway Traffic was the sole knowledge-centered institution that trained would-be urban traffic forecasters in significant numbers, and which also did some research in the domain, fifteen years later, the academic landscape for urban transportation planning in general and urban travel demand modeling in particular looked very different. Thus, a 1969 survey showed that there were at that time no fewer than twenty-nine urban transportation programs being carried out in nineteen transportation centers within the university system,[206] several of which would distinguish themselves as a home to urban traffic forecasting.[207] Unsurprisingly, the academic institutions that had cultivated traffic issues for years, some even since the interwar period, were among the organizations that displayed an early interest in urban travel demand modeling. Purdue University was one such early example. In the mid-1910s its School of Civil Engineering began to stage an annual statewide highway conference named the Purdue Road School.[208] In March 1937 an act of the state legislature officially established the Joint Highway Research Project (JHRP)[209] between Purdue University and the Indiana State Highway Commission.[210] No wonder, then, that in the 1950s and 1960s Purdue University hosted many research projects,[211] including master's and doctoral dissertations, that dealt with various aspects of the four-step model.[212] A similar kind of cooperative research program between a government agency and a knowledge-related organization took place in Texas. The partners were now the local state highway department and the Texas Transportation Institute (TTI), the original charter of which was given by the Texas A&M board of directors in 1950.[213] Though an academic offspring, TTI was able to feature among its early leaders Thomas H. McDonald, the famous "boss" of the BPR from 1919 to 1953 (chapter 1). As a result, TTI's various R&D activities would regularly benefit from financial aid provided by BPR and other federal agencies.[214] In the mid-1960s Texas A&M University was one of the very few academic institutions to have an IBM 7094 mainframe, while in the late 1960s TTI researchers had access to the IBM 360 system, one of the biggest, fastest, and most powerful computers in the world in the mid-to-late 1960s. As part of an organization well-endowed with computer resources, a number of TTI staffers, including Vergil G. Stover, a Ph.D. holder from Purdue University in 1963,[215] were to prove instrumental in the production and diffusion of original urban traffic forecasting practices within the four-step general modeling framework in the 1960s and the early 1970s.[216]

In the summer of 1947, the California State Legislature passed an act providing for the establishment of the Institute of Transportation and Traffic Engineering (ITTE) at the University of California, now a four-campus network under the name Institute of Transportation Studies.[217] Operating within the departments of engineering on the Berkeley and Los Angeles campuses, ITTE was designed as a hybrid institution pursuing a variety of goals, including formal academic instruction and research as well as the development of collaborative projects with administrative and operative agencies concerned with transportation in California. Very quickly, ITTE grew into a significant organization. In 1951 the professional staff comprised some twenty-five individuals from various disciplinary backgrounds, ranging from engineering to library science.[218] In 1969, the Berkeley branch, which was the largest, was composed of 119 people among whom there were ninety graduate students.[219] From 1964 on, the latter had access to an IBM 1620 machine, acquired to familiarize them with transportation modeling practices through the use of computer programs designed by ITTE's members.[220] Judging by the various course titles,[221] the publications of the Institute's senior members[222] as well as from students' various projects and research contributions,[223] ITTE could legitimately boast of having made itself one of the powerhouses of American urban traffic forecasting in the 1960s. Given the organization's strong commitment to travel demand modeling, it is no coincidence that Frank A. Haight (1919–2006), an ITTE staffer at the Los Angeles branch, was the first editor-in-chief of *Transportation Research,* while one of the classic textbooks in the field in the 1970s was signed by another Institute member, Adib Kanafani (born in 1942).[224]

While the decision to set up the Transportation Center at Northwestern University, currently known as the Northwestern University Transportation Center (NUTI), was made in early 1954,[225] it was only in the fall of 1956 that the newly created organization began its formal program.[226] In 1957 the center was home to fifteen faculty personnel who taught transportation in several university units, including the Department of Civil Engineering, the School of Business, and the College of Liberal Arts.[227] Among the first members of the center was Abraham Charnes (1917–1992), a leading figure in the OR field who was to publish an important research paper on the issue of traffic assignment at the end of the 1950s.[228] In autumn 1963, a fifteen-month interdepartmental program in transportation leading to the degree of master of science in transportation was inaugurated, administered by the Transportation Center in collaboration with the Graduate School of Northwestern University. Among the courses available to students some were dedicated to OR and urban transportation planning.[229] In 1973–1974, nineteen students received their master's degree, bringing the total number of diplomas awarded in ten years to 100.[230] Hosting a growing number of people—Transportation Center staff and participating faculty numbered thirty-four in 1969–1970[231]—the center would enrich the field and literature of travel demand modeling in the 1960s with several research projects, some funded by the Bureau of Public Roads,[232] as well as with many significant publications.[233]

In the early 1960s, the Department of Civil Engineering at MIT undertook a major reform of its curriculum,[234] while a newly equipped Transportation

Laboratory started offering research facilities in the area.[235] Urban traffic forecasting would greatly benefit from these developments as well as from the significant computational resources provided by the Department of Civil Engineering and, more generally, by MIT as a whole in the 1960s.[236] Already at the end of the 1950s, Alexander Jamieson Bone, then associate professor of transportation engineering, was introducing would-be MIT civil engineers to O&D traffic surveys as well as to traffic estimates and their applications to planning studies.[237] A few years later, Martin Wohl (1930–2009), an MIT alumnus and the coauthor of *Urban Transportation Problem* (1965) and *Traffic System Analysis for Engineers and Planners* (1967),[238] was to familiarize MIT students in civil engineering with "estimation and allocation traffic demand by mathematical techniques such as gravity model and linear programming."[239] It was also in the early 1960s that the Briton Brian V. Martin, a graduate of Imperial College (1957–1960), along with a certain Frederick William Memmott III would undertake a master's research project under Bone's supervision. Sponsored by the Commonwealth of Massachusetts Department of Public Works, their joint efforts resulted in a widely read report titled *Principles and Techniques of Predicting Future Demand for Urban Transportation*, which was reprinted several times after its first publication in June 1961.[240]

However, it was another MIT alumnus, Marvin Lee Manheim (1937–2000), who would prove instrumental in making research in urban travel demand modeling a central and sustainable activity within the Cambridge-based institution.[241] A graduate of the civil engineering department in 1959, Manheim was awarded a doctoral fellowship by the MIT–Harvard Joint Center for Urban Studies[242] to conduct a research project on decision-making in highway route location,[243] thanks to which he would be credited much later by the Nobel Prize laureate in economics Herbert Alexander Simon (1916–2001) with producing an "early example, but still a very good one, of incorporating design costs in the design process."[244] On completion of his military service, Manheim returned to MIT to become the first head of the newly created Transportation Systems Division (TSD) affiliated to the Department of Civil Engineering.[245] While conceiving an original academic and research program for the TSD, built on the idea that "transportation problems must be addressed as *system* problems,"[246] Manheim would develop a long-standing intellectual and professional partnership with the research engineer Earl R. Ruiter (1939–2008). An MIT graduate in 1962, and holder of a master's degree from Northwestern University, Ruiter was already an expert in urban traffic forecasting following his stint with the influential Chicago Area Transportation Study in the mid-1960s.[247] In 1968, the foundation of the Urban Systems Laboratory (USL),[248] a multidisciplinary and interdepartmental organization sponsored by six MIT units,[249] would provide Manheim and his teammates at the TSD with the opportunity to develop new projects both in research and teaching.[250] One of these projects was directly concerned with urban travel demand modeling. Named DODOTRANS (Decision-Oriented Data Organizer-TRansportation ANalysis System), it consisted of a series of computer programs that handled the different aspects of a transportation project, from predicting flows on a network, be it a highway or a mass transit system, to evaluating the project under study.[251] At the turn of the 1960s, transportation research at MIT was sufficiently well established so as not to be affected by the disbanding of the USL in 1974, following financial difficulties. In April 1973, Manheim and his collaborators had already found a new home for their modeling activities with the creation of the Center for Transportation Studies (CTS), an interdepartmental unit that had enlisted at the time of its creation the services of more than fifty people from six MIT schools.[252] A large part of the history of urban travel demand modeling in the United States from the 1970s onward would be written by CTS-related scholars, often under the intellectual guidance of Manheim, who repeatedly assumed a leadership role as an academic mentor to several Ph.D. students. In addition to being one of the four regional editors of *Transportation Research* when the journal was set up in 1967, Manheim would also provide the UTDM social world with a widely read and abundantly quoted manual published in 1979.[253]

### Setting the Four-Step Model in (IBM) Computer-Code Stone

In light of the foregoing, it is obvious that, as time went by, American urban traffic forecasting grew increasingly multipolar. As a result, at the end of the 1960s, the UTDM social world was populated by a significant number of actors ranging from state and metropolitan transportation agencies to consultancies and knowledge-related organizations. However, it was not devoid of a *center*. For years, the Bureau of Public Roads would consistently occupy a central place from where it could orchestrate the efforts and harness the energy of the various stakeholders inhabiting this specific social world. The federal agency had two major assets that allowed it to successfully fulfill its role as an orchestra conductor: A lot of money to financially back the various actors involved in efforts to produce new modeling knowledge and conceive original forecasting practices; as well as enough expertise of its own to conduct some in-house R&D work, provide technical assistance, evaluate the modeling work conducted within the field, and promote the best ways of doing to the larger community of traffic forecasters. Thanks to the various Federal-Aid Highway Acts there was money galore indeed, and the BPR, either directly[254] or through larger programs such as the National Cooperative Highway Research Program (NCHRP),[255] did pump many federal dollars into urban traffic forecasting in the 1950s and 1960s.[256] If funding travel demand modeling-related issues proved central among the tasks the BPR performed in the 1960s, it was not the only one. The existence of an increasing number of actors with the necessary resources to carry out in-house R&D work in urban traffic forecasting eventually resulted in this period in making the bureau's part in the production of innovation in this area less central that it had once been.[257] Nonetheless, thanks to its own staff who were well versed in travel demand modeling, the federal agency continued to develop its own expertise and enjoyed a strong "organizational reputation"[258] within the UTDM social world. However, because of their limited number—in 1961, the Traffic Assignment Section was equipped with six specialists only[259]—the innovation potential of the BPR was bound to decline over time. Still, in the 1960s people within this federal agency were able to put their skills and expertise to good use by scrutinizing and assessing the various models and modeling practices then circulating within the community of traffic forecasters, cherry-picking those they considered to be the best ones, codifying and standardizing them, and meticulously organizing and promoting their dissemination on a massive scale through training courses, manuals, and, even more effectively, with the help of free computer suites written by both in-house system analysts and, mostly, by many external collaborators as well.[260]

In January 1960, John T. Lynch, head of the BPR Planning Research Branch and one of the spiritual fathers of the Home Interview origin and destination survey, organized a roundtable discussion with three heavyweights of the traffic forecasting community of the time, including Voorhees, Schneider, and Brokke, the latter playing the role of advocate of the growth factor approach. As the dialogue eventually turned to a set of parallel monologues, Lynch felt forced to conclude that "it is fairly clear that there is no agreement among experts as to the best method for projecting future urban travel."[261] Faced with an absence of consensus, in the fall of 1960 the BPR Office of Research initiated an in-house research project that aimed to evaluate the various forecasting techniques available for estimating the interzonal movements within an urban area (the trip distribution, and second step of the four-step model). The Fratar method having already been assessed in the second half of the 1950s,[262] it was now the gravity model, Schneider's opportunity-model, and the so-called competing-opportunities model, originating in the Penn-Jersey Transportation Study in the early 1960s,[263] that became the target of intensive testing and evaluation. These heavy operations were conducted essentially by BPR staffers along with some help from individuals in the Pennsylvania Department of Highways throughout the 1960s.[264] Though the trip distribution step attracted much attention, procedures pertaining to the other stages of the FSM would also be put to the test by the federal agency's staffers.[265]

In addition to testing and evaluating the various traffic forecasting techniques available on the modeling market of the day, the BPR would also distinguish itself as the main channel through which the most effective operational modeling practices—in the bureau's opinion, it goes without saying—could reach the urban traffic forecasting community. To begin with, people outside Washington, DC could visit the BPR offices there and get help with their modeling tasks. To give just one example, on Wednesday June 15, 1960, Stephen George Jr., Paul Bliss, and Harry Reed, all involved in the Twin Cities Area Transportation Study (TCATS), departed for Washington via Capital Airlines Flight 118 at 5 p.m., "to direct, observe and assist with TCATS first traffic assignment." This was planned to be run on an IBM 704 machine housed at the National Bureau of Standards. During their ten-day stay in Washington, they worked closely with several BPR staffers, including Mertz and Brokke,[266] another early "computer-man" within the federal agency.[267] Another face-to-face mechanism that proved particularly effective was training. In the 1960s the BPR developed a two-week intensive course in urban transportation planning, covering in depth the technical details of all phases of the planning process and providing the participants with practical experience in using the then available modeling tools through a series of relevant workshops.[268] By the end of 1966, this extremely popular course, dating from October 1961, had been taught fourteen times across the nation, and had been attended by 471 professionals (figure 2.8).[269]

Very soon, in addition to face-to-face interactions and extended personal contact, the Bureau of Public Roads developed communication mechanisms *at a distance* as well. From 1963 to 1967 six highly popular manuals were issued at low prices, ranging from $0.65 to $2.50, most of which would be published several times. Taken together, these texts explained to the practitioner both the theoretical bases of the then existing traffic forecasting techniques while guiding them step by step in the use of the various computer suites designed for several types of mainframes.[270]

However important the aforementioned channels of dissemination may have been, none of them could match the so-called UTP as an effective mechanism in the spread of the four-step model across the UTDM social world. UTP stands for "urban transportation planning," but Urban Transportation Planning is also the name of an impressive, and extensible, collection of computer suites for the then recent IBM 360 machines, collectively released at the end of the 1960s.[271] The UTP program battery was arguably the most sophisticated—albeit not totally complete since the mode choice step would not be dealt with[272]—computer system of the day for conducting urban traffic forecasting studies with the help of the FSM (the June 1970 version of the package contained no fewer than eighty-five programs[273]). Moreover, thanks to federal largesse, the entire battery could be obtained from the Bureau of Public Roads by any member of the UTDM social world for free.[274] Unsurprisingly, such a complex and sophisticated endeavor could not be performed by one single actor. The UTP was indeed the outcome of the efforts deployed by a large "simulation code collective"[275] acting under the aegis of the BPR (figure 2.9).[276] While a number of the programs that were part of the IBM 360/battery were written with the help of BPR staffers, many other actors present on the travel demand modeling

**Figure 2.8**
Extract from the 24th Urban Transportation Planning Course, May 11–22, 1970, presented by the Staff of the Urban Division, Office of Planning, Bureau of Public Roads. *Source:* George Mason University Libraries, Special Collections Research Center, William L. Mertz Papers, C0050, Box 26, Folder 98.

11:45 Lunch **NOTES**

12:45 (2.5) TRAFFIC ASSIGNMENT TECHNIQUES

Bill Reulein

Discussion of the various types of assignment including all or nothing diversion, and capacity restraint. Description of the S/360 version of capacity restraint, procedures used in running programs, and options available.

1:30 (2.6) CODING OF NETWORKS FOR S/360 TRAFFIC ASSIGNMENT PROGRAMS - WORKSHOP INSTRUCTIONS

Dick Duemler

Part I - The details of coding a network for the urban planning traffic assignment programs will be discussed. The preparation of network maps, the use of link coding forms, types of errors, and the computerized network description will be described. Also included, will be a brief description of Sioux Falls, South Dakota, the city to be used for all future workshops.

Part II - Traffic Assignment Workshop Instructions.

2:15 (2.7) WORKSHOP - CODING OF NETWORKS

Dick Duemler, Bob Probst,
Ed Fleischman, and Gama Olvera

Each group will code a portion of a preselected traffic assignment network for Sioux Falls.

4:15 Adjourn

6:00 Class Party

NEWSLETTER OF THE
TRANSPORTATION PLANNING
COMPUTER PROGRAM EXCHANGE GROUP

U.S. DEPARTMENT OF COMMERCE
BUREAU OF PUBLIC ROADS

Volume 1, Number 1 July 15,1963





CONTRIBUTOR PAGE



Introduction to Second Printing

This is the first of a series of quarterly Exchange Group Newsletters. A second printing has been required in order to extend distribution to U. S. Bureau of Public Roads Regional and Division Engineers and to the Transportation Planning Organizations of the State highway departments.

Merton J. Rosenbaum, Jr., Secretary
Transportation Planning Computer
Program Exchange Group
U. S. Bureau of Public Roads
1717 H Street, NW., Room 850
Washington, D. C. 20235

U. S. BUREAU OF PUBLIC ROADS
Richard J. Bouchard reporting:

IBM 7090 Gravity Model Publication Now Available

The Office of Planning, of the U. S. Bureau of Public Roads, has just released a volume entitled "Calibrating and Testing a Gravity Model for Any Size Urban Area." This volume explains and illustrates the theory and the use of a system of analytical procedures and electronic computer programs for calibrating and testing a gravity model trip distribution formula for any urban area. By combining the techniques described in this volume with those concerned with traffic assignment, it is possible to complete much of the analysis and, forecasting phases of a comprehensive transportation planning study.

The analytical procedures described in this volume have general application. The electronic computer programs which mechanize these procedures, however, have been prepared for specific use on the IBM 1401 (16K) and the IBM 7090 (32K) computers. The IBM 1401 programs are used in processing basic origin-destination survey data and converting it into the required form suitable for use with the IBM 7090 computers. They are completely compatible for use on the IBM 1410 (40K) computer if it is equipped with a compatibility switch. It is the IBM 7090 programs which actually calibrate and test the gravity model trip distribution formula. These programs are completely compatible for direct use on the IBM 7094 computer.

Very briefly, the system described in this volume is designed to edit, sort, and link a set of detailed trip records that can be produced from basic travel inventories. It will build complete tables of zonal trip interchanges for any desired combination of trip purposes or travel modes. The system will compute the actual trip length frequency distribution for a given set of data, and completely develop multipurpose gravity models, so that they simulate the actual trip length frequency distribution and balance trip attractions in each zone with a very minimum of hand calculation. Finally, the system will compare zonal trip interchanges computed by the gravity models with those obtained from the basic travel inventories.

Copies of this publication are being sent to each State highway department and all Public Roads division and regional offices. Transportation study groups or other official governmental agencies should send requests to the Director of the Office of Planning, Bureau of Public Roads, through their local Public Roads division office. Others desiring copies should write to the Director, Office of Planning, U. S. Bureau of Public Roads, Washington 25, D. C.

IBM 1401 and 1620 Gravity Model Programs

The Office of Planning, of the U. S. Bureau of Public Roads, is in the process of preparing two volumes which may be of interest to members of the group. In the final stages of preparation is a volume which explains the theory and use of a recently developed system of IBM 1401 (16K) programs designed to calibrate and test a gravity model trip distribution formula with a minimum of manual processing. This system will handle a maximum of 280 traffic zones and can be used in conjunction with the present IBM 7090 traffic assignment programs. Appropriate IBM 7090 converter programs have been prepared to aid in transferring between the two machines.

**Figure 2.9**

The first page of the *Newsletter* issued by the Transportation Planning Computer Program Exchange Group (T-PEG), which was set up in the early 1960s under the aegis of the Bureau of Public Roads (the first issue of the *Newsletter* was released on July 15, 1963). *Source: Newsletter of the Transportation Planning Computer Program Exchange Group* 1, no. 1 (July 1963): 1.

scene in the 1960s, ranging from transportation consultancies[277] to research centers[278] to computer-related firms,[279] contributed to this highly cooperative undertaking.[280]

By the early 1970s, the four-step model had gathered significant momentum, and even acquired a paradigmatic status among the urban traffic forecasters of the day. Yet, it was already assailed by several challenges, emanating from either external sources or internal tensions. These challenges prompted the then UTDM social world to introduce a series of novelties in the general FSM framework. The disaggregate approach was one of these.

# II Giving the Four-Step Model a New Lease on Life, 1970s and 1980s

# 3 Travelers Are Utility-Maximizing Rational Individuals

## The Four-Step Model under Fire

> The urban transportation planning models grew out of a need to justify new construction. . . . Take a technique here and a technique there, from physics, from thermodynamics, or from what-have-you; put them together and you have urban transportation planning . . . large, data-eating monsters that serve more to obscure real issues of transportation than to shed any light. Now and then, one of the faults of the model is recognized by somebody, and an attempt is made to patch it up.
>
> —Gerald Kraft, "Discussion of Allen-Boyce Paper," *Papers in Regional Science* 32, no. 1 (January 1974): 58–59

This was individual criticism at its most bellicose. Still, it captured a more general mood. In the early 1970s, after years of intensive innovation, great expectations, and massive use, the four-step model was about to run out of steam. Originally designed to properly size new highways from scratch, it had done its job as most of the Interstate Highway System had been completed by the early 1970s.[1] Yet the FSM seemed much less suitable to address the new environmental, economic, and social issues of the day. What adjustments should be made to transportation systems *already built* in order for them to meet and maintain air quality standards, and to conserve energy, for example? How would people react to an increase in gasoline prices or to lower transit fares? Which groups would gain and which would lose as a consequence of a particular transportation plan? Subjected to close scrutiny, central pillars of the four-step model came under attack, serious deficiencies that had long been overlooked under pressure of immediately urgent affairs were now persistently brought to light, several courses of actions to correct defects were now revealed, and efforts to catch up with new needs and unprecedented goals were made. And since necessity is the mother of invention, new practices emerged and challenged previous ones. The *disaggregate approach* was one such new modeling line.

### A New Legislative and Policy Framework Sympathetic to Mass Transit

*Stochastic Choice of Mode in Urban Travel: A Study in Binary Choice* was published in 1962. Authored by Stanley Leon Warner (1928–1992),[2] the book was based on his Ph.D. thesis conducted at the Transportation Center at Northwestern University. Warner's doctoral research was about traffic forecasting, but the modeling approach adopted contrasted sharply with the general philosophy dominating the field in the early 1960s. Instead of treating urban travel in terms of *urban zones* and *aggregate* traffic flows between them, as the FSM does, the author instead focused his analysis on the *single individual* and his or her *travel decisions*. Drawing upon a general method of analysis already tested in fields as various as archaeology and inventory control, Warner made use of origin and destination traffic survey data in order to estimate "the 'probability' that an individual chooses this or that mode of travel," through studying how differences in travel time and cost between the mode options as well as the trip-taker's income affect this probability.[3] Warner's research work in mode choice modeling remained a voice in the wilderness, since the UTDM social world would stick to the aggregate principle underlying the FSM for several more years. However, at the turn of the 1960s several research studies explicitly dealing with the modeling of the behavior and decision-making of *individual* trip-makers were conducted in the United States. This was no coincidence. In the 1960s and the early years of the following decade, the national urban transportation scene underwent a series of dramatic changes that increasingly placed the then dominant four-step model under growing pressure.

> Our National transportation policy should place its principal emphasis on the *individual*, his need for mobility, and the quality of his life. . . . Yet millions of American either without cars or unable to drive are deprived of the necessary mobility. . . . Equality and opportunity have little practical meaning in our complex society if one does not have mobility. . . . Dislocation of people and neighborhoods, air and water pollution, noise, traffic fatalities are important side-effects to our transportation investments which should now receive full attention.[4]

These lines are illustrative of the general atmosphere at the moment when Richard Nixon (1913–1994) was about to come to power as president. By the mid-1960s, the various adverse impacts of urban freeways were slowly coming to be understood. As a result, closely identified with the highway system, the staffers of the Bureau of Public Roads, dubbed the "Highwaymen," progressively came under fire both from below—the country witnessed, in fact, a series of "freeway revolts" in the late 1960s—and from above as well.[5] Having become a subunit within the Federal Highway Administration, a new federal agency within the recently created Department of Transportation (1966), the BPR was now subject to a level of administrative supervision and control it had never experienced before. It eventually disbanded in 1970, when, after a long period of growing disaffection with mass transit systems, the nation was rediscovering the many advantages associated with this transportation mode.[6] Not only was mass transit now (re)considered an efficient means for decreasing traffic congestion on urban streets, and the resulting pollution, but it also was thought that it could be put at the service of the Great Society and the many Americans suffering from lack of mobility, be they too poor to afford to own and maintain a car or too old/young to regularly operate an automobile even if one were available. Environmental, and, from the early 1970s on, energy issues also helped to make the transportation equation even more complex to solve. Throughout the 1960s and the first half of the 1970s, a spate of highway legislation, mass transit legislation, and new laws focused on environmental protection and quality reflected these new concerns.[7] Among the signs of the times, the Williamsburg Conference held in 1965 endorsed the concept of making maximum use of existing transportation facilities, instead of building new ones, through traffic management and land use controls,[8] while in August 1967 even such hard-nosed highwaymen as the engineers of the BPR backed the preferential use of highways lanes for buses. No later than 1969, exclusive bus lanes started operating along Shirley Highway in the Washington, DC, area.[9]

## Replacing Zones with Rational Individuals: The Disaggregate Approach

New historical setting, original challenges, specific tools: these developments would soon generate passionate debates among urban travel demand modelers. Quite quickly a rather large consensus emerged, especially within the university segment of the UTDM social world: whatever its (real) merits, the argument went, the FSM was plagued by several limitations, making it difficult for it to be effectively used in the new settings.[10]

The four-step model is *aggregate* in nature in the first place, since it divides the urban area under study into large zones, and then deals with global traffic data, either in the form of aggregate totals, such as trips per zone, or in the form of aggregate rates, such as trips per household and per zone. This use of aggregated data implies an *averaging* over the individual behavioral units within the traffic zone, resulting in a major loss of variability, unless all the individuals and households within the area are perfectly homogeneous. However, zones are heterogeneous entities,[11] while variability does matter a lot. In point of fact, since the "rich, poor, healthy, and handicapped are rarely homogenous," forecasting methods based on aggregated data and average conditions risks being highly ineffective in predicting the consequences of alternative policy solutions, while precluding, by way of definition, "the possibility of answering questions such as *who* benefits and *who* pays for policy changes."[12]

Based on aggregated data, the collection of which can easily consume half of the budget of a traditional urban transportation study according to a BPR staffer commenting on cost figures averaged from eight typical studies that had been completed by the end of the 1960s,[13] the FSM was also accused of embodying a "statistical" style of modeling, which puts the emphasis on statistically derived correlations between the model variables while ignoring the possible *causal* (and truly explanatory) relationships among them.[14] As a result, the FSM is seriously lacking in a sound scientific basis, and proves therefore, as its opponents hammered home, particularly susceptible to the "ecological fallacy."[15] Because of this, the statistical correlations between the FSM variables observed at the *aggregate* level can prove, in fact, to be the result of *chance*, or the existence of a hidden variable that does not explicitly feature in the model.[16] It follows, then, that even strong actual statistical associations between variables do not guarantee close correlations, and, consequently, accurate forecasts. The fact that the correlated variables of the four-step model are not linked by cause-effect relationships, for lack of an explicit theoretical framework for the trip-maker's travel behavior and decision-making, means also that the performance of the FSM is too dependent on the site from which the dataset used was generated. As a result, the FSM is lacking in geographic transferability, and proves therefore very onerous since each metropolitan area has to build its specific version of this general modeling framework on the basis of its own local data. And last but not least, for its critics, even when the FSM does succeed in producing correct forecasts, it is only because they are "self-fulfilling prophecies," for "the sizeable volumes" predicted by the model "will indicate a need for more highway investment," and "this great improvement in the transport system will affect the activity system to the extent that the large volumes that were forecasted will now be realized."[17]

Since many of the main defects of the four-step model seemed to flow from its *aggregate* nature, then we should try something radically different, argued its critics: a *disaggregate* modeling line, which substitutes the *individual travelers* for the zonal traffic flows. After all, "Zones don't travel; people travel!"[18] In addition to helping the modeler avoid the epistemological defects of the FSM, according to the advocates of the new approach at least, this focus on the individual could also benefit from the general ideological atmosphere surrounding many of the Great Society programs through which economists, psychologists, and sociologists sought to conquer poverty, inequality, and racism by carrying out changes at the level of the family and the person.[19] And since it is hard to find a more individual-centered academic discipline than modern microeconomics, it is not surprising that among the first proponents of the disaggregate approach in urban traffic forecasting one can find several economists. No wonder also that, given the then growing interest in mass transit, the issue of *choice of mode* by a rational individual faced with several transportation options proved the most popular theme of the day.[20]

## A Bunch of Ph.D. Degrees and Other Research Studies

Like Warner, Thomas Edward Lisco (born in 1939) was a Chicago-based Ph.D. candidate in economics who was also interested in the mode choice issue. To put value "on the time spent by commuters during their daily travels to and from work," he constructed a model of mode choice consisting of seven variables that "may be directly or indirectly associated with time, cost, income, age, sex, and two-family structure situations differing from 'normal.'" Based on the quantitative results he was able to produce with his disaggregate binary choice model, Lisco could proudly contend that his doctoral dissertation was about to debunk "two of the most prevalent myths concerning commuters and their commuting habits": first, that "commuters are to a great extent irrational in their choice of transportation mode, and second that once a commuter is lost to mass transportation, he is lost for good."[21] Thanks to Warner and Lisco, Chicago could boast that it was at the vanguard of the search for original travel demand forecasting practices in the 1960s. The move to Illinois of two British scholars keen on disaggregate modeling would reinforce this feeling. Peter L. Watson (born in 1944) was appointed assistant professor of economics at Northwestern University in 1970,[22] two years after Peter Robert Stopher (born in 1943) had been hired by the Department of Civil Engineering. Awarded a B.Sc. in civil engineering from University College London in 1964, Stopher also received his Ph.D. in his native country in 1967.[23] While drawing upon the existing American literature on the subject—from the early work carried out by Warren T. Adams of the Bureau of Public Roads[24] to the various contributions of the Traffic Research Corporation and Warner's pioneering research study—Stopher was able to enrich the existing body of knowledge regarding the mode choice step with a disaggregate model of his own.[25] Upon his arrival in the United States,[26] Stopher would enthusiastically preach the gospel of the new approach. Witness his various, and thorough, lecture notes for undergraduate and graduate students,[27] apparently immediately spotted by other people involved in the field,[28] the subject matters of several master's theses and doctoral dissertations conducted under his supervision,[29] the various grants he was awarded from the DOT[30] as well as his many publications, including a summary of his Ph.D. study in 1969,[31] a manifesto article in favor of the disaggregate approach coauthored with Lisco in 1970,[32] and his 1975 textbook on urban transportation planning and modeling crafted with the Cornell University professor Arnim H. Meyburg (born in 1939), a 1965 German Fulbright exchange student and Ph.D. holder from Northwestern University in 1971,[33] in which the new line of modeling covers several chapters (figure 3.1).[34]

It was in 1968 that Charles Arthur Lave (1938–2008) defended his own doctoral dissertation in economics at Stanford University. The distance between the West Coast and Illinois did not prevent him from acknowledging his intellectual debts to Warner and Lisco. Like the work of his Evanston predecessors, Lave's research study was distinguished from the then dominant aggregate approach to the mode choice problem by "its explicit behavioral approach to the factors which motivate the commuter's choice."[35] A respect to a function G(x) as a symmetrical sigmoid curve called the logistic curve. This model may be written mathematically as:

$$p_1 = \frac{e^{G(x)}}{1 + e^{G(x)}} \qquad (7.08)$$

This model is applied to mode choice by defining the "event" mentioned above as being the choice of one mode in preference to the other.

The two models developed on this basis are somewhat different in the form of the function G(x). Thiel uses a multiplicative function of cost ratio, time ratio, income, age and sex, in the following form:

$$e^{G(x)} = e^{\alpha} \left(\frac{T_1}{T_2}\right)^{\beta} \left(\frac{c_1}{c_2}\right)^{\gamma} \prod_{i=1}^{3} \left(Q_i^{\delta_i}\right) \qquad (7.09)$$

The calibration technique which he then uses is based on taking logs of the above expression, after noting that

$$e^{G(x)} = \frac{p}{1-p} \qquad (7.10)$$

The log version of (7.09) then becomes:

$$\log \left(\frac{p}{1-p}\right) = \alpha + \beta \text{ lo } (T_1/T_2) + \gamma \log (c_1/c_2) + \sum_{i=1}^{3} \delta_i \log Q_i \qquad (7.11)$$

Linear regression can be applied to this model to produce estimates of the parameters $\alpha$, $\beta$, $\gamma$ and the $\delta_i$'s. However, as we have noted on previous occasions, when we are interested in predicting p/1-p rather than its log, a set of regression estimates of this form have somewhat dubious validity.

The model which Stopher developed was based on a G(x) function employing travel costs, travel times and income in the following form:

$$G(x) = \alpha_1 (c_1 - c_2) + \alpha_2 (t_1 - t_2) + \alpha_3 \qquad (7.12)$$

where $\alpha_1$, $\alpha_2$ and $\alpha_3$ are each functions of income, and the c's and t's are costs and times. Although several calibration methods are possible, the best one here appears

**Figure 3.1**

Extract from "Individual Behaviour Theory," a chapter prepared by Peter Stopher for his students at Northwestern University. Stopher's "Lecture Notes of Urban Models" constitute one of the earliest examples of the teaching of disaggregate modeling techniques in the United States. *Source:* Peter R. Stopher, "Lecture Notes of Urban Models, 1970/71," typescript, copyright © P. R. Stopher 1970, 1973, 151 (Courtesy Northwestern University Transportation Library).

similar individual-based approach could also be found in Robert Gordon McGillivray's (born in 1938) own Ph.D. work at Berkeley.[36] Still in California, in the early 1960s researchers from the Stanford Research Institute (SRI) started a multiyear study aiming to estimate the monetary value of time for motorists. Leading to a series of disaggregate models,[37] the research was initially commissioned by the Bureau of Public Roads, for which the conversion of time savings in hours to dollar value was of critical importance in the economic analyses of the highway projects conducted under its aegis in the 1960s. However, the decisive "site" for the making of disaggregate modeling into both a full-fledged academic subject and a widespread professional practice was Massachusetts, the transportation scene of which in the late 1960s was filled with anti-freeway protestors fighting current highway projects,[38] and think tanks looking to "post-highway" transportation policies, including the Cambridge Task Force on Transportation Policy set up at the end of the 1960s and chaired by MIT faculty member Marvin Manheim.[39] Here, at the turn of the 1960s, several productive marriages between the world of consultancy and academia were made.

Gerald Kraft was born in Detroit in 1935 and obtained his master of arts degree from Harvard in 1957. There he had been a student and even a collaborator of John Meyer,[40] and embarked as a member of the famous Harvard Transport Research Program on a Ph.D. dissertation provisionally entitled "Elasticity of Demand for Freight and Passenger Transportation in Underdeveloped Countries."[41] However, it was the Northeast Corridor Transportation Project that provided Kraft, then a staffer at Systems Analysis and Research Corporation (SARC) from 1961–1964, with the opportunity to develop an original traffic forecasting model, immediately dubbed "direct demand model."[42] Unlike the then dominant modeling practices, in the multimodal model that Kraft helped to devise, the generation, distribution, and mode choice steps—the three separate and individual FSM stages—were instead *combined* in a single overall framework, and carried out *simultaneously*. Kraft, who joined the Cambridge Task Force on Transportation Policy,[43] imported to Charles River Associates Incorporated (CRA) both his critical stance toward the four-step model, and the point of view of the economist seeing urban transportation as just another commodity. A tiny firm when it was set up in the Boston area by the author and another collaborator in 1965,[44] CRA would grow rapidly to employ approximately a hundred persons fifteen years later.[45] Shortly after its creation, CRA along with Peat, Marwick, Livingston & Co. (chapter 2) were charged by the San Francisco Bay Toll Crossing Agency with the task of designing an urban travel demand model for the San Francisco Bay Crossing.[46] Like the intercity Kraft-SARC model, of which it was an urban outgrowth, the new CRA modeling work was also economics-based. Derived from the theory of consumer demand,[47] it was carried out by Kraft himself, the French citizen Jean-Paul Valette, a Ph.D. holder from the University of Colorado in 1962, and Thomas A. Domencich (1933–2020), a graduate of Ripon College who had begun his career as a consultant with Arthur D. Little, Inc. Though the calibration of the model proved a rather daunting and expensive task,[48] very soon the company that Kraft founded would prove an excellent partner to operational urban traffic forecasting.

Charles River Associates started as a Harvard-related consultancy, as among its early members, in addition to Kraft himself, were the economist Franklin M. Fisher (1934–2019), a former Ph.D. Harvard student of Meyer, and Paul Osborne Roberts (born in 1933), a Ph.D. holder from Northwestern and a member of the Harvard Transport Research Program in the 1960s. Yet it was through its collaboration with MIT that the firm would definitively associate its name with the advent of operational disaggregate travel demand forecasting techniques. In 1970, CRA was awarded a $110,000 grant (around $715,000 as of 2018) from the Federal Highway Administration to develop a "Behavioral Modal Split Model" for the mode choice step.[49] The project leader for this study was Domencich, while Peter A. Diamond (born in 1940) and Robert E. Hall (born in 1943) were primary consultants on the project. The two young MIT scholars solicited in turn the assistance of another economist of their generation from Berkeley, who was visiting MIT in 1970–1971. The colleague in question happened to be Daniel McFadden (born in 1937), the future winner of the 2000 Nobel Memorial Prize in Economic Sciences.[50] When he arrived at MIT, McFadden had not published any scientific articles dealing with transportation issues. However, such issues were not totally foreign to him. While teaching on the West Coast, he had been contacted by Phoebe Humphrey Cottingham, then a Ph.D. candidate who was seeking advice on how to process the information she had gathered on the decisions made by the California Division of Highways about the selection of freeway projects.[51] In order to help her, McFadden developed a probabilistic model of choice behavior of his own, based on the so-called Luce's choice axiom.[52] Expressed in no mathematical terms, this axiom states that the probability of selecting one alternative over another from a set of alternatives is affected neither by the presence nor absence of any additional alternatives in the set. Selection of this sort has therefore independence of irrelevant alternatives (IIA), which means that if *A* is preferred to *B* in the presence of *C*, it would still preferred to *B* in the absence of *C* as well.[53] Armed with Luce's choice axiom, McFadden launched into building an empirical econometric model, which he named the "conditional logit model" because of the logistic form (or S curve) to which it is reduced in the case where there are only two alternatives (figure 3.2).

McFadden even wrote a calibration computer program of his own to estimate the model's parameters.[54] As a result, for the implementation of the work done by the MIT scholars, McFadden could rely on his already built "conditional logit machinery" (his words), and along with Domencich he set to work on testing the proposed model. Since the modeling approach was based on individual observations rather than zonal aggregates, the CRA model proved very economical in respect of the data. A total sample of only about two hundred households was used for running both a mode choice model for work trips, and a complete shopping demand model for which, in addition to the choice of mode, the choices of time of day of travel, shopping destination, and trip frequency were also modeled and predicted.[55]

Appearing first in the shape of grey literature in March 1972, the work that Domencich and McFadden performed was transformed into a book in 1975,[56] thanks to which disaggregate techniques in travel demand modeling reached a much wider audience. Readers could now easily familiarize themselves with some key elements of the new modeling approach, which was *discrete*, *probabilistic*, and based on the *random-utility maximization* hypothesis. Discrete: remember that travel choices are discrete entities, since you either travel or you don't, go either by car or by train, reach either

**Figure 3.2**
An early typical binary logit curve. *Source:* Paul Inglis, De Leuw, Cather & Company of Canada, Ltd, "A Multimodal Logit Model of Modal Split for a Short Access Trip," *Highway Research Record*, no. 446 (1973): 12–20, 15, figure 1. Copyright © National Academy of Sciences. Reproduced with permission of the Transportation Research Board.

Figure 1. Typical 2-dimensional logit curve.

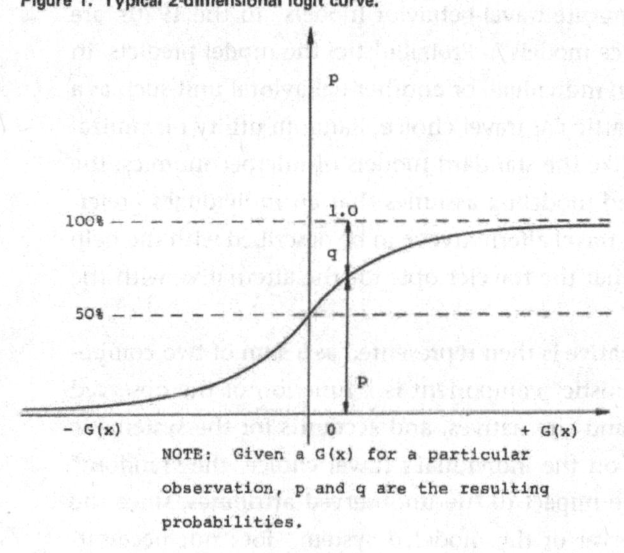


chosen. To have a model that can be realistically analyzed and that conceptually follows the formulation stated above required that the coefficient of the system variable be kept constant for each mode. Also each system variable must be related to each mode. For example, a comfort rating could not be entered into the bus mode if it were not entered in the car mode or in every other mode. On the other hand, the user characteristics need not be entered in each mode equation and should not have like coefficients unless there is a like correlation.

Several advantages are attributable to this technique, particularly in terms of theoretical assumptions. There is no assumption of normality to be met in the G(x) function, resulting in a more generally applicable model. Also the use of a probabilistic sum for aggregation gives a better conceptual idea of the true process that occurs in this mechanism.

To make this model operational requires some aggregation because collection of the information for each trip would be prohibitive. The methodology for this aggregation is still a question for review and testing. There are 2 distinct possibilities. First, if our sample is 10 percent, then each datum is a proxy for 10 other members of the population. It may be assumed that all 10 will make the same absolute choice as the proxy, i.e., the mode of highest probability according to the model. Thus, to aggregate, one multiplies the result by the sample ratio. Second, again if the sample is the same, the probabilities for the individual mode choices and not the ultimate choice are considered. To aggregate, one obtains the sums of the probabilities and multiplies those sums by 10. This better describes the behavioral process because we are dealing with human beings who can and will change their habits in this nonexact manner. The scope of this work included only a minor attempt to determine the superior methodology. That is a matter for further study.

## DATA BASE AND VARIABLES

The data used to derive and test the models have been titled the "suburban station access" data. They were collected from people using the Chicago and Northwestern Railroad suburban routes from the northwest corridor into Chicago. The trips that were modeled were not trips to the center of the city but rather shorter access trips between home and the commuter station. The 4 modes involved in the study were that destination or another (this is why this family of disaggregate models, mostly known as "disaggregate travel-behavior models" in the 1970s, are now called "discrete choice models"). Probabilistic: the model predicts, in fact, the *probability* that an individual, or another behavioral unit such as a household, will make a particular travel choice. Random utility maximization hypothesis-based[57]: like the standard models of microeconomics, the new line of travel demand modeling assumes that an individual's preference among the available travel alternatives can be described with the help of a *utility function*, and that the traveler opts for the alternative with the greatest utility.

The utility of an alternative is then represented as a sum of two components: while its "deterministic" component is a function of the observed attributes of individuals and alternatives, and accounts for the systematic effects of these variables on the individual's travel choice, the "random" component allows for the impact of the unobserved attributes, since the modeler, who is an observer of the modeled system, does not necessarily possess complete information about all the elements considered by the trip-maker.[58] The random utility-based model posits, then, that a rational individual compares the available alternatives, and has a higher probability of choosing the alternative with a higher utility than any other alternative.

After McFadden flew back to California, thanks to federal funding people at Charles River Associates continued cultivating the disaggregate approach during the 1970s and the early 1980s.[59] However, the next decisive act for disaggregate modeling would be performed in two other "sites" in Cambridge, Massachusetts: MIT and a firm closely associated with it, Cambridge Systematics. A civil engineer, Moshe Emanuel Ben-Akiva (born in 1944) headed for Cambridge after his graduation in Israel in 1968.[60] Two years after he had been awarded his master's degree, he submitted his Ph.D. research to the Department of Civil Engineering in May 1973. The dissertation was concerned with three models designed for forecasting destination and mode choices for shopping trips. The first model was the one preferred by the author of the thesis in the end, as it was considered the most realistic from a behavioral point of view and capable of leading, despite its computational costs, to sound practical applications. It had a "simultaneous structure," while the other two models were sequential in nature since the mode and destination choices were regarded as two possible separate phases (it is worth noting that the CRA's shopping model was also a sequential one).

However, despite their differences, all three were "disaggregate probabilistic models." This comes as no surprise. Among the people Ben-Akiva thanked for helping him during his doctoral research were proponents of the disaggregate approach, including McGillivray, Kraft, Domencich, and Daniel Brand, an MIT graduate and at that time a professor at Harvard.[61] Ben-Akiva also felt beholden to a number of other MIT graduate students whose names are now associated with the development of the disaggregate approach, including Frank Sanford Koppelman, an MIT civil engineer (1959) and holder of an MBA from Harvard in 1961,[62] and Steven Richard Lerman, another MIT civil engineer (1972) who got his master's and Ph.D. degrees from the same institution in 1973 and 1975 respectively.[63] Both Koppelman and Lerman specialized in transportation, and their doctoral dissertations were eventually submitted to the Department of Civil Engineering. Charles Frederick Manski (born in 1948) was also an MIT pupil, but his B.S. and Ph.D. degrees would be in economics. Although there was no transportation in his "Analysis of Qualitative Choice,"[64] which was instead concerned with the choices of colleges by high school seniors, several threads would connect the MIT economist with his fellow engineers. These links were first of all intellectual, since all these Ph.D. candidates made use of the same class of disaggregate choice models studied by McFadden and Domencich at Charles River Associates. They were also social: being nodes of larger and specific site-based networks, these young scholars knew and intensely interacted with each other during their studies while they often worked together afterward. Thus, after helping Ben-Akiva to carry out his research study by providing the latter with a calibration computer program of his own,[65] Manski would coauthor several articles with Ben-Akiva as well as with Lerman,[66] while also contributing to seminal books dealing with disaggregate modeling in the field of travel demand forecasting in the 1970s.[67]

After defending their dissertations, Ben-Akiva and Lerman remained at MIT and joined the faculty of the Department of Civil Engineering. They rapidly developed a substantial teaching program on disaggregate discrete choice modeling both at the undergraduate and, especially, graduate levels.[68] In August 1979, the MIT Center for Transportation Studies, to which both belonged, started offering a series of summer transportation programs designed for professionals in both industry and government; among them, there was a course taught by Ben-Akiva featuring the disaggregate approach on its syllabus.[69] In contrast to his fellow students, Koppelman left Cambridge in 1975 for a position as an associate professor at Northwestern University, which had already hosted pioneering work in disaggregate modeling for several years. During his long stay in Evanston, Koppelman would actively participate, both through his teaching and research activities, in the furthering and spreading of the new approach while maintaining strong bonds with his former fellow MIT disaggregate modelers.[70]

Once returned to Berkeley, McFadden launched a major research program on disaggregate modeling at the Institute of Transportation Studies.[71] At that time, the Bay Area Rapid Transit (BART) system was under construction, and McFadden decided to transform this mass transit project into a kind of outdoors experimental site for the new modeling approach, seeking to answer the question: would disaggregate behavioral tools be able to forecast travel demand accurately for an entirely new transportation mode? Coming under the motto "Zones don't travel; people travel,"[72] the project produced many results, some of which seemed quite impressive. Based on data collected in 1972, the model forecasted that BART would carry 6.3 percent of work trips; in 1975, counts showed that it was actually carrying 6.2 percent! Too good to be true? Commenting on these figures several years later, McFadden would write that "to some extent, we [he and his team] were right for the wrong reasons. We overestimated people's willingness to walk to public transport, so we thought that auto or bus access to BART was less important than it has turned out to be. We overestimated bus use in the presence of the BART system."[73] To add insult to injury, problems in forecasting were compounded by some failures regarding model transferability, both between urban and suburban residents as well as between cities.[74] However, there were also several items of good news. Though "in terms of fits to data and forecasting accuracy, the disaggregate models were comparable to aggregate models, but not significantly better," overall, the project "found that it is feasible to collect sample survey data and calibrate disaggregate behavioral models at very favorable cost levels relative to traditional aggregate methods." Policy analysis using the new modeling approach proved "feasible and cost-effective"[75] as well. Moreover, the project left a lasting legacy. Conducted by a large team, the research study succeeded in winning a number of people over to the cause of disaggregate travel demand forecasting. Kenneth E. Train, an economist who had received an A.B. from Harvard in 1974, and from 1979 a member of the

Berkeley faculty, was one of the four participants in the BART modeling project who defended their doctoral dissertations at Berkeley between 1976 and 1979.[76] A writer of calibration software for disaggregate modeling,[77] Train would also author a "masterful" book on discrete choice models in 2003[78] and put together a distance learning course on discrete choice methods, described as excellent by a seasoned travel modeler in 2008, for travel modelers with a strong quantitative background at least.[79]

Still in Berkeley, in the fall of 1978 several graduate students were about to be used as guinea pigs on a course entitled "Advanced Topics in Transportation Theory." Carlos F. Daganzo, a Spaniard with a Ph.D. from the University of Michigan in 1975 and a former MIT academic, wanted to class-test the first draft of his *Multinomial Probit: The Theory and Its Application to Demand Forecasting*, published in 1979.[80] As its title indicates, the text was essentially devoted to a specific member of the family of discrete choice models, the so-called multinomial probit (MNP) model, which had been used already in its binary version (two alternatives) by Lisco some ten years earlier in Chicago. In contrast to most proponents of the disaggregate approach in the 1970s, who ranged themselves, in the manner of McFadden himself, under the banner of the multinomial logit (MNL) model, Daganzo was frankly an advocate of the probit version. MNP and MNL models share much: both are modeling procedures based on the random-utility maximization principle, and both posit that the utility function is divided into a deterministic part and a stochastic one. However, they differ from each other in their assumptions about the properties of the *stochastic* component, a difference that entails several serious consequences. More general and flexible, the *probit* variant is unfortunately more demanding than MNL one from a mathematical and computational point of view. Yet, while easier to manipulate, the *logit* variant suffers in turn from the fact that it can lead, under certain circumstances, to counter-intuitive (and false) outcomes, especially because of the independence of irrelevant alternatives property, definitively both its "blessing" and its "curse."[81]

By the mid-1980s, several scholars in Cambridge, Massachusetts, Berkeley, and Chicago had been won over to the cause of the disaggregate approach.[82] The latter was then displaying several characteristics indicative of a young but thriving subdiscipline. It was served by a cohesive academic community whose members were adhering to the same paradigm, erected on the random utility theory of the individual's choice, and applied to travel decisions. This consensus around a core set of common hypotheses opened up a whole territory to explore. While the first generation of disaggregate models in the 1960s had solely focused on mode choice, which is the third stage of the FSM, the new approach was now applied to a series of other travel-related options, including trip destination—an issue dealt with by the trip distribution step within the FSM framework—and travel frequency, car ownership, departure time, residential location, and housing. The emphasis also progressively shifted from the general principles underlying the new approach to more technical issues—the equivalent of the famous "puzzles" characterizing Kuhnian normal science. What is the most appropriate form of the utility function? Given the probabilistic nature of disaggregate models—which rules out the use of simple calibration techniques such as linear regression analysis[83]—how can the various parameters of the model be estimated effectively, and its validity tested? What is an effective data collection survey for disaggregate forecasting techniques? These and other questions of this nature energized both graduate students and more seasoned scholars across the nation.[84] They even offered the proponents of the new approach the luxury of some controversies and internecine oppositions, often a sign of good health for an academic discipline. As has been stated, a majority of travel demand modelers opted for the logit class of the discrete-choice family and spared no effort to overcome the drawbacks with which it was saddled, those related to the IIA property in particular.[85] However, the probit model also had its faithful backers. And while scholars at MIT, including Ben-Akiva, modeled certain travel choices simultaneously, McFadden and Charles River Associates staffers sometimes dealt with the same choice set sequentially, as a proponent of the new approach remarked in a letter, adding that "dogmatism about choice structure is unwarranted," especially given that "choice hierarchies are something about which little is either known or obvious."[86]

By the mid-1980s, despite some persisting fault lines among its proponents, the existence of puzzles still waiting to be solved, and a certain dampening of the initial enthusiasm as, with the passage of time, unanticipated problems emerged and the sheer originality of the new approach was even contested,[87] disaggregate travel demand modeling, already present in nearly all popular urban transportation planning manuals of the day,[88] was both rich in content and stable enough in form to be etched into editorial stone. Having been rather long in the making,[89] *Discrete Choice Analysis:*

*Theory and Application to Travel Demand* was eventually published by the MIT Press in 1985 under the names of Ben-Akiva and Lerman.[90] Still on the MIT Press catalog, it rapidly acquired the status of a classic even outside the field of transportation.[91]

## From Scholar to Practitioner

> Although newer methods offer significant advantages over conventional methods, criticism of newer methods is widespread. Most practitioners view them as excessively mathematical and/or theoretical, cumbersome and jargon-bound, poorly packaged and disseminated, difficult to understand in lay terms, of uncertain precision and accuracy, and of questionable relevance to the practicing profession. Few real-world tests have been made of the value of many procedures, and thus their increased usefulness over existing or more traditional techniques is not clear.
>
> —David T. Hartgen, "Executive Summary," in *Travel Analysis Methods for the 1980s*, ed. Transportation Research Board (Washington, DC: Transportation Research Board, 1983), 3

The author of the preceding lines is David T. Hartgen, holder of a Ph.D. degree from Northwestern University in 1970, who was then working for the New York State Department of Transportation. He was not the only one to think that way. In the early 1980s, several academics advocating disaggregate modeling, including Lerman and Joel L. Horowitz, a Ph.D. recipient in physics (1967) then at the Department of Geography and Economics of the University of Iowa, were also aware of and regretful about the gap between the "state of the art," represented by the new approach, and that of practice, dominated by four-step modeling, in the field of urban travel demand modeling.[92] Lerman, Horowitz, and Hartgen were three of the some seventy-five experts from government, the research community, and consulting firms who gathered, thanks to federal funding, at Easton, Maryland, on October 3–7, 1982, to think and talk about "Travel Analysis Methods for the 1980s."[93] At the time when the various workshops around which the Conference was organized were unfolding, the contrast between the potentialities of disaggregate modeling and the limited number of its practical applications—the earliest of which was in all likelihood a study performed by the firm Peat, Marwick, Mitchell & Co in 1970[94]—seemed indeed very sharp.[95] However, the theory-practice gap would progressively shrink. The early teaching of disaggregate methods in a handful of elite academic institutions, including MIT, Northwestern University, and University of California, Berkeley, followed by the gradual introduction of the new approach in a growing number of universities across the nation,[96] would eventually pay off. In addition to teaching, a number of other factors at work would help disaggregate modeling spread across the UTDM social world. As a result, in the 1990s and the 2000s the new approach, essentially in its logit variant, would become the standard operational forecasting technique in the United States for travel mode choice modeling.

The 1982 Conference at Easton, geared toward bringing together academics and professionals, was part of a chain of similar events, the origin of which can be traced back ten years earlier to Williamsburg, Virginia. There, in December 1972, a five-day conference took place. Sponsored by the U.S. Department of Transportation and conducted by the Highway Research Board, it was devoted to the issue of updating the then urban travel forecasting techniques in order for them to keep pace with "changing values in society and the increasing complexity and interdependence of urbanized and industrialized society."[97] Nearly all the people on the front line of disaggregate modeling at that time participated in the conference, and so did the most important consultancies specializing in urban traffic forecasting.[98] The Williamsburg conference, widely regarded as a success by the UTDM social world, would be emulated very soon, as in July 1973 several of the attendees decided to gather again in South Berwick, Maine, to debate the issue of "Behavioral Demand Modeling and the Evaluation of Travel Time."[99] This time the conference was held by the Value of Time Task Force, an entity created in the early 1970s within the HRB. Chaired by Stopher, and populated by people aiming to bring disaggregate modeling into the mainstream of transportation planning, the Value of Time Task Force was rapidly reformed into the Committee on Traveler Behavior and Values.[100] Led in the 1970s by two pioneers of disaggregate modeling, Stopher and Hartgen,[101] and immediately welcoming several supporters of the new approach, including Koppelman, Kraft, and Lisco,[102] the Travel Behavior and Values Committee would play a significant role in spreading disaggregate forecasting techniques to practitioners. In particular, it would regularly organize during the annual meetings of the Transportation Research Board (TRB)—the new name of the Highway Research Board after 1974—specific sessions entirely devoted to the various aspects of the new approach while publishing the most important lectures given at the meetings in *Transportation Research Record*,[103] the TRB's official journal and one of the most popular periodic publications with urban travel modelers.[104] In the 1970s and 1980s, following in the footsteps of the Williamsburg and South Berwick events, the Travel Behavior and Values Committee continued promoting the cause of disaggregate travel behavior modeling through a series of regular international conferences across the world.[105] In the 1970s, two of these, which took place in 1975 and 1979 respectively, held workshops explicitly dedicated to the practical issue of the implementation of operational disaggregate models.[106] And while at the beginning of the conference series in 1973 54 percent of the participants were from universities, in 1985 over 60 percent were practitioners from governmental organizations (51 percent) and consultancies.[107] All these conferences and meetings, to which should be added the now famous seminar organized by the Department of Transportation in December 1976 in Daytona Beach, Florida,[108] eventually paid off. Leading to rich proceedings and other heavy publications, they did a lot to spread disaggregate modeling to increasingly wider circles of practitioners. Supplementary channels also greatly helped the new approach penetrate the sphere of practice.

In 1977, the Federal Highway Administration published a report entirely devoted to disaggregate modeling.[109] Authored by Bruce Spear, a former Ph.D. supervisee of Stopher, the report was essentially concerned with the presentation of three areas of transportation planning, where the new approach had already been applied. About a decade later, the Urban Mass Transportation Administration,[110] the FHWA's counterpart for mass transit systems, promoted a "self-instructing course" intended for transportation practitioners wanting to familiarize themselves with disaggregate forecasting techniques. Horowitz, Koppelman, and Lerman, three academics heavily involved in the development of this modeling approach, therefore set themselves the goal of crafting a complete course in order to explain "what disaggregate mode choice models are, how they work, and how they can be applied to practical problems, and to do this with a minimum of mathematics and jargon."[111] In 2006, Koppelman allied himself with his former Ph.D. student Chandra Bhat in order to seriously refurbish this course by producing a new text that was more rigorous in the mathematical details than its 1986 ancestor, while heavily capitalizing on the "increased awareness and application of discrete-choice models over the past decade."[112] Just several months after its release, the course would be hailed as one of the (seven) best books available on transportation demand modeling.[113] But let us stay in the 1970s a little longer.

As has been amply seen in chapter 2, throughout the 1960s the various travel forecasting techniques forming the FSM were able to be diffused to the UTDM social world, in large part due to the development of (free) software under the aegis of the Bureau of Public Roads. In the 1970s and 1980s the gospel of disaggregate modeling would also spread beyond the confines of the university and a handful of consulting firms in the vanguard of the profession, thanks to the design of a series of computer suites facilitating its implementation. The so-called ULOGIT module, a software for calibrating disaggregate models of the logit family according to the maximum likelihood principle, was one of these computer packages. ULOGIT was not the first logit-related software in the United States. In fact, consultancies heavily involved in the early development stages of the new approach, such as Peat, Marwick, Mitchell & Co. and Charles River Associates (CRA), as well as individual scholars, including Stopher, had already conceived proprietary computer programs for their own disaggregate modeling undertakings.[114] ULOGIT was nevertheless the first one to be available to many planners as part of a much larger, and extremely popular, program library for the IBM system 360/370, eventually known as Urban Transportation Planning System. This was developed during the 1970s under the guidance of Robert Dial by the Urban Mass Transportation Administration, with the objective of treating computationally all aspects of multi-modal transportation planning.[115] At the end of the 1980s, UTPS was the single most prevalent computer resource used in traffic forecasting by practitioners.[116] ULOGIT was essentially the product of the efforts of Christoph Johann Witzgall (born in 1929), a gifted German-born mathematician closely associated with the National Bureau of Standards, where he had previously worked on shortest path algorithms used in the assignment stage of the four-step model.[117] It seems that the program crafted by Witzgall, assisted by a team of people from the NBS and three staffers from the transportation firm De Leuw, Cather & Company, was so fast and produced such good results that McFadden, a connoisseur of this kind of programming since he himself had (co)designed a similar software called QUAIL,[118] asked Dial, not without a dash of jealousy, how the federal administration had managed to have a software that performed so well. It was easy, Dial replied, "I just asked Chris to do it for me."[119]

In addition to the calibration operations performed by ULOGIT, other aspects of disaggregate modeling were also incorporated later into UTPS, as part of the UMODEL module, for example.[120] As the time went by, and especially after the advent of microcomputers, urban travel demand forecasters willing to carry out traffic studies with the help of disaggregate models could progressively resort to a number of other statistical software packages, some of which had not been specifically designed for handling transportation issues.[121] And last but not least, they also could increasingly rely on the disaggregate modeling routines designed and commercialized by all major transportation planning proprietary software firms that appeared on the traffic forecasting scene from the early 1970s on (see chapter 4).[122] Yet, the new approach would probably have encountered much more difficulty in forcing its way into practice had the practice-oriented segment of the UTDM social world not itself undergone a significant transformation in the 1970s. In the absence of a dedicated term, let me call this transformation the *academicization* of major consultancies involved in urban traffic forecasting, a term referring here to the growing interpenetration of the academic world and the universe of consulting firms. Heavily populated by academics as well as by Ph.D. and master's degree holders, cultivating strong links with knowledge-related organizations on a systematic basis, these new firms eventually made themselves into the essential repository of urban forecasting traffic expertise in the United States (see part III).

Employing some 280 people, around fifty of whom specialized in urban travel demand modeling in the early 2010s, Cambridge Systematics has been one of the major transportation consulting firms in the United States since the 1970s.[123] Founded in 1972, the consultancy's first office was located less than half a mile from the heart of MIT, of which it was in large part an outgrowth. As a matter of fact, among the consultancy's founding principals were Manheim, who had already performed consulting activities for Barton-Aschman Associates from 1969 to 1972; Paul Osborne Roberts (born in 1938), a Ph.D. holder from Northwestern University and a former Charles River Associates collaborator; Wayne M. Pecknold (1940–2000); and Albert Scheffer Lang (1927–2003), all of whom were then members of MIT faculty. Before joining Cambridge Systematics as its fifth founder, William A. Jessiman, a graduate of and 1963 master's degree holder from

MIT, had headed the Boston branch of Peat, Marwick, Livingston & Co., formerly Traffic Research Corporation, a firm with which Manheim himself had also collaborated from 1968 to 1969.[124] The Cambridge-based university's footprint in the new firm would prove large and durable, as the founders brought with them several MIT people, including Earl Ruiter, Manheim's close collaborator on many research projects,[125] Ben-Akiva, Lerman, Koppelman, and Leonard Sherman, also a Ph.D. holder under Manheim's supervision.[126] And when Cambridge Systematics opened a second office on the West Coast in 1976 to draw upon the resources of the Bay Area, McFadden and his student Train were among the people who staffed it. With personnel consisting of such highly competent persons in disaggregate modeling, it comes as no surprise that Cambridge Systematics has long associated its name with the new approach. Among other things, in the second half of the 1970s the firm would be instrumental in the design and implementation of the modeling system for the Metropolitan Transportation Commission (MTC), the planning organization for the San Francisco Area, which is now considered to be the first large-scale operational disaggregate modeling suite in the United States. This undertaking took about two years (1975–1977) and cost nearly $250,000 (around a million dollars in 2018), while the resulting system proved rather expensive in terms of both time and money, since a complete running required about one week of elapsed time and cost over $6,000 in computer charges. As a result, the customer was not initially entirely pleased with the goods delivered by the consultants. Already well versed in four-step (aggregate) modeling in the recent past,[127] MTC staffers expressed, in particular, the concern that "the models may not do the job we need or may not do it within reasonable time and resource constraints."[128] Be that as it may, the pioneering work done by Ben-Akiva and his Cambridge Systematics fellow modelers did become the starting point for many other disaggregate modeling projects across the nation, in which the Massachusetts-based firm was often present. It is worth noting that the originality of Cambridge Systematics within the realm of transportation consultancies in the 1970s did not elude the attention of such a seasoned practitioner as Thomas Deen.[129] In 1977, the then head of Alan M. Voorhees & Associates, while commenting on several of his own company's professional rivals, would jealously note: "Others are favorably tied to universities where products from academic federally financed research can be sold commercially and where the brightest and best of each academic class can be funneled to the firm" (figure 3.3).[130]

The work conducted by Cambridge Systematics and its partners for the Metropolitan Transportation Commission seemed to have served as a catalyst and ushered in an era of the growing spread of disaggregate modeling to practitioners.[131] By the early 1990s, a survey conducted on travel demand forecasting processes used by ten large metropolitan planning organizations with populations of over two million[132] showed that eight of them had already resorted to disaggregate modeling for the mode choice step within their overall FSM framework, with the logit variant being used by seven agencies.[133] Both a sign and a factor in accelerating this growing use of the new approach, the 1998 (and updated) version of a reference and widely read manual on urban travel modeling contained several passages dealing with the disaggregate approach.[134] In the 2000s, the cause of disaggregate modeling had been definitively espoused by the American urban traffic forecasting community. In 2007 a web-based survey of more than two hundred MPOs, complemented by in-depth interviews with a smaller sample of agencies, reported that "almost all mode choice models" used were "either multinomial logit or nested logit."[135] Utility-based disaggregate modeling is now commonplace in the United States. However, its massive adoption by the UTDM social world neither amounts to a clear victory of its advocates nor to an utter defeat of their opponents. As a matter of fact, the passage of time blurred its initial sharp outlines of radicalism, and far from completely dethroning aggregate approaches from urban traffic forecasting as its more enthusiastic followers in the late 1960s and the early 1970s intensely wanted and fiercely fought for, the new approach has simply been smuggled into the general four-step modeling framework to essentially occupy the "box" of the mode choice stage only, while living in symbiosis with the various aggregate models that have been developed for the other steps. In point of fact, in the second half of the 2000s, trip distribution models (second step of the FSM) were still overwhelmingly based on the good old gravity model, albeit in an updated version,[136] and only some large MPOs reported using logit destination choice models.[137]

So then, by eventually monopolizing the mode choice step, has disaggregate modeling simply locally altered the FSM, with the latter managing to preserve its original four-part structure and essentially remaining aggregate

## FIGURE 6

### STAFF BY LOCATION AS OF 10/1/77
(Excludes Subsidaries)

| Location | Department | Staff |
|---|---|---|
| **AMV** | | |
| McLean | Transportation Operations | 11 |
| | Transportation Planning | 11 |
| | PEI | 5 |
| | Rail Systems | 2 |
| | Administration | 4 |
| | F&A | 17 |
| | International | 2 |
| | Technical Services | 13 |
| | Subtotal | 68 |
| Orlando | | - |
| San Francisco | | 10 |
| Honolulu | | 6 |
| Los Angeles | | - |
| Denver | | 2 |
| St. Louis | | 7 |
| Boston | | 14 |
| Caracas | | 7 |
| Detroit | | 2 |
| Dallas | | 1 |
| Tehran | | - |
| Phoenix | | 3 |
| Pittsburgh | | 1 |
| Melbourne | | 2 |
| Chicago | | 1 |
| | Subtotal | 56 |
| AMV - Grand Total | | 124 |
| **RDSA** | | |
| Lake Success | | 22 |
| Minneapolis | | 4 |
| Atlanta | | 8 |
| Los Angèles | | 7 |
| New York | | 4 |
| Washington, D. C. | | 3 |
| RDSA Grand Total | | 48 |
| TOTAL RDSA/AMV | | 172 |

on the whole? On the face of it, the answer seems to be an unqualified yes. On second thought, however, the disaggregate approach has done more than that. As a matter of fact, the new modeling line focusing on the individual stimulated the development of original ways of capturing people's travel behavior and made itself into the main channel through which new data collection, new treatment procedures, and new actors entered urban travel demand modeling.[138]

## Data Gathering and Treatment Techniques for Disaggregate Modeling

By contrast to aggregate models, often described as data-hungry monsters, the disaggregate approach is much more economical in terms of travel information. As early as in the mid-1970s, proponents of disaggregate modeling found that samples as tiny as 300 individual trips were sufficient for statistically good fits.[139] However, the impact of the new modeling approach on the design of travel data collection procedures and instruments has gone beyond the sole sample issue. In point of fact, its increasing adoption for the modeling of the mode choice step was also reflected in a growing attention to a new kind of *travel data*, including vehicle occupancy and availability, parking costs, or information on the modes of access to and egress from public transportation.[140] And there was more. Not surprisingly given its individual-centered character, the disaggregate approach brought with it an interest in data about an *individual's subjective* evaluations of his or her travel options, which led in turn to the importation into urban traffic forecasting of a new kind of data collection technique, the stated preference (SP) approach.[141] What was all this about?

Urban traffic forecasting models need various kinds of data. For years the information required for their calibration, validation, and utilization for

**Figure 3.3**
Alan M. Voorhees & Associates staff by location as of October 1, 1977. *Source:* Thomas B. Deen, "RDSA/AMV: What and Who Are We? What Do We Want to Be? (A White Paper Prepared to Stimulate an Intra-Company Dialogue on Our Company's Current Course and Future Directions; to Be Presented to the Steering Committee, November 19, 1977)," typescript, figure no. 6. George Mason University Libraries. Special Collections Research Center, Thomas B. Deen Papers, C0106, Box 6, Folder 4.

prediction had been "objective" in nature, since, traditionally, travel surveys such as the Home Interview origin and destination traffic studies were designed to collect data describing *actual* travel behavior, be they actual modes, actual travel times, or actual destinations. This kind of information is usually referred to as revealed preference (RP) data, since decision-makers *reveal* their underlying preferences through the choices they *actually* make in the travel marketplace.[142] As their name indicates, "stated preference" surveys, a technique having been used extensively in marketing research since the mid-1970s, are, on the contrary, information gathering techniques that do not rely on real-world behavior but, instead, on *statements* made by individuals who are asked about what they *would* choose to do, or how they *would* rank or rate certain options *in a hypothetical situation*.

Although until the early 1980s urban travel demand models had exclusively consumed "objective" data, relying on actual choices and travel patterns, interest in the subjective realm of the travel-maker had developed within the UTDM social world much earlier. Thanks to public money, during the 1960s scholars at the universities of Pennsylvania and Maryland, researchers at Stanford Research Institute, and even people working for private for-profit firms—including Chilton Research Services, a division of Chilton Company—carried out a series of studies in an attempt to gain a better understanding of the attitudes of trip-makers toward the various attributes, such as cost and time but also reliability and the state of the vehicle, associated with the various transportation modes and their relative importance in the traveler's decisions and choices.[143] In 1973, just a few years after these pioneering works saw the light of day, two papers, now hailed as seminal, seriously alerted the transportation modeling community to the appeal of a family of methods for evaluating an individual's preferences when the practitioner faces new situations that not only do not yet exist but of which the people whose behavior is to be modeled have little or no experience at all.[144] For Jordan J. Louviere—who gained a Ph.D. from the University of Iowa in 1973[145] and who worked as an academic geographer till 1978 when he was recruited to the marketing department of his university and changed careers—it would be the first contribution in a long series of highly influential papers on SP techniques and their utilization in transportation, culminating in the now classic book coauthored with two of his colleagues in 2000.[146] In 1978 Louviere would ally himself with Steven Lerman, the then young MIT proponent of the disaggregate approach, to co-sign an important article exploring the potential of a synthesis of RP techniques and SP methods in the estimation of a travel demand model.[147] A few years later, the same Louviere would introduce to SP techniques four academics affiliated with Dartmouth College, and the future authors of the first national guide to travel demand modeling not based on actual behavior.[148] However, despite this early interest in stated preference methods, especially by people and organizations involved in disaggregate modeling,[149] SP techniques truly began to be accepted among transportation professionals in the United States only during the 1990s.[150]

This renewed interest in collecting SP data was not coincidental.[151] Two distinct but largely concurrent issues seemed to be the driving forces that helped SP techniques become increasingly popular with the UTDM social world from the 1990s on. In the late 1980s, the concept of a high-speed rail system, already a tangible reality in Europe and in Japan, eventually reached U.S. territory.[152] As there was no existing operating intercity service akin to high-speed rail, SP techniques started to be regarded as the most appropriate method to determine *potential* patronage for the new system. Thus, by the early 1990s, the consultancy CRA had already used its SP data collection and analysis method, called value perception analysis (VPA), in large studies to model and forecast ridership and revenue on proposed high-speed rail lines in Canada and a number of U.S. states and areas, including Texas, California, Florida, Nevada, and the Northeast Corridor (Boston-New York-Washington, DC). VPA surveys were administered to travelers on each existing intercity mode (air, auto, bus, conventional rail), and the latter were asked to choose between the mode used and the proposed new rail mode alternative, where each mode was described by their travel times and costs.[153]

This period also witnessed the emergence of new and original options for handling urban transportation problems, including travel demand management that encouraged the use of ridesharing modes, such as carpool, vanpool, and transit, and/or discouraged the use of drive-alone auto, including carsharing,[154] and pricing strategies (tolls) (more in part III). Facing the question of how the traveling public would respond to such original strategies, traffic forecasters resorted, as they did in the case of high-speed rail systems, to SP methods in order to circumvent the lack of data based on actual travel behavior. At the turn of the 1990s, despite persisting concerns over the validity of the results obtained by SP techniques and their transferability from a hypothetical context to the real world, the new data-gathering approach, when applied, alone or in conjunction with more traditional RP methods,[155] to urban traffic forecasting could boast both a rich theoretical heritage[156] and a—though still short—operational past. As a matter of fact, by 1996, Portland, Oregon, and the state of New Hampshire had already conducted a series of stated preference studies in conjunction with their household travel surveys, as did Dallas for its travel demand forecasting process. San Francisco had also used a combined SP/RP methodology in its 1996 survey conducted as part of a congestion pricing experiment on the San Francisco—Oakland Bay Bridge.[157] No wonder that by the end of the 1990s SP techniques had obtained pride of place in popular travel survey manuals and other widely read travel forecasting-related publications,[158] and had become the subject matter of specific courses delivered by SP specialists, including Louviere himself.[159]

The previous paragraphs have provided the reader with a series of milestones in the development and use of the SP method in urban traffic forecasting up to the 1990s, when this specific survey technique became a permanent fixture within this modeling field. Though it is impossible to know with certainty what would have happened if some events in the past had played out differently, one can plausibly contend that the incorporation of stated preference procedures into urban travel forecasting—and the same applies to other innovations in travel survey procedures since then—would probably never have occurred had the community of those involved in travel data collection and treatment not progressively undergone its own profound metamorphosis. Despite the fact that the traditional Home Interview O&D traffic study had largely benefited from the high-quality expertise possessed by the Bureau of the Census in data gathering in the early 1940s,[160] it was only at the end of the 1970s and the early 1980s that the field of household travel surveys began to seriously engage in a dialogue with surveys in other fields, as transportation practitioners became more and more cognizant of a vast and increasingly international literature on conducting surveys of various types in domains other than transportation.[161] Both the federal administration and Transportation Research Board would play a decisive role in this process, through the organization of a series of largely attended conferences[162] as well as via the funding of large R&D programs often leading to widely read publications.[163] The importance of the part taken by such public stakeholders as the FHWA and the TRB, especially through its Subcommittee on Survey Methods set up in 1984 and which became a full-fledged committee six years later, in the spread of original survey practices to the UTDM social world should not obscure the heavy involvement of the private sector. Indeed, from the early 1980s an increasing number of market research consultancies saw their competencies and skills being systematically solicited by transportation agencies.

Anderson, Niebuhr & Associates was set up in 1974 by John F. Anderson and Douglas R. Berdie, then members of the University Measurement Services Center in Minnesota, and the authors of an early publication that quickly became a standard reference and training text for people using survey research techniques.[164] Founded in 1981, Schulman, Ronca & Bucuvalas, Inc. (SRBI) is another consultancy created by Ph.D. holders, which has been involved in many household travel surveys up to the present.[165] Starting with a modest $1,000 investment in 1982, the ETC Institute, established and owned by mathematician Elaine Johnson Tatham, who earned a doctor of education (EdD) degree in educational and psychological research in 1971, has also progressively grown into a nationwide firm that frequently takes part in travel surveys. Another market research firm that has been transformed into a powerhouse in the field has been NuStats, set up in Austin, Texas, in 1984.[166] It was founded by Carlos Humberto Arce, a Ph.D. holder from the University of Michigan, where he was a study director at the Survey Research Center,[167] and Johanna Zmud, a Ph.D. holder in communication research in 1992,[168] and currently (2022) principal consultant at Resource Systems Group (on the latter, see chapter 6). Established by Werner Brög, who studied sociology, social psychology, and communication sciences at the University of Munich, in 1972, *Socialdata*, while locating in Europe, would nevertheless play a significant part in the modernization of the household travel surveys in the United States from the early 1980s on.[169] In the mid-1990s, these and some other private firms[170] had already colonized the American travel survey territory, since by then the most common procedure followed by MPOs and the states was not only to enlist the services of a consultancy but also to task it with the primary responsibility for travel survey administration as well.[171] The increasing usage of GPS-based techniques in travel surveys from the early 2000s on[172] would further strengthen the presence of the private sector by progressively adding several new items to the then existing list of major travel survey practitioners and experts. Some of them were new knowledge-intensive firms specializing in travel surveys, such as GeoStats, a small but highly innovative company connected to Georgia Tech and closely associated with the name of Jean Wolf, a Ph.D. holder from the Georgia-based engineering school in 2000.[173] Others were existing transportation consultancies long involved in traffic forecasting, including Cambridge Systematics, and, especially, Resource Systems Group.

Both utility-based disaggregate modeling and its main accoutrements, especially stated preference techniques for travel data gathering and treatment, clearly revamped the four-step model, enhanced its effectiveness, and undoubtedly contributed to its longevity. The same applies to another set of original modeling practices taking place within the fourth step of the FSM during the 1970s and 1980s, which went down in history as the user equilibrium assignment procedures, and which is the core subject matter of the following chapter.

# 4 Seeking Equilibrium on the Transportation Network

## 1973: The Annus Mirabilis of the Traffic Equilibrium Problem

It was in August 1973 that Larry Joseph LeBlanc (born in 1947), a young mathematician from Loyola University of New Orleans, submitted his Ph.D. dissertation, dealing with the assignment of traffic flows in an urban network, the last stage of the four-step model, to Northwestern University.[1] As the research project leading to the doctorate was supported in part by UMTA, the author felt obliged to send a copy of his final report to Robert Dial. Dial was not only chiefly responsible for the comprehensive computer program library that the federal administration was developing in the 1970s under the name of Urban Transportation Planning System, but he was also a modeler specializing in traffic assignment techniques.[2] Upon leafing through the report placed on his desk, Dial was won over to LeBlanc's work on the spot. He therefore set himself to designing the code for LeBlanc's algorithm, while immediately asking his friend Geoffrey J. H. Brown, then a consultant with Creighton Hamburg & Associates,[3] to implement it in the federal government-sponsored program library.[4] This was an easy task for Brown. A member of the informal group of programmers gathering around William L. Mertz at 499 Pennsylvania Avenue, Washington, DC, in the early 1960s,[5] he was already the author of many codes, including a gravity model program.[6] Following the joint action of Dial and Brown, from the second half of the 1970s LeBlanc's research study would be featured as an option in UROAD, the UTPS module dedicated to the assignment step (figure 4.1)[7]

In hindsight, Dial's intellectual excitement in the face of LeBlanc's work was perfectly understandable. The thesis submitted by the young operations

IDENTIFICATION

TITLE: HIGHWAY TRAFFIC ASSIGNMENT PROGRAM (UROAD)

WRITTEN BY: ROBERT B. DIAL (U.M.T.A.)
GEOFFREY J. H. BROWN (CREIGHTON, HAMBURG, INC.)
STEVEN C. GIBSON (VOORHEES & ASSOC. INC.)

SPONSORS: FEDERAL HIGHWAY ADMINISTRATION
URBAN MASS TRANSPORTATION ADMINISTRATION

SUMMARY

UROAD PERFORMS THREE FUNCTIONS:

(1) PATHFINDING (TREE OR BUSH BUILDING)
(2) PATH SKIMMING, AND
(3) TRAFFIC ASSIGNMENT, WITH OR WITHOUT CAPACITY RESTRAINT, USING
 (A) ALL-OR-NOTHING ASSIGNMENT <14.1>
 (B) ALL SHORTEST PATHS ASSIGNMENT <14.2>
 (C) PROBABILISTIC MULTIPATH ASSIGNMENT <14.3>

ANY OR ALL OF THESE FUNCTIONS CAN BE EFFECTED IN ONE RUN OF UROAD FOR SELECTED OR ALL ORIGIN ZONES.

OUTPUTS OPTIONALLY INCLUDE:

(1) PRINTED DESCRIPTIONS OF PATHS
(2) BINARY FILE OF INTERZONAL MATRICES OF COST, TIME, AND/OR DISTANCE
(3) LINK AND TURN VOLUME REPORTS AND/OR
(4) SUMMARY STATISTICS RELATED TO TRAFFIC ASSIGNMENT
(5) AN UPDATED BINARY NETWORK DESCRIPTION CONTAINING ASSIGNED LINK (AND TURN) VOLUMES AND ASSOCIATED SPEEDS

UROAD READS A HIGHWAY NETWORK FILE IN THE FHWA HISTORICAL RECORD FORMAT AND A (UMTA FORMAT) TRIP TABLE, AND, AT THE OPTION OF THE USER PERFORMS EITHER AN ALL-OR-NOTHING, ALL-SHORTEST-PATHS, OR A PROBABILISTIC MULTIPATH ASSIGNMENT. THE ASSIGNMENT MAY BE ITER- -ATIVE USING A DEFAULT CAPACITY RESTRAINT TECHNIQUE.
THE TYPE OF ASSIGNMENT PRO-CEDURE EMPLOYED BY UROAD DEPENDS UPON THE VALUES OF THE PARAMETERS THETA. <1.0>

UPON COMPLETING THE ASSIGNMENT, UROAD PRINTS A REPORT OF ALL CALCULATED LINK AND TURN VOLUMES AND OPTIONALLY WRITES AN UPDATED NETWORK FILE, APPENDING THE LINK VOLUMES TO EACH LINK RECORD.

CAPACITY RESTRAINT

THROUGH USE OF THE "THETA" PARAMETERS, UROAD CAN BE MADE TO PER- FORM CAPACITY-RESTRAINED ASSIGNMENTS FOR UP TO TEN ITERATIONS. A DEFAULT CAPACITY RESTRAINT MODEL IS INCORPORATED IN SUBROUTINE ROA23. THE USER MAY SUPPLY FORTRAN CODE (THROUGH SUBROUTINE ROA23) TO OVERRIDE THIS MODEL AND TO MODIFY LINK SPEEDS EITHER DURING OR AT THE END OF EACH ITERATION. THE CAPACITY RESTRAINT PROCEDURE research specialist contained, in fact, the first efficient operational answer to a critical question that had intrigued the UTDM social world for years. This was the so-called "equilibrium assignment problem," about which Dial was himself at that time in contact with Professor Marvin Manheim and his team at MIT.[8] What was this question about, and why was it a pressing issue for travel demand modelers? A short reminder as well as a brief history of traffic assignment techniques follows.

## Going from One Point to Another via the Shortest Path

Take a road network—a system consisting of intersection and interchange points called nodes, joined by individual routes, referred to as links, in UTDM parlance (figure 4.2)[9]—and a group of motorists wanting to reach their destinations through this network. Suppose that the drivers are rational beings who do not like wasting their time unnecessarily. They are therefore very likely to take the "shorter path" from origin to destination, the one that minimizes their own travel time.[10] This minimum time assumption underlay the majority of assignment techniques that were developed by the early 1960s for both highway and mass transit networks.[11] Thus, the so-called "all or nothing" assignment procedure, the most basic model of this family of techniques developed in the 1950s,[12] defines for each link a travel time based on its capacity, and assumes that all of the trips between an origin-destination pair are assigned to the shortest path (and nothing on other routes). A rather natural extension of the "all or nothing" principle was to recognize the effect of congestion on travel time. As a matter of fact, if the number of travelers for a given origin-destination pair is large enough to exceed the capacities of the various links composing the

**Figure 4.1**

Extract from the UROAD's identification form. UROAD, an essential part of the UTPS, the comprehensive transportation computer program library developed by the federal government in the 1970s, was dedicated to the assignment step. *Source:* Urban Mass Transportation Administration/U.S. DOT, *UMTA Transportation Planning System Reference Manual* (Washington, DC: n.p., April 1974), [UROAD (01FEB74) 1].

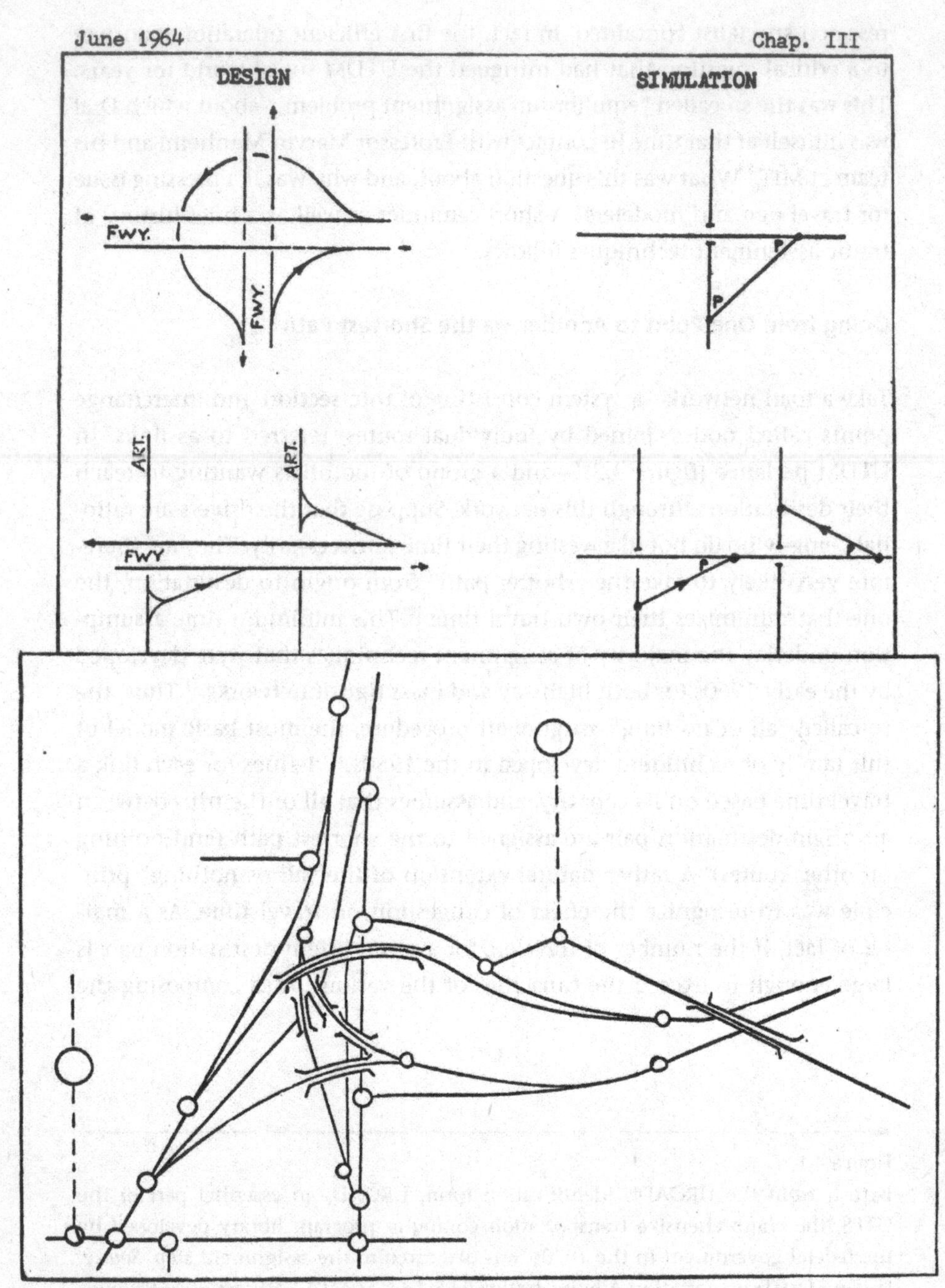


Figure III-12.--Examples of directional coding.

III-29

shortest path, the latter may cease, because of the congestion produced, to be the more interesting route among the available paths. As has been seen in chapter 2, the "capacity restraint" assignment, an original procedure first devised in Chicago, was able to take into account the effects of congestion on travel times, and therefore on drivers' choices.[13] Many other modelers, be they academics or working for consultancies and other planning agencies, immediately followed in the footsteps of Carroll's team in Chicago and crafted various assignment techniques embodying, each in their own way, the general "capacity restraint" principle.[14]

In reaction to this proliferation of assignment techniques, the Bureau of Public Roads and other funding agencies decided to support a series of research projects aiming to compare the then available products on the assignment "market." In the early 1960s, while at MIT, the Briton Brian V. Martin, from Imperial College, London, conducted such a comparison program by coding and running several minimum path algorithms on different networks and computers.[15] However, the most systematic and in-depth comparison study was probably the $99,675 research project conducted by a team from the Yale Bureau of Highway Traffic from February 1, 1964 to November 30, 1966.[16]

Four different "capacity-restraint" assignment methods in use in the first half of the 1960s as well as the older "all-or-nothing" technique were therefore incorporated into a unique computer program to permit a direct comparison on a common network and data inputs, while the results of each procedure were compared with actual ground counts from two road networks, located in the Pittsburgh area and the Raleigh region in North Carolina, respectively. Although the authors of the study felt the necessity to state that "the results or answers from a traffic assignment program still require judgment in interpretation and application," the team after

**Figure 4.2**
Examples of directional coding of a highway network (in directional coding, turning movements and weaving movements may be specified as links in the transportation system). *Source:* Bureau of Public Roads/U.S. Department of Commerce, *Traffic Assignment Manual for Application with a Large, High Speed Computer* (Washington, DC: U.S. Government Printing Office, June 1964), III-29.

"weighting adequacy of results with computation costs" eventually declared that the method devised by Morton Schneider, the gifted member of the famed Carroll team in Chicago, was "the most desirable one to use."[17]

Despite the aforementioned efforts, no consensus emerged on the "best" assignment method in the 1960s, and several members of the UTDM social world continued to explore new avenues. One of them was the *stochastic assignment*, soon associated with the name of Robert Dial. Moving away from the minimum time assumption underlining the then existing assignment procedures,[18] stochastic assignment contended that a motorist in real life does not necessarily always choose the shortest path. It follows that if "there are two ways to go from i to j and if the path times are not much different, then there is a non-zero probability that each will be used by trip makers."[19] For every origin-destination pairing, a non-zero probability of use is therefore assigned to each reasonable itinerary, with the minimum time path having the higher probability of use. Though Dial was neither the first travel modeler to underscore the various problems plaguing the shortest path-based assignment techniques nor the only traffic forecaster to try something new in the second half of the 1960s,[20] he would definitively be the one who on American soil luckily proposed a workable alternative, or at least one largely perceived as such.[21] As the discerning reader will have noticed, Dial, a huge admirer of Voorhees and Schneider, imported into his approach the idea "that a trip has non-zero probability of terminating at any reachable point, regardless of its point of origin,"[22] an assumption that his two heroes had already put into practice in their own models dealing with the trip distribution stage of the four-step model. The reader may have also already discerned that Dial's probabilistic model was in line with another important innovation at that time within urban travel demand modeling, the disaggregate approach, as it was especially applied to the mode choice step (chapter 3). As a matter of fact, just as the proponent of the disaggregate approach gives each mode choice decision by a trip-maker a probability that is dependent upon the utility the latter ascribes to various available alternatives, so every path picked by a driver is granted by Dial a probability of use depending on the route's length. It was not by chance that Manheim along with his colleagues and students at MIT in the mid-1970s, all heavily involved in the development of the then recent disaggregate approach, sought to establish links between Dial's work and the area of modeling they were about to cultivate and promote (figure 4.3).[23]

Thus, by the early 1970s, the community of urban traffic forecasters could resort to a variety of assignment techniques. Moreover, those regarded as the most efficient had been gradually incorporated into the comprehensive computer libraries on transportation planning built by the federal administration in the 1960s and early 1970s as well in some university-produced suites, including the DODOTRANS package developed at MIT,[24] and commercial products, mostly offered by computer companies such as Control Data Corporation (CDC).[25] However, the search for new assignment techniques continued unabated, since the then available models were still regarded by both the academic and the more practice-oriented segments of the UTDM social world as not wholly satisfactory, albeit often for different reasons. Thus, though capable of dealing with the effects of congestion on travel times, the various "capacity-restraint" techniques were not able to guarantee the desired convergence to an equilibrium (stable) solution, even after several iterations. While reflecting much better than its competitors the actual behavior of drivers and being computationally efficient, as it requires only a modestly larger "cost" than the cheapest "all-or-nothing" procedure, Dial's approach nevertheless had difficulty in taking account of the effects of congestion on travel times, and also proved less efficient when alternative routes overlap. What was to be done?

## Addressing the Equilibrium Problem Scientifically

This was what the assignment modeling scene looked like when LeBlanc started to struggle with the traffic assignment problem in his dissertation at Northwestern University. In hindsight, the battle was well worth waging. LeBlanc's research study would immediately be hailed as a groundbreaking and long-awaited work. It made operational an assignment principle that—while articulated several years earlier—had remained a no starter for lack of adequate computational techniques. Let us take another trip back in time and start with the general problem to which LeBlanc's contribution was an effective solution.

In the mid-1920s, Frank Hyneman Knight (1885–1972),[26] an economist now celebrated as one of the founders of the controversial Chicago School,

Appendix A. Relation of Aggregate and Disaggregate Forms

We can show the relationship between disaggregate and aggregate forms (under certain assumptions) as follows:

Consider the space of choices (F,D,M,R) open to the traveller t residing in zone i:

(1) $(f,d,m,r) \in (F,D,M,R)$

where f = specific trip frequency

d = specific destination

m = specific mode

r = specific route

and F,D,M,R = set of all combinations of f, d, m, and r.

We assume the multinomial logit form of disaggregate demand model, to express the probability that individual t chooses any particular choice (f,d,m,r):

$$(2) \quad P_t(f,d,m,r : F,D,M,R) = \frac{e^{X_{fdmr}\Theta_{fdmr}}}{\sum_{f,d,m,r \in \{F,D,M,R\}} e^{X_{fdmr}\Theta_{fdmr}}}$$

where $X_{fdmr}$ = attributes characterizing an alternative trip and individual t; and

$\Theta_{fdmr}$ = parameters

As any joint probability function, this simultaneous form can be written also as a sequential form:

$$(3) \quad P_t(fdmr:FDMR) = P_t(f:F)P_t(d:D_f)P_t(m:M_{fd})P_t(r:R_{fdm})$$

where F = the set of all relevant trip frequencies,

$D_f$ = the set of all relevant destinations for trip frequency f

$M_{fd}$ = set of all relevant travel modes for trip frequency f and destination d

was embroiled in a debate with the Briton Arthur Cecil Pigou (1877–1959) about the effects of competition on the investment of resources in industries of increasing cost. While not dealing with transportation matters in particular, both Pigou and Knight, the latter even more explicitly, articulated the first formulations of the notion of *equilibrium* in a transportation network. Their example consisted of two roads sharing the same origin and destination but having different characteristics and used by an increasing number of vehicles. In Knight's words, "as more trucks use the narrower and better road, congestion develops, until at a certain point it becomes *equally* profitable to use the broader but poorer highway."[27] Some thirty year later, the aforementioned passage would be reproduced by three scholars: the German-born Martin Joseph Beckmann (1924–2017),[28] C. Bartlett McGuire (1925–2006), and Christopher B. Winsten (1923–2005). After quoting Knight, the three coauthors of *Studies in the Economics of Transportation*,[29] proceeded to a reformulation of the latter's insights, by redefining equilibrium and articulating the mechanisms leading to it, first for a single road, then an entire network.[30] Much better versed in mathematics than Knight was three decades earlier,[31] Beckmann, McGuire, and Winsten were able to transform their prose into numbers. They wrote down the equations expressing the state of equilibrium in a network, explored in detail their properties, and gave two hypothetical examples "intended to show a possible method of computation and to exhibit the process of convergence toward a stable equilibrium."[32] By laying out and mathematically dealing with the conditions that underlie the equilibrium state in a transportation network, Beckmann and his two fellow Cowles Commission members were about to rediscover, apparently without fully realizing it,[33] what every travel demand modeler now knows as "Wardrop's first principle," which the

**Figure 4.3**

Extract of a September 24, 1973 memorandum from Marvin L. Manheim (1937–2000) to Robert Dial (born in 1934) further amplifying their telephone conversation on August 17, 1973. *Source:* Memorandum from Marvin L. Manheim to Robert Dial, "Proposed Direct Equilibrium Approach Using Exponential Functions," typescript, dated September 24, 1973, 6. Massachusetts Institute of Technology, Libraries, Department of Distinctive Collections, Marvin L. Manheim Papers, MC-0330, Box 8, Folder 16.

British mathematician and OR specialist John Glen Wardrop (1922–1989)[34] had already announced in 1952, albeit rather as an aside. According to this principle, or criterion in Wardrop's parlance, when a network is in a state of equilibrium the "journey times on all the routes actually used are equal, and less than those which would be experienced by a single vehicle on any unused route."[35] To put it another way, under equilibrium conditions an individual driver traveling in a congested network *cannot* reduce his path travel by *switching* routes. Thanks to the contribution of Beckmann and his collaborators as well as the insights of Wardrop, the notion of network equilibrium was therefore clearly articulated in the mid-1950s. However, it was not until LeBlanc's undertakings in the early 1970s that a workable algorithm was devised to assign traffic in a *real* network in a way that satisfies the network equilibrium conditions. Yet, LeBlanc's contribution did not come out of the blue. The twenty or so years that had elapsed between Wardrop's article and LeBlanc's dissertation had been full of developments both in practice and in research.

Although in the 1950s and 1960s a number of practitioners were aware of the works by Wardrop and Beckmann and colleagues,[36] unsurprisingly it was academics familiar with advanced operations research techniques who would prove more receptive to the principles posited and the analyses contained in the *Studies in the Economics of Transportation* and in Wardrop's "Some Theoretical Aspects of Road Traffic Research." As early as 1963, Niels O. Jorgensen, then a graduate student from Denmark at the Institute of Transportation and Traffic Engineering at Berkeley, produced a research paper in which he explicitly referred to Wardrop (but not to Beckmann et al.).[37] Jorgensen's work would be among the starting points for the research carried out by a young Greek by the name of Stella C. Dafermos, née Theodorakopoulos (1940–1990), one of the first women in the United States to obtain, in 1968, the title of doctor in the very masculine field of operations research.[38] In 1969, Dafermos cosigned an article based mainly on her doctorate with Frederick Tomlinson Sparrow, her dissertation supervisor and Ph.D. holder in economics and operations research.[39] The article appeared in the in-house journal of the National Bureau of Standards, which had financially supported Dafermos's doctoral research,[40] in expectation of potential operational dividends. In spite of its practical motivation, Dafermos's thesis was a decidedly theoretical and mathematized piece of work. Highly theoretical as they may be, both the dissertation and the 1969 article would nevertheless prove a milestone in the long search for a *practical* solution to the traffic assignment problem, rigorously satisfying the equilibrium conditions set by Wardrop. First off, the article grandly helped to clarify and stabilize the vocabulary, somewhat floating until then, in use in the field. Thus, for networks the travel patterns of which are set up by "egoist" tripmakers tending to optimize their own route by each choosing the shortest path, Dafermos and her coauthor suggested the term "user optimized"—today we speak more simply of user equilibrium (UE). But as Wardrop had himself announced in the 1952 article, one can envision a transportation network whose travel patterns stem from individual choices consistent with some *collective* optimum. In that case, individual drivers behave "altruistically," by considering the overall situation and not just what suits themselves. Dafermos and Sparrow termed such a network "system optimizing," since the resulting traffic pattern satisfies the "criterion" of minimization of the *total*, instead of the *individual* travel time.[41] In addition to their work of conceptual clarification, Dafermos and Sparrow designed a series of relevant solution algorithms for the traffic assignment problem. Though the exercise remained largely theoretical, since the networks treated were highly stylized and small in size, the computer "experiments" performed by the two OR specialists yielded rather encouraging results.

In stark contrast to the meager harvest of the 1960s, the early 1970s witnessed an increase in the pace of innovation in the field of equilibrium-seeking assignment techniques.[42] Thus, after moving to the Department of Operations Research at Cornell University, Dafermos, this time financially backed by the Bureau of Public Roads, continued to explore traffic assignment issues, as did three of her colleagues, who would build on her work to devise with the help of federal money an algorithm capable of handling larger networks.[43] However, as has been seen, it was LeBlanc, among the OR scholars confronted with the traffic assignment problem, who managed eventually to bridge the gap between academic work and practice. Although this author was aware of the most recent academic contributions that attempted to address the issue, including the work carried out by Dafermos, in his bid to produce a solution "which is proven to converge and which closely approximates an equilibrium without excessive computational requirements,"[44] LeBlanc had to dig out a fifteen-year-old mathematical object. Going down in history as the "Frank-Wolfe" algorithm, it had been proposed in the mid-1950s by two young mathematicians working on the Princeton Logistics Research Project sponsored by the Office of Naval Research:[45] the French-born Marguerite Josephine Straus Frank (1927–), one of the first female Ph.D. students in mathematics at Harvard University, and Philip Starr Wolfe (1927–2016), now renowned for his great prowess in the field of OR.[46] To test its efficiency, LeBlanc ran his Fortran IV-coded algorithm on a large CDC 6400 machine, using as an experiment bench the (small) road network of the city of Sioux Falls, consisting of seventy-six links and twenty-four nodes. With an execution time of less than ten seconds, LeBlanc was justified in believing that his Frank-Wolfe algorithm-based solution to the traffic assignment problem was a highly promising one.

LeBlanc submitted his dissertation in August 1973. At that time, some 1,200 kilometers from Evanston, a certain Sang Nguyen, then a Ph.D. candidate working at the Center for Research on Transportation (CRT), was about to put the finishing touches to his own dissertation in the field of operations research, submitted to the University of Montreal in December 1973.[47] Although LeBlanc's work does not appear in the bibliography cited by Nguyen, the subject of the latter's research was exactly the same as that dealt with by LeBlanc in his own dissertation. Nguyen did refer to the Frank-Wolfe algorithm,[48] but the computational procedure he devised was an adaptation of the so-called convex-simplex method, a generalization of Dantzig's famous linear simplex method.[49] And, like LeBlanc, Nguyen would also make use of an experimental protocol close to that designed by LeBlanc for his own thesis. Thus, a real-life network, that of the City of Hull, Canada, consisting of 155 nodes and 376 links, served as a testing ground for the devised algorithm, run on the University of Montreal mainframe, a CDC CYBER 74 machine. Though the author recognized himself that "more tests . . . as well as comparisons of the equilibrium traffic pattern with actual ground counts must be carried out before any robust conclusion can be stated," the preliminary computer-based results obtained were nevertheless "very encouraging," while strongly indicating "that the proposed algorithm can efficiently solve moderate size assignment problems."[50]

The years following the defense of LeBlanc's and Nguyen's dissertations would witness a dramatic increase in the number of scholarly works dealing with equilibrium assignment techniques.[51] Many of them addressed sheer computational aspects, such as the speed of execution of the various equilibrium procedures that had been devised or the required computer memory; some others confronted the predictions produced by user equilibrium algorithms with ground counts.[52] As available algorithms were accumulating, comparative analyses were undertaken regarding the respective efficiency of the proposed assignment procedures.[53] Another set of academic works would go a step further by suggesting applying methods already crafted to new problems. Thus, while the first operational equilibrium algorithms were being designed to assign a *fixed* travel demand between an origin and a destination—in more technical terms, to assign a *fixed origin-destination trip matrix*—some scholars sought to find travel patterns satisfying the equilibrium conditions when travel demand is *variable*. Indeed, by experiencing congestion, the motorist might decide not just to modify his path between his origin and destination, but also to use, instead, a different mode of travel, opt for a new destination, shift the time of travel, or even forego some trips altogether. Any of these decisions results in a change in the number of trips between the different origins and destinations. In order to solve the traffic assignment problem when demand is not fixed—usually referred to as the UE problem with variable (or elastic) demand—a new class of models saw the light of day in the mid-1970s. The traditional four-step model is *sequential* in nature and partitions the traffic forecasting issue into four successive submodeling stages—trip generation, trip distribution, mode choice, and traffic assignment—with the result from each submodel used as inputs in the next one. The "combined network equilibrium models" (the first theoretical formulation of which dates back to the seminal work by Beckmann and colleagues in the mid-1950s), however, aim to treat *simultaneously* two and even three of the steps of the traditional approach. Over the 1970s and the 1980s, LeBlanc and Nguyen, the latter working often in tandem with his dissertation supervisor Michael Florian (more on him to follow), were very soon followed by an increasing number of colleagues trying hard to achieve the various assignment-related research issues mentioned in the preceding paragraphs. All these scholars collectively transformed the traffic assignment problem into a thriving subfield, enlisting the services of a significant part of the UTDM social world.[54]

At about the same time when a segment of the traffic forecasting community started to engage with the "combined equilibrium models," some other scholars who were seeking to promote original assignment techniques began to explore another path, the stochastic user equilibrium (SUE) approach, by combining the principles underlying the probabilistic multipath assignment algorithms, associated with the name of Dial, with Wardrop's user equilibrium principle. Following in Dial's footsteps, the advocates of SUE injected therefore into user equilibrium assignment the *subjectivity* of the trip-maker, by substituting *actual* travel times with *perceived* ones. Within the new theoretical framework, equilibrium was reached when no traveler "*believes* he can improve his travel time by unilaterally changing routes."[55] The SUE approach was first cultivated at Cambridge, Massachusetts, by two scholars, Carlos Daganzo[56] and his Ph.D. student Yosef Sheffi (born in 1948), a former alumnus of the Technion Institute of Technology.[57] It is not by chance that SUE first emerged at MIT. As a matter of fact, the new assignment procedure makes use of techniques also present in the disaggregate approach intensely cultivated at MIT in the 1970s (chapter 3). Not surprisingly, Sheffi's doctoral dissertation bibliography features several publications produced by scholars connected to the Cambridge-based establishment, while two young academics then at the vanguard of research on disaggregate modeling, Steven Lerman, a member of the dissertation committee, and Moshe Ben-Akiva, were both thanked by the author "for their constructive comments."[58]

A decade after LeBlanc and Nguyen defended their theses, academic research into traffic assignment equilibrium had already produced enough publications to give birth to the first manual distilling the main results and techniques of the subject. Grown out of a set of notes prepared by Sheffi, then a member of MIT faculty, for his courses, the book was also a collective undertaking as it benefited from the contributions of several other scholars involved in advanced traffic assignment techniques, including Daganzo, Dafermos, and LeBlanc, some of whom had joined Sheffi in his research while others assisted the author in the preparation of the manuscript by commenting on its early versions.[59]

As the manual composed by Sheffi testifies, by the mid-1980s, equilibrium assignment techniques and other advanced traffic assignment procedures had already constituted a significant component of the modeling activities of the university segment of the UTDM social world. Yet, the situation looked very different as far as practice was concerned. Despite the early incorporation of LeBlanc's work into the federal government-sponsored computer battery UTPS, followed by a handful of attempts made by scholars—including the Montreal-based Nguyen and Florian as well as U.S. academics David Boyce and Bruce Janson, a Ph.D. holder from the

University of Illinois at Urbana-Champaign—to apply results obtained within academia to real road networks,[60] the large majority of practitioners were still sticking to the older "heuristic" methods[61] developed in the 1960s. However, as happened with another 1970s novelty, the disaggregate approach discussed in chapter 3, the gap between academic-produced theory and operational practices would be filled progressively from the mid-1980s on, in large part thanks to a new kind of actor that had first entered the travel demand modeling scene in the 1970s never to leave it since then: the *proprietary software firm* specializing in transportation planning and which cultivated strong bonds with the university system.

## Four-Step Model Meets Microcomputers

Nature abhors a vacuum. At the beginning of the 1980s an epochal decision regarding urban traffic forecasting software would be made in the United States. As has been seen, the U.S. government had been by far the main provider of computer programs for urban traffic forecasters, through the Bureau of Public Roads in the 1960s, the Federal Highway Administration,[62] and, especially, UMTA in the following decade. Yet it now decided to bring its involvement in the production and free distribution of software for this modeling field to an abrupt end. Is this because high-ranking civil servants imposed the new presidential agenda advocating the withdrawal of the federal government from fields in which the latter used to be heavily involved in the past? And were the anticipated adaptation costs to the microcomputer revolution, which already loomed large, deemed too high? The answer to both questions seems to be yes.[63] In October 1982, speaking on behalf of the new Republican majority under President Ronald Reagan, Richard B. Robertson from the FHWA informed the traffic forecasting community that no "new software development similar to PLANPAC[64] or the Urban Transportation Planning System sponsored by the U.S. Department of Transportation (DOT)" should be expected any longer.[65] However, Robertson assured his audience that the Reagan administration was still willing to continue serving as a central clearinghouse for the dissemination of information and technology, and to support microcomputer applications through user support centers while encouraging a wide range of technology sharing.

As a matter of fact, throughout the 1980s DOT's administration did much to assist the transition from the mainframes of the past to the machines of the future, and to spread the use of microcomputer technology among travel demand modelers. To start with, "in order to maximize the utility of the federal investment in UTPS and to promote competition among commercial software developers," UTPS source code was made available to transportation professionals for free,[66] while a one-day microcomputer seminar was organized several times a year at different locations around the country.[67] In 1982 the federal government issued *Microcomputers in Transportation: Software and Source Book*, a publication updated on a yearly basis in the 1980s that contained a complete, annotated listing of the microcomputer software available for transportation planning and traffic forecasting. A few years later, in 1986, the Center for Microcomputers in Transportation (McTrans Center) was founded under a competitive contract from the Federal Highway Administration at the University of Florida.[68] Besides serving as a clearinghouse for distributing microcomputer software, McTrans Center was designed to provide various levels of support to users for the programs to be distributed by the center.[69] Thanks to these and other actions, including sponsoring "user groups,"[70] or staging and participating in various events,[71] the federal government had a role in launching the microcomputer era in transportation-related software. However, very soon it would cede the driving seat to the private sector.[72]

Without having to fear competition with the federal government anymore, a number of private firms immediately embarked on the design and distribution of traffic forecasting-related programs for the microcomputers of the time, notably for Apple and IBM PCs. Several signs seemed to indicate that the balance of power between the mainframe and the microcomputer was rapidly tilting in favor of the latter. Thus, according to a study based on the responses of 109 questionnaire participants from across the United States and Canada, most of whom had direct experience in traffic forecasting—state DOTs, MPOs, local governments, private consultants—by the very end of the 1980s more than one-half of the organizations contacted were using microcomputer hardware/software to forecast traffic, even though more than 60 percent of the respondents made use of their microcomputer only in concert with an IBM mainframe-based process, that is, UTPS or PLANPAC.[73] Rather unsurprisingly, the first producers of such microcomputer suites were mostly existing consulting firms already involved one way or another in urban traffic forecasting. Moreover, their ability to design and commercialize travel demand-related programs run on microcomputers was directly related to their familiarity with the battery of programs produced and distributed under the aegis of the U.S. government in the past.[74]

COMSIS was one of the firms that benefited from the federal withdrawal and the vacuum that created.[75] The company was set up in March 1969 as a minority business enterprise (MBE) by Clarence Lee,[76] a Taiwanese American then working for the Service Bureau Corporation, which was set up as a subsidiary of IBM in 1957.[77] From its incipient stages, COMSIS would be home to a number of seasoned transportation analysts who had previously worked for the federal government. Arthur Bruce Sosslau was one of them. A graduate of the Yale Bureau of Highway Traffic in 1958, Sosslau—Art to his friends—had been a member of the "499 Pennsylvania Avenue" club of early computer programmers while he was a staffer at the Bureau of Public Roads from 1957 to 1965.[78] In the mid-1960s, he moved to the Tri-State Transportation Commission and two years later he started working for the Service Bureau Corporation, where he in all likelihood met the founder of COMSIS. At the end of the 1970s, Sosslau was president of the company, with Larry R. Seiders as his vice president. A holder of a B.S. degree in civil engineering from Penn State University in 1959, Seiders wrote computer code for the BPR in the 1960s and, at the end of the decade, he was also working for Service Bureau Corporation.[79] Martin J. Fertal (1936–2014), a graduate of Carnegie Tech in 1958 and one of the vice presidents of the company in the mid-1980s, had also been a former BPR staffer. Throughout the 1970s and the early 1980s, COMSIS had distinguished itself as a close collaborator of the DOT in urban traffic forecasting and so the company staffers were able to continue familiarizing themselves with the development of the federally sponsored computer suites for travel modeling.[80] It is no wonder, then, that when the federal government decided not to migrate its program batteries from the mainframe system to the nascent microcomputer one, COMSIS immediately stepped into the breach to market a proprietary software product called MINUTP, a largely microcomputer-based version of the federal program libraries. In October 1982, MINUTP, which at that time included modules for the trip generation, trip distribution, and traffic assignment FSM steps—but not for the mode choice one—could already run in a demonstration mode.[81] A year later, an IBM PC-based version of the software cost $5,000 (around $12,500 in 2018), making it a rather cheap planning transportation tool. The development of a comprehensive highway or transit planning capability using the mainframe-based UTPS computer package of the day required a major investment of at least $100,000, including staff and data collection costs, for the first year only.[82]

MicroTRIPS, then marketed by PRC Engineering in the United States and MVA Systematica in Europe,[83] was largely based on the work initially done in the 1960s by staffers of the AMV consultancy founded by Alan Voorhees in 1961, including Robert Dial.[84] In early 1986, MicroTRIPS, then a comprehensive software system paralleling the federally sponsored UTPS in functional capability as it included programs for all stages of the FSM, was one of the most popular microcomputer-based urban traffic forecasting packages of the day, with over 125 installations in twenty-five countries.

The origins of TRANPLAN (TRANsporation PLANning) go back to the early 1960s, when Control Data Corporation, a pioneer in the design of supercomputers in the 1960s and an IBM competitor, decided to enter the field of computerized transportation planning techniques. From the mid-1960s on, CDC benefited from technical assistance from AMV as well as from other consultancies for the programming of its traffic forecasting software package on the CDC 3600 mainframe model. Thanks to a close collaboration between the computer firm and the large transportation consultancy De Leuw, Cather & Company, at the turn of the 1960s TRANPLAN had grown into a comprehensive multimodal transportation planning program battery for the CDC 6600 computer system.[85] In 1971, the CDC system could be accessed through public terminals available at forty-five locations.[86] A decade or so later, TRANPLAN migrated to a microcomputer environment in large part thanks to the actions of R. James W. Fennessy.[87] A graduate in civil engineering from Carleton University (Ottawa) and holder of a master's degree from University of California, Berkeley, Fennessy was an employee of Wilbur Smith & Associates in the mid-1960s. After a brief stint at the Control Data Corporation, where he worked on TRANPLAN, he joined De Leuw, Carter & Company in 1970.[88] In 1984, Fennessy, then manager at the Systems Division of DKS Associates—a transportation engineering consultancy set up in 1979 in Oakland, California, by former staffers of De Leuw, Cather & Company—was pleased to announce that the microcomputer version of TRANPLAN was already operational on UNIX systems as well as on IBM PC DOS. In 1988, the TRANPLAN microcomputer package was available at different prices, ranging from $500 for an academic license fee, to $6,500 for nonsource code licenses, to $13,500 for source code licenses.[89] At that time, Fennessy was working along with the traffic demand modeler Edward F. Granzow, holder of a B.A. degree in social ecology from the University of California at Irvine, at the newly founded (ca. 1985) Urban Analysis Group, a California-based organization that eventually acquired TRANPLAN from DeLeuw, Cather & Company in 1990.[90]

When they were originally marketed in the first half of the 1980s, MINUTP, MicroTRIPS, and TRANPLAN were, in large part, a microcomputer version of programs initially designed for the mainframes of the 1960s and 1970s. All of them stood the test of time, and through their successors (more to follow) they have been serving the urban traffic forecasting community up to the present. However, while they succeeded in progressively forcing out of the market several other microcomputer proprietary software developed in the 1980s,[91] they and their heirs had to victoriously withstand the attacks of some new entrants, all of them part of or cultivating strong bonds with the academic world from the start.

INRO was typical of these new entrants. Like Cambridge Systematics, a leader in the realm of disaggregate modeling in the 1970s and 1980s (chapter 3), INRO, located on the other side of the Canada–United States border, was also established by academics in 1976.[92] Among its founders, and indisputably the most important figure in its success story, is Michael Florian (born in 1939), the long-standing president of INRO. A native of Romania, Florian obtained his Ph.D. in the field of OR from Columbia, New York, in 1969, then immediately embarked on a thirty-five-year career at the University of Montreal. In 1973, Florian was appointed director of the newly founded Center for Research on Transportation, and, along with a handful of colleagues and graduate students, he set to work on urban traffic forecasting issues, in large part thanks to a generous grant awarded by the Ford Motor Company of Canada. As a result, CRT very soon distinguished itself in the development of sophisticated equilibrium assignment techniques. Impressed with the results obtained by the members of the CRT, the Transportation Development Agency of Transport-Canada, though a nonacademic institution, came into contact with Florian in 1976 and granted him financial assistance to develop and, above all, test on a real site, the city of Winnipeg, a multimodal equilibrium assignment model combining the steps of mode choice and traffic assignment, building on research work already carried out at the University of Montreal. Involving a total of eleven people associated with the latter, three years or so later, the project, named

EMME—Équilibre Multimodal/Multimodal Equilibrium—was deemed successful in the eyes of its authors. However, as they themselves admitted, though EMME was a success from an academic point of view, the new travel demand modeling suite lacked one essential element to become truly useful to the practitioner: a *user-friendly interface* summarizing and visualizing EMME's many analytical results. This would make it possible for the planner and the transportation analyst to interpret voluminous amounts of data and effectively search for, explore, and evaluate a large number of alternative designs and courses of action in a short period of time.[93] Soon after the original project came to an end, the arrival of Heinz Spiess (born in 1954) at the University of Montreal as a graduate student of Florian would prove instrumental in the further extension of EMME. A graduate of the University of Basel, Switzerland, specializing in theoretical physics, Spiess now had to wade deeply into the waters of urban traffic forecasting. More precisely, the supervisee was asked to develop for his Ph.D. research an original assignment model for mass transit systems and, above all, to devise in collaboration with other CRT members the much sought-after "interactive graphical tool for urban transportation network planning."[94]

Depicted as brilliant by his supervisor, Spiess was well equipped intellectually for crafting such a graphical tool. Before exchanging Europe for the New World, he had spent the years 1973 to 1978 working as a part-time systems analyst for a Swiss transportation consulting firm directed by Matthias H. Rapp. Rapp had been a pioneer in the use of interactive graphic methods in transportation since his doctoral research conducted in the United States at the end of the 1960s and in the early 1970s under the supervision of Jerry B. Schneider, one of the then great specialists in the field.[95] Thanks to the work carried out by Spiess and a team of scholars at CRT, an enhanced version of EMME, called EMME/2, was released in the early 1980s. In August 1983, the Transportation Planning Department of the Metropolitan Service District (METRO) of Portland, Oregon, a loyal client of the mainframe-based UTPS package since 1976, became the first operational transportation agency in the United States to acquire EMME/2 for $20,000. Described by the buyer as "the first interactive graphic multimodal microcomputer-based turnkey system that has reached the market place,"[96] the INRO product pleased many of the METRO staffers, who publicly congratulated themselves on their decision to acquire EMME/2. Obviously, they were not the only ones to have been satisfied with Florian's team software. In October 1986, there were sixteen users of the package in North America and a further ten in Europe, while in May 1989, EMME/2, now developed and marketed by INRO after a royalty agreement between the firm and the University of Montreal, already had sixty-one users in North America and fifty-eight additional customers in Europe. This commercial success, which has continued unabated despite the software's higher price compared to other popular traffic forecasting software packages,[97] comes as no surprise as EMME/2 and its successors—Emme 3 and now Emme 4[98]—have continued over the years to include scientific as well as marketing work. INRO has regularly been joined by knowledge workers and among its thirty or so employees at the start of 2012, nine had a master's degree and six were Ph.D. holders. INRO has also developed a large network of distributors all over the world (seventeen in 1998) and offers its clients a large range of services. These include a detailed and regularly updated user's manual; free technical support;[99] a newsletter called *EMME/2 News* that was published from 1986 to 1998; and training courses, conferences, and workshops subject to a fee,[100] while the company also hosts online discussion forums on its website.

The example of INRO, acquired by the infrastructure engineering software company Bentley Systems in 2021, is not unique. Founded in 1983 in the larger Boston area, Caliper Corporation shares several characteristics with its Canadian competitor.[101] While not an academic, Caliper's founder and president Howard L. Slavin spent many years of his life within the university system and has cultivated strong bonds. After obtaining an A.B. degree in mathematics and urban studies from Yale College, followed by a master's degree in urban and regional planning from Harvard, Slavin moved to the University of Cambridge in the United Kingdom, where he received his Ph.D. in 1979.[102] Once back in Massachusetts, he worked for a while for the federal government at DOT's Transportation Systems Center then located in Cambridge, a facility where more than six hundred persons were employed.[103] There, he managed a research group evaluating innovative experiments in urban transportation and related issues, including travel modeling, and came to be familiar with the faculty and students at MIT as well as with Massachusetts-based transportation firms. Following Reagan's victory in the 1980 presidential election, Slavin left the Volpe Center in 1981, and went to work at Charles River Associates (CRA). In his capacity as the firm's first director of marketing research, he helped in the commercialization of the use of discrete choice models, while working closely with seasoned modelers such as Gerald Kraft and Daniel Brand (see chapters 2 and 3). After a rather brief stint at CRA, Slavin founded Caliper Corporation in Newton, Massachusetts, in fall 1983. The product that enabled the firm to enter the transportation modeling software market in the 1980s is the now world-known TransCAD. Like INRO, Caliper has been a knowledge-intensive firm[104] since its beginnings. As a matter of fact, the design and release of TransCAD was the outcome of the joint efforts of a small number of highly qualified people. Eric Adam Ziering, the cofounder of Caliper and former CRA staffer was a civil engineering graduate of MIT, where he earned his master's degree in 1979.[105] Zvi Tarem, another CRA staffer in the early 1980s, and the first person Slavin and Ziering hired, was another MIT master's degree holder.[106] A "brilliant programmer" (Slavin's words), Tarem had been Yossi Sheffi's research assistant at MIT and had written codes for stochastic user equilibrium procedures among other things. Andres Regueros (now Rabinowicz), a Ph.D. holder from Technion–Israel Institute of Technology in 1992, joined the team in early 1993 and was another pivotal character in Caliper's early history. Thanks to the collective work done by this small group, TransCAD, first released in 1988, was the first computer software package to provide the UTDM social world with a tool integrating a geographic information system (GIS)[107] and traffic forecasting modeling.[108] Like Emme, TransCAD was a major success. In 2006, according to information disclosed by the company, there were more than eight hundred clients worldwide using Caliper's proprietary software, around half of which were inside the United States.[109] Like Emme's success across the world, TransCAD's popularity with traffic urban forecasters can largely be attributed to the ability of Caliper's staffers—among whom there were five Ph.D. holders and six with a master's degree in 2018, most of them from MIT[110]—to regularly inject into the software the latest advances in urban travel demand modeling (see also chapters 6 and 7). And, as in the case with INRO and its other competitors, Caliper has been providing TransCAD's users with a wide range of services, ranging from manuals and technical support, to training courses and hosting a User Center for its customers.

Another popular and particularly cheap[111] traffic forecasting package that illustrates the growing importance of academic research for practice in urban traffic forecasting is Quick Response System II (QRS II), developed and marketed by AJH Associates, a company of just two people formed in

June 1987.[112] Though much of its success has been due to the continuous efforts deployed by academic Alan J. Horowitz, a Ph.D.-holder from the University of California at Los Angeles in urban planning in 1974, QRS II has its roots outside the university system. In many respects a light version of the traditional four-step model, and consisting of a set of essentially manual traffic forecasting techniques, the initial QRS was the outcome of a federally sponsored research project performed by COMSIS over the period 1974 to 1978.[113] QRS techniques proved highly popular with practitioners,[114] and in order to facilitate and expand their utilization, COMSIS developed a microcomputer version, named QRS I, which was released in the public domain in February 1984.[115] Since it was introduced in the late 1980s, QRS II, first developed at the University of Wisconsin-Milwaukee for the Urban Mass Transportation Administration, has been regularly updated by Horowitz with the help of other forecasters, including the users of the software.[116]

Emme, TransCAD, and QRS II can all be regarded as examples of a more general trend toward increasing interactions between forces emanating from the university system and the private sector. These academic forays into the traffic forecasting software market would not leave unaffected the trajectory of the other proprietary microcomputer packages that had seen the light of day in the early 1980s, such as MINUTP, MicroTRIPS (later renamed TRIPS), and TRANPLAN. Following a series of acquisitions and mergers, more recent versions of these initial UTPS microcomputer clones would eventually be integrated, along with some other similar and related computer suites, into a comprehensive software package called CUBE. This has been developed and marketed by Citilabs, a company set up in 2001, and acquired, in 2019, by Bentley Systems, a large global provider of comprehensive software and digital twin cloud services for advancing the design, construction, and operations of infrastructure.[117] Like its various competitors, such as INRO (also now part of Bentley Systems) and Caliper, Citilabs has been home to several knowledge workers, and regularly injected university-produced knowledge into its own traffic forecasting software, including the traffic assignment component.[118] The same applies to Planung Transport Verkehr (PTV) AG, a large German engineering firm set up in 1979 and closely linked to the University of Karlsruhe, the flagship product of which—a software package for urban travel demand modeling called VISUM—has also made its presence increasingly felt in the U.S. market through PTV Group America, the firm's North America wholly-owned subsidiary established in 2004.[119]

COMSIS, Urban Analysis Group, INRO, Caliper, AJH Associates, and, more recently, Citilabs, and PTV America: it was in large part thanks to these proprietary software companies that user equilibrium assignment techniques based on or making use of Wardrop's first principle progressively left the academic sphere to effectively spread to practitioners. EMME/2 seems to be the first microcomputer package to provide rigorous Wardrop-based assignment techniques in the 1980s, although a few years later TRANPLAN and MINUTP followed suit.[120] From the 1990s on, all the major transportation planning software vendors present on American soil have offered equilibrium assignment techniques in their catalogs, and what initially had been the preserve of a few grew into a commonplace tool taking pride of place within the last stage of the FSM. In fact, while in the early 1980s state and local transportation agencies across the national territory—all then resorting to federally sponsored computer suites—seemed to be split over using the old "all-or-nothing" or "capacity-restraint" traffic assignment procedures, with only few of them making use of stochastic assignment,[121] by the mid-2000s equilibrium assignment procedures were used by 76 percent of all metropolitan planning organizations in the United States, and this percentage rose even to an impressive 91 percent for the larger ones.[122] However, despite the fact that as time went on, the current "Big Four"—Caliper, Citilabs, INRO, and PTV—have regularly incorporated into their products a variety of new and more effective assignment methods,[123] in the early 2010s most MPOs kept resorting to assignment techniques still computed with the old Frank-Wolfe algorithm.[124] Seemingly, it's hard to teach an old dog new tricks, as the saying goes.

Taken together, disaggregate modeling (chapter 3) and equilibrium assignment techniques (chapter 4) did breathe new life into the FSM. However, the latter continued to face a number of problems, some of which emanated from external sources while others stemmed from persistent internal tensions. As a result, from the 1990s on the UTDM social world embarked on a series of attempts to supersede the rejuvenated FSM by newer and alternative modeling lines. First experimented with locally, and later deployed on a large scale across the nation, the new traffic forecasting practices have brought, in several respects, a profound rupture with the past, thus making it possible to talk about a "paradigm shift" in urban travel demand modeling.

# III Urban Travel Demand Modeling Enters the Post-Four-Step Model Era, 1990s–2010s

# 5 Traffic Forecasting's Attempted Manhattan Project

## Original Challenges and New Resources for Traffic Forecasting

By the early 1990s, two decades had passed since the utility theory-based disaggregate approach (chapter 3) and rigorous user equilibrium assignment techniques (chapter 4) first appeared on the American travel demand modeling stage. With the passage of time, the then novelties had long ceased to be new, and a nearly thirty-year-old four-step model, with its major procedures having not been significantly revised for a decade or so, had to confront new, and serious, challenges.

Air pollution was one such challenge, and probably the most pressing one.[1] When the Sierra Club Legal Defense Fund, a conservation organization, took the San Francisco Bay Area transportation planning agency to court at the turn of the 1980s, it did so by invoking violations of the provisions of the Clean Air Act Amendments enacted in 1977.[2] Thirteen years later, the Clean Air Act Amendments (CAAA) of 1990—immediately followed by the Intermodal Surface Transportation Efficiency Act (ISTEA) of 1991 that emphasized the need for a truly *multimodal* approach to mobility[3] that included biking and walking—made air pollution reduction the driving force behind transportation planning for the next thirty years.[4] As a result, since they were a significant component of the latter, travel forecasting procedures found themselves under great strain. Both the 1990 CAAA and 1991 ISTEA, along with the various post-ISTEA acts, forced the UTDM social world to scrutinize well-established practices, and strongly prompted, even compelled its members to explore original modeling avenues more in line with the new legislative agenda.

Yet, they were not the only forces that would help shape the post-1990 urban travel demand modeling landscape. If pollution was a rather new challenge, congestion, an old urban scourge, was also back with a vengeance while wearing a new (suburban) face (see also chapter 7). And there was more. The post-1990 CAAA and post-1991 ISTEA years also witnessed increasing concerns about environmental justice, social equity, and discrimination issues,[5] while experiencing a series of important mobility-related societal changes, such as telecommuting[6] and the shift in travel from peak to off-peak times. Technological changes, including the advent of ITSs and the availability of new powerful computational resources, also helped build a favorable context for the deployment of original travel demand modeling practices across the country.

## Revamping the Four-Step Model through Feedback

The first reaction of the UTDM social world to the new legislative setting was rather quick. Thus, three years after President George H. Bush enacted the 1990 CAAA, a *Manual of Regional Transportation Modeling Practice for Air Quality Analysis* saw the light of day.[7] Prepared by DHS, a consultancy set up in 1987, with Grieg Harvey (1950–1997), an MIT graduate in civil engineering, and University of California, Berkeley, faculty member Elizabeth Deakin (and Harvey's spouse) leading the effort, the manual greatly benefited from the expertise of many organizations and individuals belonging to the UTDM social world. Though several other more or less similar "travel forecasting guides" were published in the first half of the 1990s,[8] it was the Travel Model Improvement Program (TMIP) that would probably most symbolize the nation's desire to improve its urban travel demand modeling practices. TMIP was jointly established in the fall of 1991 by the Department of Transportation, the Environmental Protection Agency, and the Department of Energy. Going beyond developing new forecasting procedures only, TMIP consisted of four major tracks collectively aiming to assure that transportation practitioners could have, essentially through a continuing program of training and technical assistance, access to what were deemed the best transportation planning methods of the day.[9]

Unsurprisingly, one of the first large-scale actions undertaken within the TMIP framework was the revamping of the FSM, still the most common procedure for producing urban travel forecasts across the country in the early 1990s. Shortly after TMIP became fully operational in 1994, COMSIS, the well-known proprietary software firm (chapter 4), was commissioned by the Federal Highway Administration to conduct a research study on the incorporation of feedback loops between the various steps of the FSM.[10] While the exploration of methods for introducing such loops was not new,[11] by the early 1990s most practitioners had been reluctant to implement feedback in their models.[12] Because of the degree to which vehicle emissions rates are affected by speed, the 1990 CAAA, as well as a number of guidelines issued in the years following the introduction of the act, required that the speeds used in the process be realistic in comparison to what might be observed on the road, and reasonably consistent—as a matter of fact, it had not been unusual to find different speeds and travel times used in different parts of the four-step forecasting procedure.[13] After selecting two major metropolitan areas, Memphis, Tennessee, and Salt Lake City, Utah, as test sites, COMSIS staffers, with the help of several experts, including the academic David Boyce, Robert Dial, and the consultant Gordon Schultz from Parsons Brinkerhoff, proceeded to assess alternative methods for implementing feedback into the FSM framework. Following COMSIS's undertaking and other studies, which showed that feedback mechanisms could significantly affect the forecasting results when there was congestion in the network, by the mid-2000s over 80 percent of large MPOs had adopted such mechanisms in their models and forecasting practices.[14] However, though the feedback issue remained an important one within the UTDM social world for several years,[15] very soon the bulk of efforts deployed by both academics and practitioners with a strong theoretical background would be directed toward the development of new and innovative travel demand modeling practices able to replace the old FSM paradigm altogether.

Nothing better illustrates than TRansportation ANalysis and SIMulation System (TRANSIMS) the huge swaths of time and money spent as well as the vast amounts of effort deployed by the UTDM social world in its bid to substitute the FSM with a new set of practices capable of meeting the new challenges and demands that emerged in the 1990s. At the same time, nothing better embodies than TRANSIMS the difficulties encountered in building a global modeling approach that radically departs from longstanding practices.

## Los Alamos after the Cold War Ended: From Military to Civilian Science

> Traffic is particles with motive. I think it's cool as hell.
>
> —Chris Barrett, TRANSIMS project leader, quoted in Alan Sipress, "Lab Studying Science behind Traffic Patterns," *Washington Post*, August 5, 1999

TRANSIMS was the outcome of the improbable encounter of Big History with traffic forecasting. In the early 1990s, the Cold War was over and so was the corresponding "cold war science."[16] Suddenly, the major national research centers across the country had to adapt to new political conditions, which required less research for the development of original weapons of mass destruction. Like other federal organizations of its kind, the Los Alamos National Laboratory (LANL) had to sign a new contract with American society, as it was now urgently summoned to recycle the competences it had honed during the Cold War years in order to produce knowledge and practices for (hopefully) more peaceful times.[17] Transportation therefore presented itself as an opportunity to be seized, especially after the passage of the 1990 CAAA and 1991 ISTEA, both of which posed a serious challenge to travel demand modeling practices of the day.[18] In October 1991, LANL decided to join forces with the Sandia National Laboratories, the University of New Mexico, New Mexico State University, and the New Mexico State Highway and Transportation Department (NMSHTD) to form the Alliance for Transportation Research (ATR), with David Albright as its first president.[19] Only several months after ART was established, a large team of scientists from LANL drew up the "TRANSIMS White Paper."[20] Given that TRANSMIS was about transportation modeling, the authors of the white paper had to familiarize themselves with what would be their new job. Two specialists in urban traffic forecasting, John Hamburg, the former CATS staffer (chapter 2), and Rick Donnelly, both then working for the local branch of the nationwide consultancy Barton-Aschman Associates, spent two months on site to hand on their knowledge and expertise to Los Alamos researchers.[21] However, the TRANSIMS project was thought of as something totally different from anything done before in transportation planning and modeling. Its promoters were, in fact, displaying the ambition of building a (mega)modeling suite capable of simulating the second-by-second movements of every person and every vehicle throughout the transportation network of a large metropolitan area. Judged by the standards of the UTDM social world of the day, TRANSIMS was certainly a huge (and original) project, and the people involved were aware of that. Still, they were optimistic and confident. They had already successfully conducted large-scale simulations of complex systems for military purposes, and while TRANSIMS was far "beyond the hardware/software capability that currently exists in the transportation community,"[22] their organization was endowed with the computing machines necessary for carrying out the new project.

It was Thomas Deen, the comrade-in-modeling-arms of Alan Voorhees in the 1960s and the executive director of the Transportation Research Board since 1980, who first reacted to the "TRANSMIS White Paper," following a letter received from David Albright on May 29, 1992[23] (figure 5.1).

While the paper exuded ardor and hope, Deen's own missive, though polite, was clearly lacking in enthusiasm.[24] Deen did concede that if the Los Alamos project were to eventually succeed, it would "allow better estimates of the performance of the highway system under different proposals, and this in turn will enable better estimates of air quality impacts." Yet, he obviously had serious doubts about TRANSIMS's feasibility and cost-effectiveness. In his opinion, if "TRANSIMS has gone to great length to describe the linkage of traffic simulation with planning models,[25] . . . it seems to ask the reader to take on faith the ability of the authors to make real advances on other fronts such as trip generation, distribution or modal split." And there was more. Given the complexity of each submodel within the overall TRANSIMS structure, Deen wondered whether there was a single other site outside Los Alamos that had the necessary IT resources for the operational use of TRANSIMS. As for the cost of the project, in the order of $50 million, it was certainly too high for a civilian research program. Obviously, Deen was rather skeptical about the possible outcomes of LANL's proposal and expressed serious doubts as to whether the TRANSIMS project could live up to expectations and deliver on its promises. However, his personal reluctance to share the fervor of his interlocutors was not strong enough to block the project. The latter was clearly appealing to the federal government, which LANL approached in 1992. "In 1993, defense technology came to the highway community,"[26] a number of federal agency staffers proudly announced. In fact, mandated to assist in the implementation of the new transportation agenda following national legislation enacted in

June 10, 1992

David Albright
President
Alliance for Transportation Research
University Research Park
851 University Blvd., S.E., Suite 100
Albuquerque, New Mexico 87106

Dear David:

I have read the draft you sent describing the Transportation Analysis and Simulation System (TRANSIMS). As you indicated it is an impressive piece of work and your team has clearly proposed an ambitious effort to improve our ability to make better transportation decisions.

You must understand that I have been out of the planning business for some years, and even when I was in it I didn't consider myself an expert in transportation models. Nevertheless, I did use them extensively in my career and even have some experience in developing new models.

As I understand it, the major innovation in TRANSIMS is to integrate improved microlevel traffic simulation models with improved macrolevel travel forecasting models. Successful achievement of this will permit better estimates of the performance of the highway system under different proposals, and this in turn will enable better estimates of air quality impacts. The document goes to great lengths in describing how this might be done, and it is very impressive in the detail to which this aspect has been thought through. I believe that this would be an achievement of considerable magnitude and could be of great benefit.

However, I personally am persuaded that the great deficiencies of our planning models for today's practice are as follows:

1. They are data hungry and our efforts to develop and run them with inadequate data simply serves to reduce their credibility.

2. Trip generation models show little sensitivity to congestion and inadequate supply. Maybe they are accurate in this respect but it is counterintuitive in supersaturated conditions. How bad do things have to get to discourage tripmaking?

3. They almost never deal with freight movement and since one

**Figure 5.1**

Extract of a letter in which Thomas Deen (born in 1928) comments on the "TRANSMIS White Paper." *Source:* Letter from Deen to David Albright, dated June 10, 1992, George Mason University Libraries, Special Collections Research Center, Thomas B. Deen Papers, C0106, Box 50, Folder 8.

1990 and 1991, the DOT and its administration pinned much hope and faith on TRANSIMS.

The unveiling of the TRANSIMS proposal to the wider travel demand modeling community of the day seems to have taken place in January 1994, during a well-attended session at the annual TRB meeting. There, the Federal Highway Administration surprised many people by inviting LANL to present their white paper and announcing that FHWA had chosen to financially support TRANSIMS instead of any of the work proposed in the other white papers presented in that session (further discussion follows).[27] As a matter of fact, since the creation of the Travel Model Improvement Program in the early 1990s, its successive managers, starting with Frederick W. Ducca, Jr.,[28] put TRANSIMS on Track C, the one that "involves major research to redesign travel and land use forecasting procedures to respond to the greater information needs required by ISTEA and CAAA," while for a decade or so TRANSIMS would be, and by far, "the largest project in the program."[29] No wonder then that when TMIP staged its first workshop conference in Fort Worth, Texas, in August 1994, an event attended by all the then heavyweights in urban traffic forecasting,[30] the LANL project had pride of place. Despite the skepticism understandably expressed by several members of the UTDM social world about the effectiveness of a practice-oriented research project crafted by people whose scientific work thus far had been primarily in areas other than transportation modeling,[31] TRANSIMS would remain for many years extremely popular both with federal administration staffers and lawmakers as well. The project constantly benefited, in fact, from substantial financial backing, since from 1992 to 2003 no less than $38 million dollars was spent on the project, about three-quarters of which went to Los Alamos for basic research and development. Passed in July 2005,[32] the act called SAFETEA-LU (Safe, Accountable, Flexible, Efficient Transportation Equity Act: A Legacy for Users) allocated $2 million annually to the project.[33] Part of the money was destined to finance the implementation of TRANSIMS by MPOs interested in the tool. TRANSIMS also succeeded in capturing additional money from other sources, around $4,130,000 by 2007. We only need compare the aforementioned figures with the modest sum of $500,000 per year that TMIP had been funded for all activities other than development of TRANSIMS in the first half of the 2000s[34] to understand why, several years after the LANL project launched, a significant number of seasoned travel demand modelers and other transportation professionals would still feel they had been unfairly treated by their usual funders.[35] But let us travel back to New Mexico in 1992, when the intensity of the hopes placed in TRANSIMS was probably at its highest.

## Carrying out the TRANSIMS Project

Comprised of a substantial group of about twenty-five persons during the period 1992 to 1995,[36] the large majority of those partaking in the TRANSIMS adventure were mathematicians, physicists, and computer scientists (remember that the "traditional" travel demand modeler had been closely related to the figure of civil engineer).[37] LaRon L. Smith, appointed project coordinator for TRANSIMS, was a mathematician with a Ph.D. in nuclear engineering in 1971. Christopher Barrett, another project leader, and the person who convinced LANL that traffic was a serious and grave matter for federal scientists,[38] was a Caltech alumnus with a Ph.D. in bioinformation systems (1985), specializing in control theory as well as in the study of nonlinear dynamical systems. It was also Barrett who seems to have proposed a highly disaggregate approach tracking individuals and vehicles second by second. A member of LANL since 1971, after being awarded a doctorate in statistics in 1969 Richard J. Beckman was appointed TRANSMIS project leader for the period 1997 to 2003 and was responsible for the development of an important module in the overall modeling suite, the "synthetic population generator" (more to follow). Not only was the TRANSIMS team packed with Ph.Ds, but it was also cosmopolitan. Steen Rasmussen, a physicist interested in self-organizing processes in natural and artificial systems, was Danish while Madhav Marathe, a Ph.D. computer scientist, was native of India. Another foreigner whose contribution would prove decisive for the design of TRANSMIS was Kai Nagel. Born in 1965 in Cologne, Nagel obtained a French postgraduate degree in oceanology and meteorology, followed by a *Diplom*[39] in computational physics models of cloud formation in 1991.[40] In 1994, Nagel defended his doctoral work on fast microscopic traffic simulations. But even before completing his Ph.D. in computer science, Nagel had already made a name for himself as a modeler able to run "simulations faster than ever before." And as the LANL team badly needed "someone who could design simulations fast enough to be practical,"[41] it enlisted Nagel's services. Nagel first took part in the development of TRANSIMS during the period from July 1993 to April 1994, when he was appointed research assistant and a member of the Simulation Application Group. After he was awarded his Ph.D. in Germany, he returned to Los Alamos, where he remained until 1999. From 1998 to 1999, he would even occupy the position of research team leader. During his American stay, Nagel would work closely with Marcus Rickert, a compatriot of his and also a physicist specializing in simulation techniques.

Given their initial disciplinary backgrounds, which bore little or no relationship to transportation, the members of the TRANSIMS group had first to familiarize themselves with the literature on urban traffic forecasting.[42] For the same reason, more or less regular interactions with seasoned members of the then UTDM social world were being organized throughout the project. Thus, after initiating the LANL team into the world of transportation modeling, Rick Donnell would continue providing advice to the TRANSIMS group in the 1990s, and even coauthored a document with several of its members.[43] Another specialist in urban travel demand modeling whose services had been enlisted by the Los Alamos team was the academic Ryuichi Kitamura.[44] He was among the most fervent promoters of the activity-based approach (chapter 6), a general travel demand modeling line being developed in the United States at that time, and which the researchers at Los Alamos would also embrace in order to build certain of TRANSIMS's modules (see box 5.1). On September 4 and 5, 1997, a brainstorming workshop session was held in Los Alamos, hosting representatives of the TRANSIMS project team and several specialists in advanced urban traffic forecasting, including Koppelman, Eric Pas, Konstadinos Goulias, Mark Bradley, Keith Lawton, and Ram Pendyala (chapter 6). This provided an opportunity for LANL to be informed about the cutting-edge travel modeling knowledge and practices of the day.[45] However, despite all this intellectual trade between the travel demand modeling community at large and the TRANSMIS group, the latter sought to break with existing traffic forecasting practices by drawing inspiration from approaches and methods mostly developed outside the realm of transportation.

In the mid-1990s, some three years after it was first set in motion, TRANSIMS had undoubtedly reached an important milestone in its development process.[46] Its overall architecture was more or less stabilized, while two of its most original components were then regarded as mature enough to be tested on real sites. TRANSIMS was now a modular system comprised of four basic blocks. In addition to each module providing outputs serving as

**Box 5.1**
TRANSIMS (circa 1995)

"The TRANSIMS Project objective is to develop a set of mutually supporting realistic simulations, models, and data bases that employ advanced computational and analytical techniques to create an integrated regional transportation systems analysis environment. . . . The integrated results from the detailed simulations will support transportation planners, engineers, and others who must address environmental pollution, energy consumption, traffic congestion, land use planning, traffic safety, intelligent vehicle efficacies, and the transportation infrastructure effect on the quality of life, productivity, and economy" (on 1).

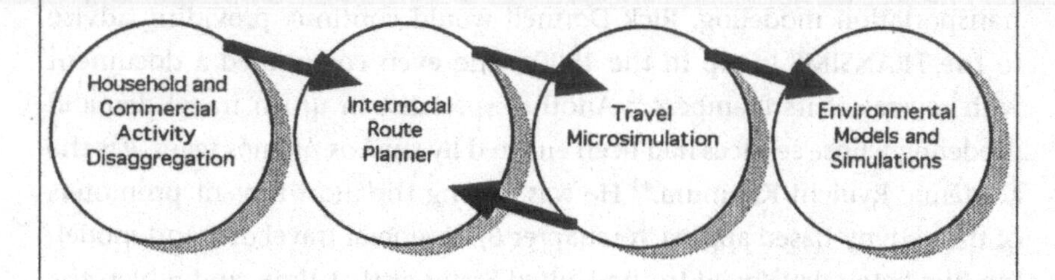


"The previous figure illustrates the TRANSIMS architecture. The TRANSIMS methods deal with individual behavioral units and proceed through several steps to estimate travel. TRANSIMS predicts trips for individual households, residents, freight loads, and vehicles rather than for zonal aggregations of households. The Household and Commercial Activity Disaggregation Module creates regional synthetic populations from census and other data. Using activity-based methods and other techniques, it produces a travel representation of each household and traveler" (on 2).

"The Synthetic Populations submodule creates a regional population imitation whose demographics closely match that of the real population. The imitation's households also are distributed spatially to approximate the regional population distribution. The synthetic population's demographics are provided to the Activity Demand submodule to derive individual and household activities requiring travel. The household locations determine travel origins and destinations" (on 2).

"The purpose of the Household and Commercial Activity Disaggregation's Activity Demand submodule is to generate household activities, activity

**Box 5.1 (continued)**

priorities, activity locations, activity times, and mode and travel preferences. The activities and preferences are functions of the household demographics created by the Synthetic Population submodule. The Intermodal Route Planner uses the activities and preferences to determine individual's and load's trip plans for the region" (on 4).

"The Intermodal Route Planner involves using a demographically defined travel cost decision model particular to each traveler. Vehicle and mode availability are represented and mode choice decisions are made during route plan generation. The method estimates desired trips not made, induced travel, and peak load spreading. This allows evaluation of different transportation control measures and travel demand measures on trip planning behaviors" (on 2).

"The Travel Microsimulation executes the generated trips on the transportation network to predict the performance of individual vehicles and the transportation system. It attempts to execute every individual's travel itinerary in the region. For example, every passenger vehicle has a driver whose driving logic attempts to execute the plan, accelerates or decelerates the car, or passes as appropriate in traffic on the roadway network" (on 2).

"The Travel Microsimulation produces traffic information for the Environmental Models and Simulations to estimate motor vehicle fuel use, emissions, dispersion, transport, air chemistry, meteorology, visibility, and resultant air quality. The emissions model accounts for both moving and stationary vehicles" (on 2).

*Source:* LaRon Smith, Richard Beckman, Keith Baggerly, Doug Anson, Michael Williams, *TRANSIMS: TRansportation Analysis and SIMulation System* (Los Alamos: LANL, ca. 1995); see also *TMIP Connection—The Travel Model Improvement Program Newsletter*, no. 2 (January 1995).

If we compare the structure of TRANSIMS with the four-step modeling architecture, we can see that, taken together, the Household and Commercial Activity Disaggregation module and Intermodal Route Planner perform functions equivalent to the first three steps of FSM (trip generation, trip distribution, and mode choice), while the Intermodal Route Planner and Traffic Microsimulation module are charged with tasks performed by the FSM assignment step. It goes without saying that TRANSIMS does more than the FSM. For a comparison of the two modeling structures, see especially Laurence Rilett, "Transportation Planning and TRANSIMS Microsimulation Model: Preparing for the Transition," *Transportation Research Record*, no. 1777 (2001): 84–92.

inputs to the next module, two of TRANSIMS constituents were also connected through a feedback loop. In the mid-1990s, among the four components, the first, and especially its submodule dedicated to the creation of a "synthetic population," as well as the "traffic microsimulation" ingredient represented the most elaborate parts of TRANSIMS. Unsurprisingly, these were the parts of the entire system that required advanced expertise and high skills in mathematical techniques and computation methods, two domains in which Los Alamos excelled.

A *synthetic population* is a mathematical construct, derived through statistical techniques from real population census data. It is both a simplified representation of the actual population—only features deemed to be relevant for travel demand modeling are taken into account—and a microscopic representation of the latter, since every household and every person in the area under consideration is represented on an individual basis. Though not identical to the actual population by definition, the synthetic population nevertheless seeks to imitate it by matching its various statistical distributions. But why resort to a virtual population and not directly use, as the FSM does, actual data obtained from real travel surveys? As a matter of fact, the general principles and objectives underlying the overall TRANSIMS modeling approach require an original tool such as synthetic population. Remember that the proponents of the project aimed to model the entire number of trips of all individuals in a metropolitan region on a second-by-second basis throughout the entire day. Highly individual-centered, TRANSIMS could not therefore rely upon the set of average and aggregated zonal data used in traditional travel demand modeling. However, an exhaustive survey of all households and individuals in a large area with a view to capturing their specific characteristics—to be associated afterward with specific travel patterns—is materially impossible, if only for economic and privacy reasons. Hence the idea developed of simulating the characteristics of each household and its individual members through mining census data and other statistics. Composed of virtual households (described by characteristics such as household size, income, number of cars, and address) and virtual individuals (represented by attributes such as age, sex, or work location), a synthetic population created with the help of a simulator is therefore *statistically* equivalent to the real population. Not surprisingly, it was the mathematicians of the Los Alamos group, led by Beckman himself, who crafted the mathematical procedures to generate synthetic populations tailored to the data needs of TRANSIMS project, on the basis of sample and census data available in the United States. To do so, the LANL group built upon existing techniques already crafted by statisticians, especially the so-called iterative proportional fitting (IPF) method. Dating back to the 1940s,[47] it was introduced in transportation literature in the mid-1970s by Daniel McFadden and his collaborators in their San Francisco area study (chapter 3), for which the authors came to "synthesize" household survey data.[48] Beckman and his teammates would present the outcomes of their efforts in several papers,[49] while presenting their findings at several conferences sponsored by the Travel Model Improvement Program.[50] Thanks to these dissemination conduits, synthetic population definitively entered the UTDM social world (see especially chapter 6).[51]

Another TRANSMIS component regarded as mature enough by the mid-1990s to be tested in real-life conditions was the *traffic microsimulation* module,[52] for the development of which the European segment of the LANL team, Nagel in particular, proved decisive.[53] With its traffic microsimulation component, the LANL team demonstrated once more its willingness to break with long-standing practices in travel demand modeling through resorting to cellular automata (CAs), a theoretical framework with which the UTDM social world had been little familiar so far. In his capacity as a physicist, Nagel was aware of this approach, originally proposed by Johann von Neumann (1903–1957) in the shape of formal models of self-reproducing organisms, a topic that had emerged from investigations in cybernetics some years after the end of World War II.[54] Following in Neumann's footsteps, physicists as well as biologists would resort to the cellular automata approach in order to study the behavior of complex natural systems on the basis of interactions between their (many) elementary parts. After a period during which research in the field petered out, from the early 1980s on a steadily increasing number of papers started appearing on cellular automata.[55] Highway traffic networks, regarded as complex systems composed of many elementary "particles," that is, vehicles, were among the beneficiaries of this growing interest in CA-based modeling techniques.[56] It was in 1992 that Nagel and a colleague of his published a paper titled "A Cellular Automaton Model for Freeway Traffic," where they first expounded what is now known as the Nagel-Schreckenberg model, dubbed the "NaSch model," for a single-lane freeway.[57] Once settled in Los Alamos, Nagel would combine his own considerable expertise in traffic analysis with the mathematical and scientific skills of his teammates—those of Barrett, Stephen Eubank (a physicist with a Ph.D. from the University of Texas at Austin in 1986), who specialized in nonlinear dynamical systems, and Rasmussen in particular—to collectively devise the traffic microsimulation module. Unlike the original NaSch model, the TRANSIMS module considers a multilane freeway, in which vehicles can change lanes to pass other automobiles. Each lane is therefore sectioned into an array of cells of approximately 7.5 meters long, which is the average distance between vehicles caught in a traffic jam. At each time step, the vehicle moves from one cell to another on the basis of a certain number of driving rules,[58] and traffic at a macroscopic level *emerges* from the interaction of all the vehicles occupying successively the various cells of which the modeled transportation network is made. Unsurprisingly, this kind of *microsimulation* (box 5.2) operating with each vehicle on the network, while easy to grasp intuitively, poses formidable computation problems when it comes to predicting the evolution of traffic on a large network hosting thousands of vehicles. To provide computational speed as well as an operation capability available to end users less well endowed with computer resources than LANL,[59] the TRANSIMS team, after testing different computational supports and techniques, decided to run the traffic microsimulation module on a network of SUN workstations.[60] Once completed, following in the footsteps of the synthetic population component, the work carried out by Nagel and his teammates would be made widely available to the UTDM social world through various dissemination channels.[61]

By the mid-1990s, the Los Alamos group had designed the overall architecture of TRANSIMS, and already crafted two of its main components. Heavily backed by the federal administration and well publicized, especially thanks to TMIP, the LANL product was nevertheless still regarded by a large part of the UTDM social world of the day with suspicion. A great (and expensive) intellectual adventure breaking with long-standing practices for sure, would TRANSIMS be able to deliver on its high promises when confronted with real-life situations? After several years of conceptual developments, it was testing time. While their product was still under development, several members of the TRANSMIS team visited six MPOs.[62] After considering various factors, ranging from the staff and management interest in the project to the organization's capabilities to data availability, the group decided to work with the North Central Texas Council of Governments

**Box 5.2**

Simulation/Microsimulation

"*Simulation* generally refers to an approach to modeling systems which possess the following two key characteristics."

1. The system is a *dynamic* one, whose behavior must be explicitly modeled over time.
2. The system's behavior is *complex*. In addition to the dynamic nature of the system (which generally in itself introduces complexity) this complexity typically has many possible sources, including:
   (a) complex decision rules for the individual actors within the system;
   (b) many different types of actors interacting in complex ways;
   (c) system processes that are path dependent (i.e., the future system state depends both on the current system state and explicitly on how the system evolves from this current state over time);
   (d) the system is generally an "open" one in which exogenous "forces" operate on the system over time, thereby affecting the internal behavior of the system; and/or
   (e) significant probabilistic elements (uncertainties) exist in the system, with respect to random variations in exogenous inputs to the system and/or the stochastic nature of endogenous processes at work within the system.

Note that in speaking of complexity, we are not merely referring to the difficulty in dealing with very large models with large datasets defined over many attributes for hundreds if not thousands of zones. Rather, we are referring to the more fundamental notion of the difficulty in estimating likely future system states given the inherently complex nature of the system's behavioral processes.

Given the system's complexity, closed-form analytical representations of the system are generally not possible, in which case numerical, computer-based algorithms are the only feasible method for generating estimates of future system states. Similarly, given the system's path dependencies and openness to time-varying exogenous factors, system equilibrium generally is not achieved, hence rendering equilibrium-based models inappropriate. In the absence of explicit equilibrium conditions, the future state of the system again generally can only be estimated by explicitly tracing the evolutionary path of the system over time, beginning with current known conditions. Such numerical, computer-based models which trace a system's evolution over time are what we generally refer to as simulation models.

**Box 5.2 (continued)**

Note that conventional four-stage travel demand models most clearly are *not* simulation models under this definition. Conventional four-stage models are static equilibrium models which predict a path-independent future year end state without concern for either the initial (current) system state or the path traveled by the system from the current to the future year state. Thus in adopting a simulation approach to modeling activity and travel behavior, one is explicitly rejecting the conventional static equilibrium view of urban systems in favor of a dynamic representation of such systems—a very significant decision, both conceptually and practically.

*Source:* Eric J. Miller, "Microsimulation and Activity-Based Forecasting," in *Activity-Based Travel Forecasting Conference, June 2–5, 1996, Summary, Recommendations and Compendium of Papers*, ed. Texas Transportation Institute (prepared by) (Arlington: Texas Transportation Institute, 1997), 152; emphasis in original.

(NCTCOG) for the first real-life study of the traffic microsimulaton module.[63] In 1996, a twenty-five-square-mile area centered on the Galleria shopping mall, infamous for its serious congestion problem, therefore became the site of the first experiment using a major component of TRANSIMS. The case study simulated approximately two hundred thousand vehicle trips into, out, and through the heavily congested area during four peak hours of the morning. Microsimulations were run for two different network improvement scenarios and were then compared to a microsimulation of the existing situation. The team also tested the traffic model's behavior, especially its sensitivity to variations in the input data. Although the MPO specialists found that the results provided by the TRANSIMS component under study were a little problematic, in particular regarding the roads acting as traffic collectors in the area under study, a consensus on the overall good performance of the traffic microsimulaton component, including the associated running times, would be established by the end of the study.[64] Thanks to the Dallas "experiment," TRANSIMS became a showcase for the LANL: When President Clinton visited Los Alamos in February 1998 to give a speech, he was shown a computer-generated animation of the results of the Dallas-Fort Worth case study.[65]

At the end of the first large-scale experimental study carried out by the LANL team, TRANSIMS Version 1.0 of the software was installed on the Sun

UNIX operating system of NCTCOG, whose staff could grasp the arcane technicalities of the modeling suite with the help of a manual of more than three hundred and fifty pages.[66] At the same time, while already diffusing the traffic microsimulation package, somewhat revised based on the Dallas experience, to a number of MPOs across the country for additional checks and feedback, the LANL team was actively preparing itself to put TRANSIMS to a second and much more comprehensive test. As a matter of fact, the planned second case study, which had in the meantime undergone a number of modifications,[67] would be concerned with the entire modeling suite, the feedback mechanisms included. In addition, while the Dallas case study was a vehicle-based simulation only, the new one was designed to be multimodal. And, last but not least, there was also a major shift regarding the spatial extent of the "experiment," since the new test would concern an entire metropolitan region, much larger than the Dallas twenty-five-square mile area.[68] It was METRO, the MPO for Portland, Oregon, that would eventually welcome the LANL team in 1998. METRO's involvement should come as no surprise, since in the 1990s this MPO was among the few agencies in the United States equipped with significant internal transportation modeling resources.[69] A short while before it had begun to collaborate with the Los Alamos group, METRO, under the intellectual guidance of T. Keith Lawton, the holder of a master's degree in civil and environmental engineering from Duke University in 1975, had done pioneering work in household travel survey techniques in the mid-1990s.[70] It had also already started to work with John L. Bowman, an MIT doctoral student, and seasoned travel modeler Mark Bradley to operationalize a new forecasting approach that Bowman had prototyped for his 1995 master's thesis, aiming to replace the traditional four-step model (see chapter 6). Given its scale, the new TRANSIMS operation required significant supplementary human resources for its deployment. As a result, the circle of partners in the project broadened to include new stakeholders.[71] In addition to Bradley (more on him in chapter 6), Paul L. Speckman, a Ph.D. statistician from the recently founded National Institute of Statistical Sciences (NISS) was invited to offer his services as well. The NISS staffer was commissioned to work in particular on the modeling of the daily activities of travelers, an issue Speckman was already familiar with following his earlier collaboration with Eric Pas, an academic well versed in advanced travel demand modeling practices (see chapter 6).

At the time when the TRANSIMS experiment in Portland was about to kick off, the entire Los Alamos project was entering a new phase in its development, not so much in terms of its very content as in respect of its legal status. Though the Dallas case study certainly helped dissipate some of the skepticism the federally sponsored project had been confronted with since its birth, TRANSIMS, entirely designed and developed by a research center, was nevertheless still regarded by most of its potential end users, including the many MPOs and state transportation departments across the country, as a sophisticated and complex tool. As a result, it encountered serious difficulties in spontaneously stirring their interest.[72] Among other things, the user interface of the original software seemed to have left much to be desired, despite the fact that the overall software architecture had been entrusted to a team of specialists, who had sought to make sure that "TRANSIMS remains flexible, expandable, portable, and maintainable throughout its lifetime."[73] At the end of the 1990s, the LANL team's efforts were focused on completing the development of the basic TRANSIMS core package, and it was hoped that the Portland case study would be finished during the summer of 2000. The proponents of the project were also actively looking for a contractor able to craft user-friendly TRANSIMS software. Organizations interested in participating in the commercialization and deployment of TRANSIMS were therefore invited to attend the TRANSIMS Opportunity Forum, scheduled for June 28–30, 1999 in Santa Fe, New Mexico. At the same time, the federal administration was calling for the implementation of TRANSIMS in a wide range of operational settings, and to this end, it was willing to provide substantial financial support to a number of local planning agencies to help them shift from their existing travel demand procedures to the Los Alamos product.[74]

### Trying Hard to Make TRANSIMS Popular

The Santa Fe Opportunity Forum seemed to be a success, for TRANSMIS proponents at least. The source code for the TRANSIMS software was made available under a nondisclosure agreement to interested parties, and the organizers managed to attract about fifty people not directly associated with the Los Alamos project. A group of nine specialists, most from LANL and METRO, interacted with the attendees. Of particular note, Chris Barrett spoke personally and privately with any significant potential bidders for the TRANSIMS commercialization process.[75] A request for proposals (RFP) was issued by LANL in August 1999, and a pre-bid conference was held in September 1999 to answer any questions on the RFP.[76] It was Pricewater-houseCoopers (PwC), the multinational consulting colossus, which eventually won the contract and was therefore commissioned to "further develop the user interface . . . to create a version more useable for transportation planning applications."[77] For its services the firm would receive approximately seven million dollars from the federal government. By September 2001, PwC had defined the overall architecture of TRANSIMS's commercial version, including a Linux server, client workstations, and a cluster of compute nodes the number of which could be adjusted to meet the computing needs of the end user. PwC had also begun to design a new user interface. The first release of the commercial version was expected for spring 2002.[78]

While PcW was stepping onto the TRANSIMS stage, the LANL team continued to deploy significant efforts to persuade the various segments of the UTDM social world to adopt its product. Thus, during the 2000 and 2001 annual meetings of the Transportation Research Board, special TRANSIMS sessions were organized.[79] Several courses of actions aimed to seduce academics were also undertaken, including inexpensive specific "research licenses," thanks to which, the argument went, the members of the academic community would be able to "spread the TRANSIMS methods and technology into university research, develop competencies in the TRANSIMS sciences, and help prepare tomorrow's transportation planners for exercising the TRANSIMS capability to its fullest."[80] In March 2001, TRANSIMS Version 1.1 was available to qualified educational organizations willing to pay a modest $500 (noncommercial) license fee.[81] Several academics were also approached with a view to strengthening the dissemination of TRANSIMS through education. Thus, in the very early 2000s, Antoine Hobeika, then professor at the Civil Engineering Department of Virginia Tech, started working with the Federal Highway Administration and the TMIP to establish a textbook and course manual on TRANSIMS as well as to provide assistance in testing and refining the software.[82] Thanks to Hobeika's involvement, Virginia Tech in all likelihood was the first American university to make itself into an active research and educational center for the TRANSIMS project.[83] Laurence R. Rilett, a civil engineer with a Ph.D. from Queen's University (Ontario) in 1992, was another academic who showed an early interest in TRANSIMS. An obvious firm believer in the latter's operational potential, Rilett, after reviewing the many differences between the old four-step model and the brand-new Los Alamos product, called upon the UTDM social world to prepare "for the transition."[84] More generally, judging from the number of research licenses for the early versions of the software,[85] the American university did not turn its back on TRANSIMS.[86]

However, Rilett's expectations did not materialize, and TRANSIMS would never make itself into a commercial and commonly used operational software. At the end of the 2010s there was no transportation planning agency in the United States that had adopted TRANSIMS as its production model. This does not mean that between the early 2000s and the end of the 2010s the Los Alamos product just melted in thin air. For several years the TRANSIMS community continued to evolve with participants from government, academia, and consulting, while since the mid-2000s the source code of the software has been freely available under an established open-source license. To see how TRANSIMS, essentially through its many byproducts, including the now emblematic "synthetic population" idea, succeeded in being part of the current travel demand modeling landscape despite its failure to supersede the traditional four-step model, let us turn the clock back to the late 1990s, when its second, and more ambitious, case study was about to start in Portland, thanks, among other things, to a powerful new computer system acquired with federal money.[87] After an enthusiastic beginning, as time went by the problems to be overcome accumulated. The calibration of the entire package proved much more difficult than that of its isolated modules, largely because of the multiple feedback loops. In the autumn of 2002, the TRANSIMS Working Group (TWG) was established, in which academics joined forces with more practice-oriented modelers to support METRO.[88] In addition to TWG, the Portland case study would also benefit from the expertise of the IBM commercialization team[89] as well as from the modeling skills of two large transportation firms. The first of these, Parsons Brinkerhoff (PB) was already familiar with the TRANSIMS project since a company staffer, Rick Donnelly, had long cooperated with the LANL team in the 1990s. The second firm involved, AECOM,[90] had just entered into a contract with the Federal Highway Administration to study the impact of using TRANSIMS for estimating vehicle emissions.[91] Following its involvement in the Portland case study, AECOM in the 2000s and 2010s has grown into one of the most important members of the TRANSIMS community, in large part because of the dedication of David B. Roden, an ex-staffer at JHK & Associates with a master's degree in engineering from UC Berkeley; he joined AECOM in 2000. However, despite the creation of such a powerful task force, work in Portland stalled. At about the same time, the Los Alamos team disbanded in 2005 as it was believed to have successfully accomplished its mission.[92] Due to funding issues and unexpected heavy staff demands regarding transit proposals, METRO eventually turned definitively away from TRANSIMS in the mid-2000s.[93] However, AECOM carried on working on the project, and continued to build upon data from Portland.[94] At the end of the decade the firm even issued TRANSIMS Version 4.0 while rewriting several of the TRANSMIS modules so that they could be executed on the Linux and Microsoft Windows operating systems.[95]

No longer supported by Los Alamos National Laboratory, TRANSIMS's career would undergo a major change in the mid-2000s. Probably considering that the prospects for successful commercialization were not very bright in the immediate future,[96] the Federal Highway Administration decided, first in May 2005 and afterward in July 2006, to make the source code of TRANSIMS Version 3.1.1 available under the NASA Open Source Agreement Version 1.3.[97] By taking this decision, which contrasted with its previous policy that aimed at bringing the Los Alamos product to market, the federal administration nevertheless continued to participate in the TRANSMIS adventure by financially backing further developments of the project and by sponsoring the deployment of new case studies.[98] And there was more. As TRANSMIS became a free software, the federal administration set to work on cultivating, strengthening, and maintaining an independent community of TRANSIMS users, researchers, and developers. To this end, the federal administration engaged Mitretek Systems[99] to help design a dedicated website providing access to a number of resources, including software release packages, documentation, TRANSIMS example cases, presentations and archived web seminars, and access to the TRANSIMS wiki and TRANSIMS forum. As of April 2008, the latter numbered 195 users, while 626 articles had been posted thus far.[100]

From 2006 until the early 2010s, the friends of TRANSIMS could also benefit from another, even larger federal initiative, the so-called TRACC project. It was on October 1, 2006 that the famous Argonne National Laboratory[101] initiated a multiyear program with the DOT to establish the Transportation Research and Analysis Computing Center (TRACC). Among the objectives pursued by the TRACC project was the creation of a national high-performance computer user facility—which was fully operational in March 2008—and the provision of technical support for its use by DOT staffers as well as their university and private sector contractors. Hardly had TRACC come into being when DOT asked the newly created research unit within Argonne to advance the use of TRANSIMS within the transportation community.[102] From 2007 to the early 2010s, a group of TRACC staffers headed by Hubert Ley, a 1994 Ph.D. holder in nuclear engineering and specialist in the development of advanced user interfaces and web-based technology, worked intensely, often in tandem with university partners,[103] government planning agencies,[104] and large transporation and consulting firms, including AECOM and Resource Systems Group (RSG), to help design new tools that would make TRANSIMS more suitable for large-scale deployment.

As the amounts of data resulting from TRANSIMS runs could not be easily interpreted without the help of powerful visualization tools, large parts of the work done at TRACC emphasized the development of network editors, visualization devices, and graphical user interfaces. The TRANSIMS Studio Graphical User Interface was therefore released in the early 2010s, to be soon downloaded close to 250 times.[105] Given the computational demands on the TRANSIMS computer package for large metropolitan applications, work on the parallelization of the software started at TRACC in 2007. While the various LANL versions of TRANSMIS had been crafted to run exclusively across multiple processors on a cluster computer,[106] version 4.0 of the software was designed, in contrast, to optimize performance for a single core. Developed under the leadership of AECOM and released in the early 2010s, TRANSIMS Version 5.0 could be run in a parallel capacity.[107] Version 5.0 was at the heart of the Regional Transportation Simulation Tool for Evacuation Planning (RTSTEP), a $2,000,000 research project started in December 2010, performed by the Argonne National Laboratory with the help of AECOM and a number of other organizations, and funded by the U.S. Department of Homeland Security.[108] In addition to participating in the furthering of TRANSIMS software and its implementation, especially in the Chicago Area, TRACC would also form the mainstay of a series of regular courses and workshops on TRANSIMS till the early 2010s. The first of these training sessions took place in the second quarter of 2007, while on April 14–15, 2011, TRACC celebrated its thirtieth training course, held at South Carolina State University.[109]

Relying on these resources, the TRANSIMS community remained very active and even continued to grow throughout the 2000s, while a whole range of quantitative as well qualitative indicators could testify to its good shape. Thus, TRANSIMS-L, a list hosted on the TMIP website, and which allowed users to post issues or questions related to using TRANSIMS software, had an estimated 179 and 206 users at the end of fiscal years 2009 and 2011 respectively.[110] As for the TMIP website, between May and September 2012 and between February and September 2013, the TRANSIMS directory page experienced the most page views with the exception of the main landing page.[111] After TRANSIMS had been transformed into an open-source software in the mid-2000s, the number of (rather small) projects[112] in which the software was involved also went up. By fall 2009 there were around thirty TRANSIMS-related operations, completed or under way, dealing with issues ranging from multimodal evacuation to congestion and emission studies to congestion pricing proof-of-concept.[113] Unsurprisingly, this increase in the number of projects was accompanied by the widening of the circle of actors showing an interest in TRANSIMS. Throughout the 2000s and the early 2010s, many heavyweights of the UTDM social world—be they transportation consultancies and individual practitioners, proprietary software providers, or government planning agencies—familiarized themselves one way or another with parts, if not the entirety, of TRANSIMS. The cases of AECOM and Parsons Brinkerhoff have already been mentioned. Another large consultancy, Resource Systems Group, also took part in two (small) projects involving the LANL product, before leading a $1.9 million research study,[114] in which parts of the TRANSIMS modeling software, the Router module and the Microsimulator one, were implemented in the regions of Burlington, Vermont, and Jacksonville, Florida.[115] Citilabs, the well-known proprietary transportation software vendor,[116] also headed a team that was awarded a contract at the end of 2007 to use TRANSIMS in a before-and-after study of the Hiawatha light rail line—rebranded METRO Blue line—in Minneapolis.[117] Among the large government planning agencies that showed interest in and made use of TRANSIMS, we should mention the Chicago Metropolitan Agency for Planning and the Atlanta Regional Commission. Last but not least, several members of the university system, including academics doing cutting-edge research into urban travel demand modeling, such as Ram M. Pendyala, Chandra Bhat, and Konstadinos G. Goulias, to name just three (more in chapter 6), also participated in some TRANSIMS initiatives.[118]

### A Productive Failure?

And then, all of the sudden, the enthusiasm seemed to fade away.[119] Despite all these projects, TRANSIMS, now in version 7.4,[120] has not been adopted for operational use by any transportation planning agency across the nation, with perhaps the exception of the Virginia DOT, which enlisted the services of AECOM in a series of projects carried out with the help of the LANL product.[121] In all likelihood, the high computation times needed for running TRANSIMS frightened more than one member of the UTDM social world.[122] Moreover, from the early 2010s on, they could rely on a series of efficient concurrent traffic forecasting practices that progressively emerged across the nation,[123] and may have felt that TRANSIMS was less appealing. On a smaller scale, even the TRACC (micro)community that had supported TRANSIMS for several years seemed to turn its back on the federal project. From the early 2010s onward, building on their familiarity with TRANSIMS but also on the modeling work previously conducted by Joshua Auld, a newcomer to Argonne in 2012,[124] Ley and his team started working on another large research program called POLARIS (Planning and Operations Language for Agent-based Regional Integrated Simulation). Also an open-source modeling framework designed, like TRANSIMS, to simulate large transportation systems, POLARIS contained a highly integrated modeling suite in which all the aspects of travel decisions, including departure time, destination and route choice, and planning and rescheduling could be modeled simultaneously.[125]

In light of the latest developments concerning TRANSIMS, the skeptic would probably shrug his shoulders and wonder whether the Los Alamos project has been so far worth the forty million dollars or so spent on its development since it was first conceived (money that could have fueled more effective modeling practices, TRANSIMS's critics might add). In contrast, the optimist would insist that TRANSIMS is still alive, although under proxy lives. The LANL project would indeed leave its stamp on the UTDM social world, although not to the degree its creators and key financial backers had originally intended or expected. To start with, some specific operational outcomes of TRANSIMS, especially the "population synthesizer" are now an integral part of the standard (advanced) tools available to travel demand modelers (see chapters 6 and 7). In addition, the current POLARIS modeling suite has in several respects been an outgrowth of the LANL project. One could also argue that TRANSIMS was part of a series of general shifts in urban traffic forecasting over the last two decades as it helped push the entire field into devising increasingly dynamic, disaggregate, and microsimulation-based modeling techniques. It also incited vendors to include microsimulation techniques in their software packages and was probably the earliest successful example of large, open-source software development in urban travel demand modeling. Through its various federal government–backed implementations as well as via the controversies it generated, the TRANSIMS project has also helped the UTDM social world as a whole envision, criticize, and adopt new and original ways of modeling transportation systems. It is no wonder that its specter hovers over the next two chapters dedicated to a number of innovative practices that, taken together, are in all likelihood on the verge of replacing the old FSM for good.

# 6 Travelers Are Social Beings

## Disaggregate but Not Utility-Based? What Activity-Based Modeling Is About

"Zones do not travel, people travel," so the saying goes (chapter 3). But who are these people? For the proponents of disaggregate modeling, and the promoters of the motto, they are separate individuals acting like the proverbial *homo economicus* portrayed by the neoclassical microeconomic theory, men and women relentlessly seeking to maximize the welfare derived from their travels, and invariably choosing the alternative that provides them with the highest utility among the available options. Such an individual is a highly problematic theoretical abstraction, argue the adepts of the activity-based modeling, since flesh-and-blood creatures are neither autonomous nor self-centered maximizing entities, not all the time at least. Should we return therefore to the old aggregate principle upon which the traditional four-step model has been erected? However critical of the utility-based disaggregate models they were, the early proponents of ABM nevertheless yearned as much as their intellectual enemies to substitute FSM with new modeling practices more focused on the *individual*, on his or her travel behavior and decision-making. But their individual was *homo sociologicus*, a creature embedded in manifold social relations.

The activity-based approach is about scrutinizing and modeling *homo sociologicus*'s travel behavior. It builds upon the general, and easy to grasp, idea that travel is not a good that is desired for itself. As a matter of fact, more often than not, travel is seldom an activity undertaken for its own sake. On the contrary, it is derived from the various activities in which people participate in order to meet their needs and satisfy their desires. In some cases, these activities may occur within their own home. But they can also be located in other places, hence the need to travel. Activity-based models therefore try to depict and model an individual's entire day of activities, including the *derived* travel for activity participation,[1] in the presence of *time* and *space* constraints,[2] while emphasizing the many *linkages* between activities and travel both for an individual person and across multiple people in the same household—think, for example, of the importance of young children, who radically transform the activity pattern and travel constraints of the adult members of household.[3] In a nutshell, activity-based modeling seeks to predict "which activities are conducted when, where, for how long, for and with whom, and the travel choices they will make to complete them."[4] Producing as its final output tables of all movements of individuals in the area under consideration by mode while providing for each movement its duration as well as its start and end times, ABM aims to replace the *first three stages* of the FSM: trip generation, trip distribution, and mode choice.

Laying out a series of general principles is one thing. Implementing them in practical settings is quite another. As the first proponents of the new approach would only too poignantly come to realize, activity-based modeling took a great deal of time to take off from a practical point of view because its much greater behavioral realism compared to other modeling lines has translated into serious computational complexity. While originating within academia in the 1970s, the activity approach would gain operational status only some three decades later. And, in an ironic twist, in order to become operational ABM also had to incorporate important components of the utility-based disaggregate approach predicated on the idea of *homo economicus*.

## British Beginnings

On July 9–13, 1973, the Briton Ian Graeme Heggie (born in 1936), a transportation economist,[5] then head of the Transport Studies Unit (TSU) at Oxford University,[6] was in South Berwick, Maine, as part of a select team of specialists gathering there to discuss "Issues in Behavioral Demand Modeling and the Evaluation of Travel Time" (chapter 3). Participants divided into small working groups in order to address specific topics related to the general subject matter of the conference. Charged with the task of examining the mathematical structure of what was then the nascent utility-based disaggregate modeling approach, Heggie and his teammates found themselves "drawn into a detailed consideration of the metaphysical basis of behavioural models as well as the empirical problem of parameter estimation." The main conclusion they drew from their discussion was that nearly everything remained to be done, since "we know practically nothing about the way in which people arrive at decisions to make a journey and the way in which they choose to accomplish it."[7] Convinced that the much-vaunted new approaches aiming to import *homo economicus* into transportation forecasting "merely caricature behaviour," Heggie, once back home, seriously envisioned an alternative modeling line thanks to which "models of behavior" would "become more human and less mathematical" while giving "realistic insights into the way in which individuals arrange their activities in time and space."[8] Fortunately, at the TSU Heggie was surrounded by a handful of people who thought like him. Peter Jones was a geographer with a B.Sc. from the University of Bristol,[9] Martin Dix had studied psychology, while Mike Clarke was a physicist by training with strong mathematical skills. Around the mid-1970s, the Oxford Group set to work on an original project that aimed to "develop a new way of describing travel behavior in terms of a model that has an explicit behavioural basis."[10] It was probably the first systematic excursion into a new travel demand modeling territory, now known as activity-based modeling.

As original as the TSU's project was, it did not appear out of thin air. Like the famous metaphor of dwarfs standing on the shoulders of giants, Heggie and his teammates could greatly benefit from work done by a series of scholars in the 1960s and early 1970s. The American F. Stuart Chapin Jr. (1916–2016), an academic long identified with planning education and research at the University of North Carolina at Chapel Hill,[11] was such a "giant," and so was the Swedish geographer Torsten Hägerstrand (1916–2004),[12] both of them unanimously recognized and revered today as the intellectual fathers of ABM.[13] Building on Chapin's attempts to explain differences in activity patterns through person and household characteristics—to this end, the author would coin the notion of "stage in the family life cycle"—and on Hägerstrand's insights into spatial and temporal constraints, the Oxford Group succeeded in producing a series of qualitative results about travel behavior, especially the importance of the spatial and temporal constraints and linkages between household members. Moreover,

Heggie and his teammates were also able to craft a series of more practice-oriented tools, including the so-called Household Activity-Travel Simulator (HATS), essentially a gaming simulation examining both the direct and secondary effects of policies on different types of households, and CARLA, an activity-scheduling model capable of predicting trips as a result of activity decisions.[14]

Hardly had the Oxford Group completed its innovative project when it organized an international conference on "Travel Demand Analysis: Activity-Based and Other New Approaches."[15] Held in Oxford in 1981, the event was later described by an ABM specialist as a "milestone in the growth of the field of activity-based travel analysis,"[16] and went a long way toward publicizing the new approach.[17] But even before the Oxford conference took place, the original work carried out by Heggie and his collaborators in the 1970s had already been circulating among transportation academics and practitioners through several channels. Leading journals in the field of urban traffic forecasting, including the widely read (American) *Transportation Research Record*, welcomed in its columns several articles authored by TSU members,[18] who also had the opportunity to preach the gospel of ABM in international forums as well. Thus, during the third international conference staged by the TRB's Committee on Traveler Behavior and Values in Tanuda, South Australia, in April 1977, the activity-based approach took up the space of an entire workshop. Three resource papers addressing "new approaches to understanding traveler behavior (including nature of travel and the structures of journeys)," followed by interactions with the audience, were thus delivered in front of a public composed of sixteen delegates.[19] Jones himself was the author of one of the resource papers as was Patricia (Pat) Burnett, also a participant in the South Berwick gathering during July 1973. A little later, Burnett and her teammates at Oklahoma University would produce a lengthy text sponsored by the U.S. government for two conferences organized on American soil this time, in which they sharply criticized utility-based disaggregate modeling in light of the ABM approach.[20] Through these forums and channels, the studies carried out by the Oxford Group reached American shores by the end of the 1970s. A decade later, all major transportation centers in the United States, from East to West coast and from Chicago to Texas, had already taken an interest in the ABM research program.

## Activity-Based Modeling Enters American Campuses

In the second half of the 1970s, Moshe Ben-Akiva and Steven Lerman, two young MIT scholars with whom the reader is already familiar, started supervising a number of postgraduate research studies aimed at incorporating insights from the activity approach into the newly developed, and increasingly popular utility-based disaggregate modeling. The earliest efforts, focused on the question of activity choice and time allocation, would be deployed by Joseph Henry Bain, who even benefited from the cooperation of Chapin himself by obtaining a part of the data used in his master's thesis.[21] Bain's study was followed by a number of Ph.D. dissertations, including those by David Damm, the author of an important early review of the ABM literature in 1983,[22] Jesse Jacobson, and Ilan Salomon, all three referring, in one way or another, to the project carried out by the TSU members in Oxford.[23]

In 1980, Eric Ivan Pas, born in Cape Town in 1948, defended his Ph.D. at Northwestern University, thus becoming the first doctoral student of Frank Koppelman,[24] an early specialist in disaggregate modeling (chapter 3). Though Pas left Evanston, Illinois, for a position at Duke University, where he would distinguish himself as a tireless and eloquent advocate of ABM to the UTDM social world until his premature death in 1997,[25] Koppelman continued to cultivate, often along with his supervisees, the new approach at the Transportation Center in Evanston. After Massachusetts and Illinois, Texas would also become an important site for activity-based modeling. Born in 1956, Hani Sobhi Mahmassani started studying civil engineering at the American University of Beirut. After receiving a master's degree from Purdue University in transportation, he defended his Ph.D. thesis at MIT under the supervision of Yosef Sheffi.[26] Mahmassani arrived at the Center for Transportation Research, located at the University of Texas at Austin, in spring 1982. There, throughout the 1980s he and his collaborators would notably produce a series of analyses based on observation of departure time decisions by commuters in a simulated environment (chapter 7). When Mahmassani left Texas in 2002, activity-based modeling was a well-established domain in Austin, thanks in particular to two young professors. Chandra R. Bhat, born in 1964, graduated from the Indian Institute of Technology Madras to become a Ph.D. student of Koppelman,[27] while Kara Maria Kockelman, who arrived in Austin in the fall of 1998 after defending her thesis that same year, had earned all her university degrees in California.[28]

In fact, during this period, the California university system would prove to be the most fertile ground for research in ABM in the United States, thanks in large part to the many branches of the Institute of Transportation Studies located across the West Coast (chapter 2). It was at the University of California, Davis, that Ryuichi Kitamura (1949–2009) decided to settle at the end of the 1970s; Kitamura, for whom it was said that "no individual is larger than the field, but some can define and redefine their field,"[29] made UC Davis the organizational center of his crusade for the ABM approach for around fifteen years. Kitamura had graduated from Kyoto University in 1972, and after getting his master's degree from the same institution,[30] moved to the United States where he obtained his Ph.D. from the University of Michigan in 1978.[31] Influenced by the spirit of the times, Kitamura took his first steps in urban travel demand modeling within the utility-based disaggregate framework. For his thesis, he applied a "heuristic stratification procedure to the exploration of variation in tastes in tripmakers' choice of travel mode."[32] But very soon, Kitamura became keen on ABM, as can be seen in a paper of his that appeared in *Transportation Research Record* as early as 1981, in which he and his coauthors built on Hägerstrand's concepts while quoting the work conducted by the Oxford Group in the 1970s.[33] Kitamura left UC Davis for his first alma mater in 1993, but he continued to advocate and serve the cause of ABM until his death in his capacity as university professor, as chair of various organizations,[34] and as associate editor of *Transportation*. The intensity of his commitment to the activity-based approach is also reflected in the number of publications he signed either as single author or, more often than not, as a member of a team, with over 280 publications between 1975 and 2009, the bulk of which are devoted to activity-based modeling.[35] Kitamura was a gifted scholar. He was also a loved, and even funny, "guide, mentor, and coach"[36] for many well-known academics now cultivating the field of ABM, including the Greek Konstadinos Goulias[37] and the Indian Ram M. Pendyala,[38] to name just two. A mentor to young scholars, Kitamura was seemingly a good fellow as well, always willing to collaborate with his colleagues in order to advance the frontiers of ABM knowledge.

Wilfred W. Recker was one such Californian colleague, and another early proponent of ABM.[39] A Ph.D. holder from Carnegie Mellon University in the field of applied mechanics in 1968, Recker shifted to travel demand modeling in the 1970s, and joined the University of California, Irvine branch of the Institute of Transportation Studies at the end of the decade. Before collaborating with Kitamura, Recker had already worked with another early activity-based approach advocate, Thomas F. Golob,[40] as well as with Lidia P. Kostyniuk,[41] who befriended Kitamura when he was still a Ph.D. student and later collaborated with him.[42] It was under the supervision of a dissertation committee including Kitamura and Recker that Michael G. McNally completed his pioneering Ph.D. work in 1986 at Irvine.[43] Supported (in part) for his research by a grant from the DOT, McNally crafted the so-called STARCHILD (Simulation of Travel/Activity Responses to Complex Household Interactive Logistic Decisions) model system consisting of five modules. Although McNally stated in his dissertation that "the proposed theory is, at best, *incomplete*," and that the "activity program generation, incorporating household interactions, is a *major shortcoming*,"[44] STARCHILD is now regarded as the first activity-based modeling attempt that could claim operational objectives.

## Good in Theory, Lacking in Practical Results

By the end of the 1980s, whether in Cambridge, Massachusetts; Evanston, Illinois; Texas; or California, all the major transportation research centers in the United States had welcomed the activity-based approach. The latter had also started to gain increasing traction within the widely attended conferences organized by the International Association for Travel Behaviour Research (IATBR), which were bringing together many heavyweights in the field of travel demand modeling across the world.[45] The highly influential Transportation Research Board seemed also more and more interested in ABM. Four years after the 1982 Easton Conference, where ABM was the subject matter of a long survey article,[46] at the 1987 annual TRB meeting in Washington, DC, an entire session titled "Activity-Based Travel Analysis: A Retrospective Evaluation" was devoted to the activity-based approach.[47] Other influential organizations in the field of transportation modeling, be they public or private, were showing signs of growing interest in the new modeling line as well. Thus, in June 1988 Kitamura was invited to deliver a keynote lecture at a large conference held by the TRB in Washington, DC, addressing the issue of the nation's future transportation system.[48] A little earlier, thanks to a $221,250 federal grant in 1978 (around $850,000 in 2018), staffers from the consultancy Charles River Associates had sought to incorporate a series of notions, including those of "life cycle" and "household structure," as well as other insights from ABM into trip generation models, the first step of the predominant four-step modeling framework.[49] However, at the end of the 1980s, despite this growing interest in ABM within both the academic and practice-oriented segments of the UTDM social world, the activity-based approach was regarded, even in the eyes of its proponents, as a field with a rather mixed track record.

On the theoretical side of the equation, thanks to their intellectual proximity to sociology and other social sciences, the activity approach advocates were both aware—even proud—of participating more than other "research schools" of the day, including the utility-based disaggregate approach, in developing a better understanding of real people's travel practices. Thus, Susan and Perry Hanson found it legitimate and interesting to inform the readers of the journal *Transportation*, a home for research studies using sophisticated mathematical techniques, of a series of results suggesting that while full-time employment does bring about significant changes to Swedish women's mobility patterns, it has, on the contrary, very little impact on their husbands' travel behavior.[50] For Kitamura himself congestion was not just a technical problem awaiting an engineering solution, but also the symptom of a much broader picture that activity approach proponents should grasp first. As he put it, drawing from books authored by the historian James J. Flink and the specialist in American literature Cynthia Golomb Dettelbach, it was "the life-style to which Americans aspire," "the American dream . . . to live in a suburban single-family house on a half-acre lot with a three-car garage" that was lying at the heart of the congestion predicament.[51] This intellectual closeness to the social sciences would prove a hallmark of ABM specialists and their fellow travelers, and has been repeatedly reflected in both their parlance and references. Thus, in their "Synthesis of Past Activity Analysis Applications," which was published in February 1997, Kenneth Kurani and Martin Lee-Gosselin included in their bibliography works by contemporary sociologist Anthony Giddens (born in 1938), anthropologist Edmund R. Leach (1910–1989), and psychologist Abraham Maslow (1908–1970).[52] In their editorial for the special issue of *Transportation* dedicated to the memory of Eric Pas, Kitamura and Bhat wrote about the "issue of time pressure, seemingly a common affliction of *postmodern society*,"[53] while Bhat, along with his former Ph.D. supervisor Koppelman this time, did not hesitate to refer to Aristotle, Plato, Descartes, Darwin, and even Freud.[54] No wonder that traffic forecasters participating in the 2006 International Conference on Travel Behaviour Research in Kyoto had the opportunity to hear and engage in a dialogue with the late sociologist John Richard Urry (1946–2016), the author of the first keynote speech of the event.[55]

On the practical side of the equation things appeared in a much less favorable light. At the end of 1980s the new approach was still suffering, in several of its own proponents' opinion, from both an uncontrolled profusion of research directions and, above all, a serious lack of operational efficiency as a transportation planning tool.[56] While most people favorably disposed to ABM seemed to share this general assessment, they often disagreed on the specific reasons why there was such a low acceptance of ABM among practitioners. For Peter Jones, the well-known member of Oxford Group, since the activity approach had not enjoyed the same levels of research funding that had been available for the development of the four-step or utility-based disaggregate modeling frameworks, it was only to be expected that one could not see a wide application of activity-based approaches in everyday transportation planning.[57] For Lidia Kostyniuk, Kitamura's old friend and collaborator, a shortage of funds was indeed an important factor at play for at least two reasons. First, since most ABM research had not been sponsored, each academic studied what was of particular interest to him or her, which led to a fragmented body of knowledge. Moreover, as the work conducted by researchers primarily aimed to satisfy their own curiosity and not to respond to specific questions posed by practitioners, all in all little effort went into developing planning methodologies.[58] David Hartgen, a specialist of utility-based disaggregate modeling, was less welcoming to the new approach. Although in his opinion there was "potential for the use of this tool," and with more work "eventually activity analysis will find its place," for Hartgen the much-vaunted "theory," a significant part of it at least, put forward by ABM proponents was "really almost pop sociology, almost trivial." As for the "largely empirical analyses" conducted under the banner of ABM, they "have told us very little so far, not much more than what we could glean by a few minutes of thought about our own patterns and our personal lives." And last but not least, in contrast with the older utility-based disaggregate approach, the new modeling area had not succeeded so far in establishing a clear tie with practical needs.[59] However, it was Mahmassani, though favorably disposed to ABM, who probably provided the most articulate explanation of the various predicaments in which ABM had found itself, especially its rather minimal impact on practice so far. First off, the new approach seemed to be the victim of its own excessive subtlety. To illustrate this point, Mahmassani made use of two famous examples from physics. First, think of Ohm's law in electricity, he addressed the reader: Had one been "overly concerned with the detailed paths of the electrons and what these electrons were doing before crossing the resistor," it is very likely, the argument continued, that Ohm's law would never have come into being. And second, if Newton had been absorbed in the detailed growth characteristics of his proverbial apple, he would surely have missed a great opportunity to discover the principles of gravity. Given the "'search for complexity' syndrome" that had characterized ABM proponents so far, it is only to be expected, the argument went, that few operational tools had grown out of the studies conducted by activity-based approach advocates. Exchanging examples from physics for epistemological considerations reminiscent of the ideas developed by Thomas Kuhn, Mahmassani contended that the group of ABM proponents may have missed what was their intellectual enemies' strength: An intense interaction between two already highly structured disciplinary groups, economists on the one hand, and transportation engineers and analysts on the other hand. Certainly, on the one hand, the underlying theory and the models to which the utility-based disaggregate approach gave rise "may have lacked behavioral realism, and the ability to discern subtler socio-demographic changes and tackle complex interpersonal interactions may have suffered." But on the other hand, the "result of this interaction has been an easily communicable theory, a powerful methodological apparatus, practical applications, and a capability to use these tools that is not limited to a handful of academic researchers."[60]

## Operational Implementation of Activity-Based Modeling Is Drawing Near

The aforementioned critical considerations were first expressed as a series of comments on a review paper by Kitamura during the 1987 annual Transportation Research Board meeting in Washington, DC.[61] However, about a decade later, the general atmosphere surrounding ABM seemed noticeably different. "After many years of development, the activity-based approach to travel is ready for implementation at a time when the planning and policy analysis issues of the day cannot be suitably addressed by the existing, trip-based, four-step travel demand model." These lines were written by Eric Pas, who went on to describe the activity-based approach as "the only real scientific revolution or paradigm shift, in Kuhnian (1970) terms, in the history of the development of travel demand forecasting models." Pas continued: "The shift from aggregate to disaggregate models that took place starting in the 1970's was a shift in *statistical technique* rather than a shift in the *paradigm* and thus can be considered an incremental change in the approach to travel demand modelling."[62] On the face of it, the author's optimism and his confidence in ABM were all the more surprising since just a few years earlier the same Pas was questioning, and with much gravity, the state of health of the field.[63] However, in hindsight, Pas's bold assertions would prove prophetic. Taken together, several developments dating back to the early 1990s would contribute to eventually giving activity-based modeling a central place within the American UTDM social world.

As a matter of fact, the same factors that prompted the federal administration to massively back the TRANSIMS project would work also to the advantage of the ABM. As has been seen in chapter 5, following a series of significant legislative initiatives including the 1990 Clean Air Act Amendments and the 1991 Intermodal Surface Transportation Efficiency Act, the federal government launched the Travel Model Improvement Program, one of the main objectives of which was to promote new modeling practices capable of addressing the multiple points of intersection between transportation, on the one hand, and air quality, energy, economic development, land use, social and environmental justice as well as quality of life, on the other.[64] If for several years the lion's share of TMIP and other government-sponsored programs went to TRANSIMS, ABM would be also among the main beneficiaries of the continuous efforts to substitute new forecasting knowledge and practices for the older—and increasingly ineffective in the post-1990 CAAA and post-1991 ISTEA eras—four-step model. In the mid-1990s, given the new legal and political environment, ABM was no longer a purely academic matter; it was now invested with high operational hopes. Not surprisingly, the aforementioned text by Pas was his contribution to a large conference held by TMIP in New Orleans on June 2–5, 1996, the principal goal of which was to "promote use of activity-based approaches for travel forecasting." Judging by the number and the characteristics of the participants—there were about a hundred participants, two-thirds of whom worked for planning agencies, government organizations, and consultancies—in the mid-1990s both the academic and the professional segments of the UTDM social world were about to display a great interest in the ABM.

In addition to the new legal and political environment of the day, a series of developments internal to urban traffic forecasting in the 1990s would help the activity approach progressively penetrate and take hold within the world of practice. To begin with, the early 1990s witnessed a significant change in the ways of carrying out household travel surveys. Several metropolitan planning organizations, starting with the Boston and Los Angeles agencies (1990–1991),[65] quickly followed by other MPOs,[66] would adopt a new kind of transportation questionnaire,[67] the *activity diary*, in which the traditional question "where did you go?" was replaced by "what did you do?" Like another recent shift in travel survey techniques, from *retrospective* to *prospective* procedures,[68] the original aim of the move to the activity diary was to obtain more reliable and complete data on the trips made by household members. The basic idea was that while travel is incidental in a person's normal daily experiences and is therefore likely to be forgotten or regarded as too trivial to report, the activities one performs are much more likely to be remembered and reported both completely and accurately. With the activity diaries, it is only after establishing that a specific activity had been done at a different location from the preceding one that the respondents must provide details about how they reached the place where this activity was performed. Activity-based modeling would benefit from this new survey tool, yielding much richer information about the many activities of the members of a household.[69] The fact that the stated preference techniques had already reached a stage of maturity in the mid-1990s would also work to the advantage of the new modeling approach.[70]

Last but not least, the activity approach would find a powerful ally in the continued rapid developments of computer technology and computational techniques, especially microsimulation. Like TRANSIMS, which by the mid-1990s had incorporated an "activity" component albeit not a very effective one,[71] ABM would largely benefit from such developments that allowed researchers and practitioners to process large datasets and to estimate models that could not easily be calibrated in the past due to the lack of required computational resources. Increasingly coupled with geographic information systems (GISs) allowing for coding, storage, and easily manipulated large geo-referenced databases, the new computational resources at the disposal of the UTDM social world would encourage, in fact, several of its members to develop new lines of modeling, dealing, like ABM, with *point-to-point* movements rather than *zone-to-zone* traffic flows in the manner of the old FSM.[72] It was not by chance that microsimulation (box 5.2) occupied the entire space of one of the three workshops of the TMIP-sponsored conference on ABM held in New Orleans on June 2–5, 1996. At this, participants were given the opportunity to hear, among others, Eric J. Miller from the University of Toronto provide "an overview of the state of the art of microsimulation modeling applied to activity-based travel forecasting,"[73] and Richard Beckman from the Los Alamos National Laboratory describe the approaches to activity-based analysis used in TRANSIMS,[74] by far the most ambitious attempt to develop a comprehensive microsimulation travel demand modeling framework thus far.

The 1996 New Orleans conference was the first large national meeting bringing together academics and practitioners already involved or simply interested in ABM in the post-1990 CAAA and post-1991 ISTEA eras. Its proceedings amply testified to the impact made by these two acts on the American UTDM social world, its university segment initially. As a matter of fact, when in July 1992 the Federal Highway Administration issued an invitation for proposals to redesign the travel demand modeling procedure in order to meet the forecasting requirements introduced by the new legislative framework, a couple of years or so later three of the four winners would "explicitly recommend that the current 'trip-based' model framework be replaced by an 'activity-based' framework in which the demand for travel is derived from a more basic demand to engage in various activities."[75]

It comes as no surprise that among the proposals that were awarded a contract was a project collectively crafted by a team featuring Kitamura and Pas among its members. In order to respond to the federal invitation, the two academics partnered with two other government practitioners, Keith Lawton from METRO, the Portland MPO, and Paul Benson from the California Department of Transportation, as well as Clarisse Lula from Resource Decision Consultants (RDC).[76] Kitamura was already familiar with microsimulation procedures following research work previously conducted with his Ph.D. supervisee Konstadinos Goulias.[77] He and his collaborators therefore decided to rely on this technique in order to operationalize the principles underlying their Sequenced Activity-Mobility Simulator (SAMS). The latter departed from the then dominant modeling practices developed within the framework of the utility-based disaggregate approach in that it modeled the traveler as a person characterized by "bounded rationality," that is, an individual having a *limited* capability to process the available information while seeking only *satisfactory*, and not *optimal*, alternatives thanks to "heuristics" (routines) that are the fruit of experience and the outcome of a trial-and-error-learning process (unsurprisingly, the authors explicitly referred to Herbert A. Simon [1916–2001] and his models of human action).[78]

The team that revolved around Kitamura was essentially composed of people already immersed in the activity approach. The other two winning teams also preaching the gospel of ABM were, in contrast, mainly composed of scholars long involved in utility-based disaggregate modeling. Thus, Peter Stopher, at that time a professor at Louisiana State University, teamed up with David Hartgen and Yuanjun Li from the Lane Council of Governments, Eugene, OR, to propose a "paradigm shift" by "focusing not on travel but on the behaviors, needs, and roles that generate it." Their general modeling framework, called SMART (Simulation Model for Activities, Resources and Travel), was "based on an integration and synthesis of ideas in the transportation literature that deal with household activities and roles," including the contributions of Stopher himself to the design of activity diaries and activity surveys. SMART was designed to operate in a geographic information system environment, and its "heart," the Household Activity-Travel Simulator, was geared toward determining "the locations and travel patterns of household members-daily activities in 3 categories: mandatory, flexible, and optional."[79]

Like Kitamura and his teammates, the designers of SMART recommended the use of microsimulation techniques. This was not the case with the third winning team proposing a shift to ABM and led by MIT scholar Moshe Ben-Akiva.[80] As has been seen, although he was one of the heavyweights of utility-based disaggregate modeling in the 1970s and the 1980s, Ben-Akiva was aware nevertheless of the activity approach from having supervised a number of ABM-related master's and Ph.D. students from the late 1970s on. For the MIT research project, Ben-Akiva could count on two of his supervisees from the early 1990s, Dinesh Gopinath[81] and especially John L. Bowman. A mathematician by training—he received his bachelor of science degree from Marietta College, Ohio, in 1977—Bowman first worked for fourteen years in systems development, product development, and management for an insurance and financial services firm. However, in the early 1990s he decided to resume his studies, and in 1995 he received a master's degree in transportation from MIT.[82] Though the members of the MIT group clearly recognized the need to replace the existing modeling practices with new and activity-based ones, they adopted an incremental approach, building their project on the current "'best practices' of disaggregate travel demand model systems."[83] Their point of departure was the existing "tour-based" models, then in the vanguard of operational utility-based disaggregate model systems. Tour-based models saw the light of day in the late 1970s and early 1980s in the Netherlands, and by the mid-1990s they had been developed and used for practical purposes in Sweden and Italy as well. This family of disaggregate models groups separate trips—the basic travel unit for both the traditional FSM and early disaggregate modeling procedures—into larger units, called "tours."[84] While tour-based models do provide an improvement over traditional trip-based models, as they enable the forecaster to address issues such as trip chaining and the various temporal-spatial constraints within a tour, they still lack a connection among the *multiple tours* taken by the same person in a day.[85] The project put forward by the MIT team consisted therefore in extending the tour-based idea by explicitly modeling tours and their interrelationships in an individual's daily activity schedule pattern.

Thanks to three sponsors,[86] the Activity-Mobility Simulator (AMOS), the key component of the SAMS model system designed by Kitamura and his teammates, could be investigated in an applied setting in the mid-1990s.[87] The Washington, DC, study was conducted by Resource Decision Consultants, and was probably the first attempt in the United States to implement a full-fledged activity-based model for transportation planning and policy analysis, through predicting household responses to a selection of travel demand management measures, ranging from congestion pricing to incentives for using alternatives to personal vehicles. AMOS proved reliable enough, in the eyes of the actors involved at least, and the success encouraged its designers, including Kitamura and his former Ph.D. student Ram Pendyala, a faculty member at the University of South Florida from 1994 to 2006, to continue investigating the modeling venues first explored in Washington, DC. The Florida Activity MObility Simulator (FAMOS), a microsimulation activity-based multimodal project conducted with the support of the Florida Department of Transportation, was completed in 2004 with its full-fledged development and application in Southeast Florida.[88]

However, it was the path first trodden by Ben-Akiva and the MIT group[89] that proved the more promising from an *operational* point of view, and the one with the most practical impact across the nation in the 2000s and the 2010s. Things seemed to start in Portland and its Metropolitan Planning Organization (METRO) in late 1993, when an expert panel recommended an investment in ABM. METRO followed this suggestion, and began working to collect the necessary travel data.[90] Lawton, then manager of modeling at METRO and one of the coauthors of the SAMS project, had obviously been seduced by the MIT approach and suggested that the latter be tested with the help of data derived from the household travel and activity survey completed in Portland in the mid-1990s.[91] Bowman, who had already made the general modeling framework proposed by the MIT team into a prototype system for the Boston metropolitan area,[92] therefore set to work as a Ph.D. candidate on the Portland project. Carried out during 1996 and 1997, his Ph.D. investigation was funded by the Federal Highway Administration as part of the TMIP. A few years later Bowman could boast of having produced the first operational implementation of the ideas contained in the MIT proposal (figure 6.1).

The MIT postgraduate student would not be left alone in his efforts. In addition to Bowman's mentor and thesis supervisor, Ben-Akiva, and several METRO staffers, including Lawton himself, the Portland project involved a number of other seasoned modelers.[93] Thomas Rossi and Yoram Shiftan from Cambridge Systematics, both specialists in tour-based modeling,[94] contributed to the project, and so did Andrew Daly. A mathematician by training and a graduate of Oxford University, Daly had already done pioneering work in disaggregate modeling while also a collaborator of Cambridge Systematics in Europe. For the Portland study, he would in particular expeditiously increase the capacity of his widely used ALOGIT software for estimating utility-based disaggregate models.[95] But it was Mark Bradley, his "partner in research and development" according to Bowman himself, who would help Ben-Akiva's supervisee to transform his designs into a "practical production system in Portland" while providing data for model development and producing forecasts.[96] Bradley was a seasoned as well a cosmopolitan travel demand modeler. An American with studies in

Introduction and Summary 23

condition any stops occurring before or after the primary activity. At each conditional level, the probability is represented by a multinomial logit model.

Figure 1.3 shows the overall structure of the activity-based model system. Lower level choices are conditioned by decisions modeled at the higher level, and higher level decisions are informed from the lower level through expected maximum utility variables.

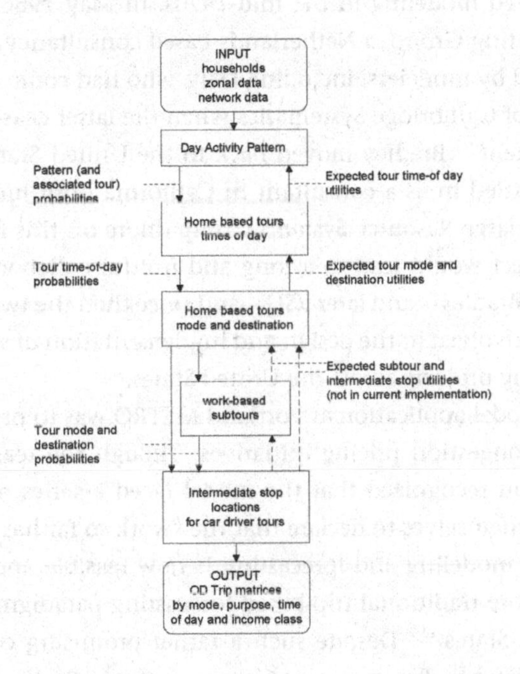


Figure 1.3 Portland day activity schedule model system

Table 1.1 shows the five main types of models included in the system, as well as the types of variables included in each of the model types. The variables include important lifestyle categories and mobility decisions, attributes of the activity and travel environment, and the expected utility variables from the conditional models. The entire system includes 633 estimated parameters, including 297 measuring the importance of lifestyle and mobility variables, 95 measuring the importance of the activity and travel environment—including

**Figure 6.1**
Portland's ABM system overview. *Source:* John L. Bowman, "The Day Activity Schedule Approach to Travel Demand Analysis" (Ph.D. diss., MIT, 1998), 23.

OR at Cornell University (1976–1980), he obtained his master's degree in systems simulation and policy design from Dartmouth College, where he was first introduced to travel demand modeling by Tom Adler, Ben-Akiva's former Ph.D. student. After working as an analyst for Cambridge Systematics from 1983 to 1985, Bradley would spend about ten years as a specialist in disaggregate modeling and stated preference (SP) techniques in Europe. He first moved to the Transport Studies Unit at Oxford University, that is, the research center where Heggie and his collaborators launched the idea of activity-based modeling in the mid-1970s. In May 1986 he joined the Hague Consulting Group, a Netherlands-based consultancy set up in 1985 and populated by modelers, including Daly, who had come from the European branch of Cambridge Systematics when the latter ceased its activities on the continent.[97] Bradley moved back to the United States in the mid-1990s, and settled in as a consultant in California until June 2012, when he joined the large Resource Systems Group (more on this firm later). The Portland project would lead to a long and fruitful collaboration between Bowman and Bradley—and later RSG—and since then the two authors have been heavily involved in the design and implementation of several activity-based modeling projects across the United States.

The first model application at Portland METRO was to provide forecasts for possible congestion pricing initiatives. Though the team involved in its construction recognized that the model faced a series of difficulties,[98] they allowed themselves to declare that the "work so far has indicated that activity-based modeling and forecasting is now feasible and can begin to replace the more traditional trip-based forecasting paradigm within MPOs in the United States."[99] Despite such a rather promising conclusion, the project carried out by Bowman and his teammates in Portland nevertheless came to a halt at the end of the 1990s for lack of funding, at a time when METRO shifted its focus to TRANSIMS thanks to federal money (chapter 5). However, in the long run the Portland project would be proved right. Dubbed the "Bowman-Bradley model,"[100] the Portland project paved the way for the implementation of activity-based modeling systems in several American metropolitan areas.

## Activity-Based Modeling Reaches Operational Status

Less than a decade after the Portland experience was completed, the number of models claiming to be built upon activity-based modeling, already operational or about to become so across the nation, was deemed by academic Chandra Bhat high enough to support the statement that a "literal revolution in the development, testing, and use of activity-based travel models"[101] was well under way in the United States. A fervent proponent of ABM at the University of Texas at Austin, Bhat was heavily involved in the design of Comprehensive Econometric Micro-simulator for Daily Activity-travel Patterns (CEMDAP), an activity-based modeling system developed for Dallas-Fort Worth.[102] Bhat made this statement during the Innovations in Travel Demand Modeling conference convened by the TRB in Austin, Texas, in May 2006, and was heard by a large audience of approximately 220 individuals from different backgrounds, ranging from the academic world to consultancies, government planning agencies, and the federal administration. Information from the conference proceedings seems to support Bhat's claim. In a review lecture in which the authors discussed the design features of activity-based models that had been conceived, and some even already implemented, at a series of urban transportation planning agencies throughout the country by the mid-2000s, Bradley and Bowman reported no fewer than eight examples.[103] Obviously the Portland experience ushered in a new era of "moving activity-based modeling into practice," to use the name of the task force set up by the TRB in 2003, and first chaired by Konstadinos Goulias, Kitamura's former Ph.D. student. As a matter of fact, from the end of the 1990s onward, the operational implementation of ABM would progress rapidly, especially in comparison with the pace of change that had characterized the trajectory of the new approach during the period 1975 to 2000.

Hardly had Bowman and Bradley left Portland when the San Francisco County Transportation Authority (SFCTA) began to develop its own activity-based forecasting procedure with the decisive help of Bradley and Cambridge Systematics. As the latter was characterized by the same basic design previously used in Portland, the shift to the new approach was rapid and the SFCTA modeling suite was designed, implemented, and calibrated within a time frame of about two years, while the initial development cost approximately $700,000 in consultancy fees. Completed in 2001, the SFCTA model became the first ABM approach to be used in practice in the United States, and by 2010 it had the largest number of applications in real-world studies across the country.[104] SFCTA is a county transportation agency. Its model had already become operational when two metropolitan transportation agencies—of New York and, a little later, Columbus—followed suit and developed in the first half of the 2000s their own activity-based models with the help of PB Consult, the then strategic consulting arm of the large international engineering firm Parsons Brinckerhoff—known as WSP USA from 2017 on.[105] These first experiences seem to have convinced even the last skeptics that the ABM concept was workable and that operational activity-based models could be constructed. Judging from the findings and recommendations put forward by the Committee for Determination of the State of the Practice in Metropolitan Area Travel Forecasting, the collective author of an important report published in 2007, by the mid-2000s ABM seemed to have been endorsed by large parts of the UTDM social world.[106] From that point on, things accelerated.

Fifteen years after the initial San Francisco model came into being, a survey conducted in March 2015 and dealing with the twenty largest MPOs in the country and three smaller ones with a strong reputation for innovation in travel forecasting clearly displayed an upward trend in both use of and interest in ABM.[107] Of the twenty-three MPOs that responded to the survey, six already had production-use ABMs,[108] another ten were developing one,[109] and one (Boston) had an ABM in a pre-development stage. Thus, by the mid-2010s, although from a sheer statistical point of view ABM was still a rather small part of the entire American travel forecasting landscape, the headway being made toward operationalizing the activity-based approach was more than significant since nearly all large MPOs had already turned away from the long-standing FSM, despite the fact that developing an operational ABM had proved a highly time-consuming and expensive task. As a matter of fact, according to the 2015 survey the average time required to translate the activity approach into practice was four and a half years, with a range of between three and six years, while for the Columbus activity-based modeling system, for example, a total of around $2.5 million was spent between 2001 and 2005.[110] A sophisticated, time- and money-consuming product, an activity-based model also has computational requirements that are usually much higher than those displayed by more conventional traffic forecasting approaches. Unsurprisingly, the initial implementation of the SFCTA ABM took thirty-six hours to run, a time span that even risked seriously jeopardizing the model's further development and application. It was only after the transportation agency started using parallel computing techniques that full runs shortened from thirty-six to twelve to fifteen hours by the early 2010s.[111]

As with the case of the novelties that emerged in urban travel demand modeling in the 1970s and 1980s (chapter 3 and 4), the part played by large consultancies in the operationalization of ABM and its spread to transportation planning agencies across the country would prove crucial. That Cambridge Systematics, highly involved in the pioneering Portland experience in the second half of the 1990s, has since then actively taken part in the design and implementation of a number of activity-based modeling systems in the United States comes as no surprise. However, although the Massachusetts-based firm succeeded in remaining a significant stakeholder in the American ABM market, it has been confronted with stiff competition, especially from two other large firms. Resource Systems Group has a history similar to that of Cambridge Systematics in several respects. The firm was founded in 1986 by three academics from Dartmouth's Resource Policy Center, including the firm's current (at this writing: 2022) president emeritus Thomas J. Adler, the former Ph.D. student of Ben-Akiva at MIT in the 1970s.[112] Within a thirty-year period, the small shop in Vermont would grow into a company of over eighty employee-owners with a national footprint. Since early 1988, when RSG built its first travel forecasting system, the firm has regularly enhanced its modeling capabilities through the recruitment of seasoned practitioners, many of whom having a postgraduate degree. As a result, at the very end of the 2010s the consultancy could boast of having on its payroll several well-known ABM specialists. In addition to Bradley, who joined the firm in 2012, one can mention Maren Outwater, then vice president of the firm and a former Cambridge Systematics staffer (1997–2007), Joel Freedman, previously at Parsons Brinckerhoff and heavily involved in the development of the first American operational ABM in San Francisco, and Kay Axhausen, a European academic and one of the most active proponents of the activity-based approach all over the world. John Gliebe, a Ph.D. supervisee of Koppelman at Northwestern University in the early 2000s,[113] as well as Joe Castiglione, a holder of a master's degree in city planning, were also among RSG staffers during the 2010s,[114] while Bowman himself has been a regular collaborator of the consultancy up to the present day. By 2019, RSG, in partnership with Bowman, had been involved in no fewer than eighteen projects involved in activity-based modeling across the country. It is also worth noting here that RSG, in addition to having grown into an ABM heavyweight, has also shown great interest in travel surveys as well.[115] At the turn of the 2000s, the firm even participated, along with Cambridge Systematics, in the GPS-only travel survey for the Cincinnati region, the first large-scale attempt in the United States to carry out a household travel survey in which paper-based travel diaries and forms would be totally eliminated.[116] Moreover RSG has recently been catapulted into the national leadership tier in the field of smartphone-based household travel surveys,[117] largely thanks to its own proprietary smartphone application called rMove™.[118] This interest in GPS-based data collection and analysis techniques displayed by transportation consultancies heavily involved in ABM[119] should come as no surprise. In fact, as they aim to provide a rigorous analytical and behavioral framework for forecasting travel demand, activity-based models call for the collection of rich and detailed information about people's travel-related activities and practices for longer periods of time than the usual single-day trip surveys, while they are more sensitive than the FSM to underreported travel and the lack of precision in the location of activities. By making it possible to *automatically* access many, various, and accurate activity and travel data items[120] at relatively low costs, GPS-based travel survey techniques greatly help the development and diffusion of ABM.[121]

Besides Cambridge Systematics and Resource Systems Group, Parsons Brinckerhoff has also succeeded in turning itself into another heavyweight, even a leader, in ABM efforts on the American traffic forecasting scene. By February 2011, PB had already been involved in more than half of the activity-based modeling projects nationwide.[122] At that time, the firm was home to some three dozen professionals specializing in urban traffic forecasting. Oddly enough, PB's expertise in ABM owes a lot to a veteran urban traffic forecaster, Gordon Wilford Schultz (1937–2006). Fresh out of Northwestern University with a master's degree in 1962, Schultz first found a job at the Tennessee Department of Highways before joining Alan M. Voorhees & Associates in the mid-1960s (chapter 2). After doing pioneering modeling work built upon the utility-based disaggregate approach for the firm Barton-Aschman,[123] Schultz worked for PB for about fifteen years before retiring. He was instrumental in the original design of the New York region's much-acclaimed activity-based model in the early 2000s, and played the role of "instigating advocate," in Bowman's words, for the activity-based approach with the MPOs for Atlanta and Columbus. Over the years, in addition to Schultz, PB has housed an experienced and large team of ABM specialists. Among those the firm relied on in the 2000s and the 2010s were Joel

Freedman, now (early 2022) senior director at RSG; Peter Vovsha, holder of a 1985 doctorate from the Moscow Technical University; Rick Donnelly, the modeler who had helped TRANSIMS team members familiarize themselves with traffic forecasting and a 2008 Ph.D. holder from the University of Melbourne; William A. Davidson, a member of the small group of transportation professionals who pioneered the development and application of advanced disaggregate models for the mode choice step; and Rosella Picado, who received her Ph.D. from University of California, Berkeley.[124]

Though the first phase of ABM systems had been developed from scratch for each metropolitan area, each one of the three major firms nationally involved in the activity approach soon developed a software platform on which their models could be adapted and improved over time as they were implemented for new regions. Thus Parsons Brinckerhoff created the open-source CT-RAMP (Coordinated Travel–Regional Activity Modeling Platform), with the first ABM of this platform's family being developed for Columbus in 2004, and adapted for the Tahoe Regional Planning Agency two years later.[125] By 2010 six different regional ABM projects throughout the nation were already sharing PB's platform, while the Southeast Florida Regional Planning Model is an example of a more recent variant of the CT-RAMP family.[126] Another major ABM platform is DaySim, first associated with the duo of Bradley and Bowman when they were working as independent consultants, and later with Resource Systems Group as well.[127] DaySim was originally developed and implemented in Sacramento in the mid-2000s and since then it has been regularly upgraded. Recently, the North Florida Transportation Planning Organization adopted DaySim for project planning in 2016.[128] A decade after it first came into being, DaySim was available as open-source software without a license fee through a consulting business model: If a client was engaging one of the copyright holders—Bowman, Bradley, or RSG—for consulting services for its implementation, then they would be given an open-source license to code. As of writing (2020), the DaySim software is open-source and is maintained online in a GitHub repository.[129] Apparently a little less popular than CT-RAMP and DaySim, TourCast, Cambridge Systematics' design and implementation software platform created in response to ABM projects in Denver and Houston, has recently been used in Baltimore as well.[130]

It is worth noting here that this overwhelming presence of these three software platforms nationwide has not prevented a series of metropolitan areas from adopting specific ABM modeling systems. Thus METRO, the Portland MPO, partnered with Portland State University, the ex-RSG staffer Gliebe, and several university students to develop the Dynamic Activity Simulator of Households (DASH). At the end of the 2000s, the Los Angeles MPO started developing an activity-based model built on university research conducted essentially by the academic triumvirate of Bhat, Pendyala, and Goulias.[131] The Future Urban Mobility Research Group, including Ben-Akiva and other MIT scholars, at the Singapore-MIT Alliance for Research and Technology has also been developing an activity-based simulation platform called SimMobility since January 2011.[132] ActivitySim is another recent open platform for ABM implemented by the consultancy Resource Systems Group on behalf of a consortium of MPOs, Departments of Transportation, and other transportation planning agencies.[133]

As noted, most of the software platforms used for the large majority of activity-based models in the United States have been governed by open-source licensing agreements. However, as the market for ABM has matured over time, companies specializing in the design of proprietary software for travel demand modeling have shown an increasing interest in the new approach. Thus, on the occasion of the release of version 5.1 of its flagship product Cube Voyageur in the very early 2010s, Citilabs asked Bradley to help add "an example of an activity-based model implemented entirely using Cube Voyageur to the Cubetown demonstration system."[134] Several years later, in July 2017, Citilabs staffer Bill Allen could proudly announce that his firm was able to provide practitioners with a "less complex, less ambitious" simplified tour-based model simplifying "ABM down to its essence," and offering "75% of the benefit of ABM for 10% of development cost."[135] By the mid-2010s, the three other major commercial software vendors on American soil—the Canadian INRO, Boston area-based Caliper, and German PTV—had all added trip chaining and activity-based modeling components to their software packages.[136] In February 2019, INRO was even pleased to announce the appointment of Peter Vovsha, one of the ABM specialsts at PB for years, as vice president of travel modeling; Vovsha would join INRO's R&D team later in the year with a view to helping, among other things, in establishing and developing ABM activities for inclusion in the company's software.

The decisive part played by the private sector in the design and implementation of operational activity-based models from the early 2000s on should not obscure the fact that both the development and the spread of

ABM to transportation planning agencies across the country also greatly benefited from the efforts deployed by other segments of the UTDM social world as well. The multiple occurrences of the term "Ph.D." in this chapter testify to the increasing popularity of ABM among academics involved in travel demand modeling, and bear witness to their efforts to serve the new approach through research projects, frequently conducted in tandem with practitioners with a scientific bent, as well as training programs.[137] Other nonprofit organizations, such as the FHWA and TRB for example, have also preached the gospel of the new approach through regularly, and sometimes generously, funding ABM-related projects and by serving as clearinghouses for their diffusion to the UTDM social world. Thus, both FHWA and TRB have funded research studies on the crucial issue of transferring activity-based model structure, software, and coefficients from one region to another in the hope of reducing development time and cost. By 2015, thanks to this transfer-and-refine process, new activity-based models had been implemented in six- to nine-month time frames, while as little as $250,000 had been spent for the transfer of an ABM system in a small region.[138] The same TRB commissioned Parsons Brinckerhoff to draw up a guide that featured ABM in a prominent place,[139] while five years later, in 2015, a *Primer*, also sponsored by the Transportation Research Board and entirely devoted to ABM this time, was published under the signature of three RSG modelers.[140] In addition to these publications, a series of webinars organized by the federal administration in the late 2000s[141] and throughout the following decade helped spread ABM to the community of practitioners,[142] while the various transportation agencies involved in the activity-based approach have regularly shared experiences—and sometimes funding[143]—through another tool supported by the TMIP: the Peer Review Program. Inaugurated in 2003, the latter has provided an MPO the opportunity to elicit recommendations for addressing its current and future modeling challenges, especially through a one- or two-day meeting with seasoned modelers: practitioners, academics, host agency planning staff, and other stakeholders.[144]

One can easily see that American ABM is served now by a rather close-knit community, characterized by both competition—for clients, funding, but also human resources, since professional mobility within the field seems to have been rather strong—and various forms of collaborative work, including common research studies and joint publications.[145] As a result, despite variations in technical details—such as on how to construct daily activity patterns, for example—the current operational activity-based modeling systems across the nation share a set of commonalities and have largely similar overall structures.[146] They all utilize synthetic population as base inputs prior to simulating travel demand, and all apply microsimulation techniques at the disaggregate level of persons and households—and one can note here the parallels with the TRANSIMS project. Today's systems are (naturally) tour-based and multimodal while they all include daily activity pattern formulation, which determines what tour purposes an individual wants to undertake during the day—the analogue to the trip generation step of the four-step model. In an ironic twist, they all belong to the econometric family of ABM (see box 6.1), since they tend to use a series

**Box 6.1**

The Various Families of Activity-Based Models

*Utility-maximizing activity-based models*: This econometric family of models is based on the premise that individuals maximize their utility when organizing their daily activity pattern.

*Constraint-based activity-based models*: This is the first generation of activity-based models. Their main goal is to examine whether an individual activity pattern is feasible within particular space-time constraints. CARLA (mentioned in this chapter), a model devised by the Oxford Group, is an early example of this family of activity-based models.

*Computational process activity-based models*: Some people, especially scholars, involved in ABM advocate the idea that travel behavior and decision-making are not governed by the utility-maximizing principle. For them, the trip-maker is characterized by "bounded rationality," which means that the individual has a limited capability to process the information he or she gets, and seeks only satisfactory, and not optimal, alternatives thanks to "heuristics" (routines) that are the fruit of experience and the outcome of a trial-and-error learning process. The SAMS model (by Kitamura and colleagues), discussed in this chapter, is an example of this family of activity-based models.

For the preceding classification, see https://tfresource.org/topics/Activity_based_models.html, accessed January 24, 2022. For a thorough, particularly clear presentation for the layperson of the various families of activity-based models, see especially Goran Jovicic, *Activity Based Travel Demand Modeling, a Literature Study* (n.p.: Danmarks TransportForskning, 2001).

of choice models[147] predicated, like the disaggregate approach of the 1970s, on the utility-maximization principle based on the assumption that people choose the option that provides them with the highest utility among available alternatives. However, unlike traditional disaggregate modeling, these ABM systems implement their choice models with the help of simulation techniques, especially the famous Monte Carlo method,[148] first used for the design of the atomic bomb during World War II.[149] Last but not least, they all interface with popular proprietary traffic forecasting software, such as Cube, TransCAD, or Emme.

By the end of the 2000s, the utility-based disaggregate approach, an intellectual curiosity for academics in the 1960s and early 1970s, had made itself into an effective operational tool widely, even exclusively, used by transportation practitioners for the modeling of the mode choice step in the four-step model. However, the remaining components of FSM, the trip generation, trip distribution, and traffic assignment steps, long preserved their original aggregate nature. In the mid-2010s, with more than 75 percent of the largest MPOs across the country about to move to the activity approach, operating at the fully disaggregate level of persons and households, the only remaining representative of the aggregate philosophy within urban travel demand modeling was the traffic assignment step—keep in mind that ABM ambitions to replace the first three steps of the FSM. But as the next and final chapter of this book will clearly show, even this ultimate bulwark of the aggregate principle in urban traffic forecasting has long been under attack, and it is highly likely to meet the same fate as the other components of the FSM in the foreseeable future.

# 7 Modeling Variable Flows, and the (Probable) End of an Era

## The Advent of ITS and the Dynamic Assignment Problem

In the early 1980s, congestion, that old urban scourge,[1] was back with a vengeance while wearing a new face. In 1983, about 55 percent of urban freeway travel during the peak hour occurred under congested conditions.[2] But what seemed so alarming about traffic congestion in the eyes of contemporaries was its pervasiveness. From a temporal point of view, congestion was no longer limited to the peak hours, while spatially it was not confined to the downtown area of cities as in the past. As a result of the soaring number of jobs—largely due to women's massive entry into the labor market[3]—and, especially, the jobs' increasing suburbanization, some of the worst traffic conditions could now be found in the suburbs.[4] In the early 1990s, in fact, suburb-to-suburb commuting constituted about one-third of all metropolitan commuting across the country.[5] Traffic congestion had become the hot issue of the day, and so the Traffic Systems Division of the Federal Highway Administration undertook a research study on it. Published in 1986, the study's alarming findings were immediately picked by the national press and received wide publicity. Congestion continued unabated nationwide in the following decade before stabilizing around 2000,[6] and has remained constantly in the limelight since then.

The California Department of Transportation (Caltrans) seems to have been the first large transportation agency to sound the alarm in the mid-1980s. Their studies for future highway development and funding requirements in the state had resulted in the unnerving result that no realistic new construction program could maintain even the present levels of congestion in the system, while a state gasoline tax increase to support such efforts was believed to be politically unacceptable. A three-day conference was therefore sponsored by the Californian transportation agency for its managers in October 1986 to consider the role of electronic technology—in contemporary parlance, ITS (intelligent transportation systems) technology—in helping make the most of the existing road infrastructure. Several outside experts and influential persons, including the FHWA executive director at the time, partook in the event, hence giving it a national significance. By the end of the decade, a broad research consensus, based on thirty-eight projects carried out over the 1980s, established that ITS can have noteworthy positive effects on traffic congestion.[7]

Later in 1990, a not-for-profit organization now called Intelligent Transportation Society of America (ITS America),[8] which brought together a variety of individuals from the public and private sector, was formally established as an advocacy group to represent the ITS technology perspective in policy formation. In December 1991, the Intermodal Surface Transportation Efficiency Act was enacted.[9] With $660 million allocated for research, development, and operational tests within an initial time frame of six years, ITS was an integral part of the new legislation. ISTEA also included a requirement for the U.S. Department of Transportation to designate a federally utilized advisory committee in the area of ITS technology. ITS America was chosen to play that role.[10] Since then, both ITS technology and ITS America have become a fixture on the American transportation landscape.[11]

But why does ITS technology matter for the content of this book? The chief reason is that it bears directly on traffic assignment procedures, the fourth and last step of the four-step model. Remember that, like the entire FSM, standard assignment procedures were initially designed to address long-term transportation planning issues. As a result, they are concerned with traffic patterns and settings in which the "demand" for travel expressed by people and the "supply" capacity provided by the network have stabilized over time, forming thus a couple in equilibrium. Such a *static* approach as this can reflect no *temporal variations* in traffic flows and conditions. The result is that standard assignment procedures, though rather efficient as a long-range planning tool geared toward sizing large capacity-expansion projects, are ill suited for analyzing traffic congestion effects at a fine-grained temporal level, while they also prove unable to assess many of the measures that can be taken to improve traffic conditions on the existing networks through ITS technology systems. It is obvious that an efficient design and implementation of ATIS (advanced traveler [or traffic] information system)-based measures, such as disseminating information to network users on traffic conditions, and the deployment of dynamic traffic management practices[12] through an advanced traffic (or transportation) management system (ATMS), are heavily dependent on the availability of timely and accurate estimates of both prevailing and emerging traffic conditions. Therefore, there is a strong need for a *dynamic* modeling of flows on the network, able to capture temporal traffic variations. In technical terms, what is needed are assignment procedures that make it possible to forecast traffic flows on the network when the latter is loaded by origin and destination traffic matrices that are not *fixed*—this is the case of static assignment—but *vary* over time. Loading variable O&D matrices to forecast traffic flows also (necessarily) varying over time is the main mission assigned to dynamic traffic assignment,[13] a unified tool for transportation planning and traffic operations,[14] the history of which forms most of the subject matter of the seventh and final chapter of this book.

As we will see, interest in DTA within the American UTDM social world grew exponentially with the advent of ITS technology.[15] Closely associated with the latter, DTA, a still-evolving technique, is more generally suitable for analyses and decision-making where travel demand is highly time dependent, and the planner and the transportation system manager have to take into account the spatial and temporal effects of *congestion* in determining route choice, time of departure, or mode choice. This includes, in addition to ITS-based measures and strategies, a series of project evaluations, including the management of special events, such as a downtown concert or the Olympic Games, work zone impacts and construction diversion, incident management response scenarios, evacuation strategies, or managed lanes and tolling projects.[16] More recently, the rise to preeminence of activity-based modeling (see chapter 6) as a valid alternative to the first three steps—trip generation, trip distribution, and mode choice—of the traditional FSM have prompted several research studies that aim to develop *integrated* traffic models with detailed temporal and spatial resolution thanks to the marriage of ABM (estimating the activities and the resulting travel of individuals) and DTA (representing the effects of this travel on the transportation system). If this integration proves operationally successful, then we will witness the end of a (very) long cycle encompassing the emergence, the rise to power, and the demise of the FSM framework.

But before the demise, the birth. Let us go back to the early 1970s, when DTA was the preserve of a handful of academics.

## Coping with Variable Traffic Flows within Academia

Samuel Yagar gained his master's degree from the University of Toronto in 1965. Financially backed by a fellowship from the National Research Council of Canada, he moved to Berkeley in the second half of 1960s, where he joined the local branch of the Institute of Transportation and Traffic Engineering (chapter 2). As a Ph.D. student, Yagar sought to extend traffic assignment procedures of the period, especially those devised by ITTE staffer Wolfgang S. Homburger (1926–2010), to allow for the *queuing* phenomenon caused by congestion bottleneck. The basic idea behind Yagar's original approach is that *dynamic* equilibrium could be readily described in terms of a *sequence* of short-term *static* equilibrium states, each corresponding to a portion of the entire time period under study. Building on this hypothesis, Yagar proposed a computerized method for achieving dynamic equilibrium flows according to Wardrop's first principle (chapter 4).[17] In November 1974, Yagar presented his graduate work as well as an extension of his method along with results from the Ottawa Queensway corridor before a gathering of specialists in assignment techniques in Montreal.[18] Among the participants of the event were two scholars from Cornell University. One was Deepak K. Merchant, who had just completed his Ph.D.[19] under the supervision of George Lann Nemhauser (born in 1937). Nemhauser was a gifted operations research expert,[20] and a colleague of Stella Dafermos in the early 1970s, upon whose work he had built an assignment algorithm capable of handling rather large transportation networks.[21] Unsurprisingly, Merchant's dissertation makes heavy use of OR techniques while drawing upon Dafermos's distinction between user optimizing and system optimizing (chapter 4). Merchant does not mention Yagar in his dissertation, but like Yagar he divided the duration over which traffic assignment is to be studied into equal time intervals of suitably small length. However, unlike the Canadian, who contributed a heuristic approach to the DTA problem while applying his method to a real network, Merchant formulated his model for minimizing total cost (system optimizing) in the form of a nonlinear and nonconvex mathematical programming problem and worked on a *fictive* network having multiple entry origin points but limited to a single destination. Portrayed by his dissertation supervisor as a "very bright and energetic young man with an outstanding future," Merchant, born in India in 1947, was cut down in his prime in a cable car accident in April 1978.[22] Fortunately, his innovative contribution would reach the American UTDM social world thanks to his participation in the Montreal symposium,[23] and, especially, through two articles jointly authored with Nemhauser, both posthumously published in *Transportation Science* in 1978.[24]

The pioneering research work by Yagar and by Merchant and Nemhauser in the early 1970s as well as another original article by the Canadian OR specialist Pierre Robillard, who also tragically died in 1975 in a car accident,[25] came at a time when substantial progress was being made in static assignment techniques (chapter 4). Following in the footsteps of these trailblazers, some scholars sought to *extend* knowledge and practices initially developed within the full-fledged static assignment field to the still-fledgling DTA domain.[26] As a result, mathematical techniques, including the so-called variational inequalities method introduced by Dafermos,[27] were imported into DTA, and the same applies for the two Wardrop's equilibrium principles (chapter 4).[28] Although the first inroads into dynamic traffic assignment were concerned with small, even idealized transportation networks, attempts to model more complex road systems were also gradually emerging. At the turn of the 1980s, Bruce N. Janson, already a static assignment specialist (chapter 4) and a member of the narrow circle of academics trying to adapt Wardrop's user equilibrium principle to a DTA framework, suggested a heuristic algorithm of his own, designed to produce an approximate solution to the dynamic user equilibrium (DUE) problem.[29] To this end, he performed test cases on two real roadway systems, the network of Sioux Falls, the city where Larry LeBlanc had already tested his famous static assignment algorithm in the early 1970s (chapter 4), and a larger Pittsburgh network—but still, nevertheless, very small compared to networks that static assignment procedures were able to efficiently handle at that time.[30]

While some members of the UTDM social world were about to investigate the venues first explored by Merchant and Yagar, other colleagues were also setting to work on DTA, focusing less on the mathematical and computational aspects of the problem—in other words, the transformation of O&D traffic matrices varying over time into flows on the network—and more on the *behavior* of motorists, whose decisions and interactions collectively, albeit unpredictably, determine the very formation and evolution of these flows. More precisely, these scholars directed their efforts toward analyzing changes in *trip departure times* and sought to assess their impact on congestion. In their assignment models, the travelers not only choose a particular route to their destination—as they do in static assignment procedures—but they also determine the time of departure. The basic idea behind this interest in the latter was that one could more efficiently combat peak hour travel congestion through judiciously modifying departure time with the help of a series of management measures, including time-varying tolls, information provided to travelers, and staggered work hours. The first wave of research on this crucial issue at the turn of the 1970s was typically undertaken with the help of idealized networks, often consisting of a single arc with a congestion bottleneck. This research program[31] from the start captured the interest of a number of universities offering transportation courses, such as Carnegie-Mellon University and MIT.[32] At the latter, Moshe Ben-Akiva started collaborating with several colleagues and students—including the Belgian André de Palma (born in 1952), a physicist transitioning into an economist and who was about to do pioneering work on modeling peak period traffic congestion—to extend the utility-based disaggregate approach to the departure time issue. Unsurprisingly, the MIT model assumes a "rational" trip-maker seeking to maximize his or her utility function by choosing a departure time on the basis of a trade-off between the travel time spent on the road and the so-called schedule delay, that is, the difference between actual and desired arrival times, a notion dating back to the 1960s.[33] In order to render their modeling approach more comprehensive and dynamic, the authors added a component addressing the *day to day* adjustment of the distribution of traffic.[34] Simulation experiments were also performed to predict the impact of various policy measures on peak period congestion, including alternative pricing policies and preferential treatment of high-occupancy vehicles.

Around the same time, Hani Mahmassani of the University of Texas at Austin[35] was also intrigued by the departure time issue. To address this, he allied himself with an older Texas colleague who became his mentor, Robert Herman (1914–1997).[36] Herman was a physicist by training, who had specialized in OR techniques during World War II, and was one of the founding fathers of the postwar modern "traffic science."[37] The two teammates published the first results of their common work in an article that appeared in *Transportation Science* in 1984. Like the bulk of research on dynamic traffic assignment at that time, their study was performed on highly stylized situations. The authors began by addressing the case of a pool of commuters going from one origin to a single destination along a unique route, the range of trip-maker's options being limited to the sole choice of departure time. They then extended their analysis by introducing the option of a second route, a gesture that allowed them to rearticulate Wardrop's first principle as follows: "User equilibrium conditions for this situation are such that no user can reduce his/her travel cost by unilaterally changing routes *and/or departure times*."[38]

Mahmassani would continue exploring "the dynamics of interaction between user decisions and system performance." However, to further investigate this interaction, he had to tackle an entire range of questions addressed to the human sciences. For example, how do people decide in a particular context and how does one learn from his or her past experiences?[39] Following in the footsteps of ABM proponents with whom he would be in constant dialogue for years, Mahmassani therefore made himself into a psychologist willing to dissect a trip-maker's behavior in real-world settings by resorting if necessary to laboratory-based experimental practices. The results of his delving into the motorist's psyche would soon go public. Partnering with Herman again, and with the close collaboration of his student Gang-Len Chang,[40] Mahmassani started tackling day-to-day dynamics, focusing on questions that had been absent from the 1984 article: How do road users caught in a traffic jam transform this unpleasant experience into a learning process? And through what mechanisms do they update their departure time patterns in order "to 'beat' congestion from one day to the next?"[41] Like Kitamura and some other proponents of ABM at that time, Mahmassani was convinced that excessive schematization plagues the paradigm of the "rational" traveler in the quest to maximize his or her utility function (a theoretical principle explicitly endorsed by the advocates of disaggregate modeling in the 1970s but also implicitly espoused by Wardrop with his first principle). In order to develop a more realistic modeling framework the Austin professor turned for help to another economist, the Nobel Prize winner Herbert Alexander Simon (1916–2001), whose thinking he was already familiar with thanks to his Ph.D. work at MIT.[42] For Mahmassani, following in Simon's footsteps, the trip-maker in the street, often lacking relevant information, displays in his or her behavior a "bounded rationality," and is content with selecting "satisficing" instead of "optimal" solutions when faced with problems. Building on the "satisficing" principle, Mahmassani and Chang therefore performed a number of computer-based experiments involving virtual commuters located along a highway facility, whose behavior was governed by simple decision rules and heuristics, including explicit mechanisms to incorporate the experience accumulated through repeated use of the highway.

The distance to cover from virtual commuters to flesh-and-blood trip-makers proved rather short. A new experimental setting was devised by the Texan team, in which a traffic flow simulation model was used to predict congestion patterns on a highway facility resulting from the *aggregation* of the departure-time choices of *actual people*, who updated their decisions daily in response to information on the arrival time calculated by the traffic flow simulation model.[43] Mahmassani was apparently happy with these first computer-based experiments, whether they were carried out with "virtual" human beings modeled in the light of Simon's theory of bounded rationality, or with real commuters. Firmly believing in their high potential, he initiated a new series of simulation experiments, based on the same general protocol, in order to study the impact of "planned traffic disruptions," such as road works on the network as well as the "stability of the boundedly-rational equilibria that arise as a result of the individual behavior rules."[44] The accumulated results prompted Mahmassani to use this approach to explore a third and new dimension of DTA, involving *real-time* operations and management schemes this time: What effects does real-time traveler information have on the performance of a transportation system, such a corridor? The issue was of great practical interest as it was hoped that the conclusions drawn from the new experiments could help design and efficiently use "real time in-vehicle information systems" to alleviate congestion.[45] In addition to a variety of results about the dynamics of interaction between user decisions and transportation system performance, the simulation experiments performed by Mahmassani and his teammates would eventually enrich the vocabulary of urban travel demand modeling with a new term: BRUE, for boundedly rational user equilibrium.[46] As the authors themselves put it, a "boundedly rational user equilibrium (BRUE) is achieved in a transportation system when all users *are satisfied* with their current choices," and therefore none of them "intend to switch."[47]

Mahmassani's interest in "real time in-vehicle information systems" was in the spirit of the times. As has been seen,[48] at the end of the 1980s the subject was relatively new but highly topical. As a matter of fact, given the congestion problems facing American cities, the issue was about to mobilize several transportation specialists, including urban traffic forecasters. But if guidance systems based on real-time information had already been the subject of systematic research on their material aspects, their effectiveness in combating congestion was still to be demonstrated. It was even suspected that disseminating information about current traffic conditions could produce adverse effects, as the users of the system were likely to massively switch to what seemed the "best path" from their individual point of view, thus potentially generating higher levels of congestion due to the (over)concentration of traffic on the supposed effective routes.[49] These and other dynamic traffic assignment-related issues would dominate Urban Traffic Networks: Dynamic Control and Flow Equilibrium, an international seminar held in Italy in June 1989 under the joint aegis of the U.S. National Science Foundation and the Consiglio Nazionale delle Ricerche.[50] Ben-Akiva, Mahmassani, and several of their respective collaborators were among the attendees. On the face of it, the two lectures delivered respectively by the MIT and Texan teams displayed a series of striking similarities as to the objectives pursued, the general approach adopted, and the conclusions drawn by the two university research groups. The two teams proposed treating real-time traveler information systems with the help of a two-tier general framework, consisting of a dynamic network performance model coupled with a driver behavior model. They also envisioned a series of issues regarding the design and the efficient use of real-time information systems and their effects on the performance of the network. And both concluded that further theoretical and empirical research was needed. Yet, crucial differences lurked nonetheless beneath these similarities. Both the traffic simulation and the user decision components of the overall modeling framework put forward by the two groups of scholars differed from each other in several important respects. In the very early 1990s, the FHWA would give Ben-Akiva, Mahmassani and their collaborators the opportunity to continue their intellectual contest—this time on American soil.

## A Dynamic Traffic Assignment Contest: MIT against Texas University

In 1988, the U.S. Department of Transportation initiated the University Transportation Centers (UTC) Program.[51] After a nationwide competition, the federal administration awarded grants to create ten regional university transportation centers, and MIT became the lead institution for the "Region One consortium." Soon after, MIT began receiving $1 million per year from the federal government to fund a series of research programs, with the consortium having to find another $1 million from nonfederal sources. Among the themes high on Region One's agenda was ITS technology.[52] This was no coincidence. At the end of the 1980s, MIT in general, especially through Joseph M. Sussman,[53] and Ben-Akiva and his team at the Center for Transportation Studies in particular, were strong supporters of ITS and had already initiated in-house research programs focused on it and its application to transportation. Thus, during the 1990–1991 academic year, Ben-Akiva was appointed program director of a major research program called the Cooperative Research Program on Intelligent Vehicle-Highway Systems. This involved, in addition to CTS, several other MIT laboratories, including the famous Lincoln Laboratory established in 1951 to build the first air defense system of the United States following the explosion of the Soviet atomic bomb in August 1949.[54] Endowed with its own financial resources, thanks partly to MIT's participation in the Region One UTC, the in-house intelligent vehicle-highway systems (IVHS) program—renamed the intelligent transportation systems (ITS) program in the mid-1990s—would immediately benefit from the enactment of ISTEA in 1991. Indeed, at the end of the 1992–1993 academic year, Ben-Akiva was appointed co-principal investigator for a research program on the application of ITS technology for the large-scale Central Artery/Tunnel project of the City of Boston.[55] Since MIT was granted no less than $2.5 million for its participation in the program,[56] it was no wonder that the university's official magazine featured the latter on the front cover of the July 1992 issue.[57]

With all this money pouring into the research program, Ben-Akiva and his many collaborators, including the Greek Haris N. Koutsopoulos and the Belgian Michel Bierlaire, set up a master's and Ph.D. dissertation "factory" in the field of ITS. For 1993 alone Ben-Akiva and Koutsopoulos supervised, either individually or in tandem, at least seven master's theses.[58] Building on the previous research studies carried out in the second half of the 1980s,

Ben-Akiva's team set to work developing a general traffic modeling framework for the ITS era, capable of achieving the two following objectives: to generate pertinent information to be sent to motorists either before they embark on their journey or during their travel; and to predict the resulting traffic flows while assessing the impact of the various travel management strategies and other control measures on the network performance.[59] Among the many master's degree students supervised by Ben-Akiva and his collaborators in the early 1990s, several would continue their postgraduate education at MIT, and embark on a series of complementary Ph.D. projects.

This was especially the case for Kalidas Ashok, Amalia Polydoropoulou, and Qi Yang. In 1996, four years after getting his master's degree, Ashok, a graduate of the Indian Institute of Technology Madras, defended his Ph.D. research on the "Estimation and Prediction of Time-Dependent Origin-Destination Flows."[60] While most research done in DTA had thus far been concerned with loading the network with time-varying and *already known* origin and destination traffic matrices in order to estimate the resulting traffic flows and their evolution over time, Ashok was asked to devise mathematical techniques for *predicting* these very matrices. Thanks to previous research[61] it was then widely believed, and rightly so, that without projection of traffic conditions into the future, real-time control and route guidance strategies could often prove to be irrelevant because they are outdated by the very time they take effect—hence the need for prediction. At the time when Ashok was busy with integrating multiple sources of information, including historical data, in order to tackle the dynamic O&D estimation and prediction problem, Amalia Polydoropoulou, a graduate of the National Technical University of Athens in 1990, was charged with the task of developing and demonstrating a behavioral framework for user response to advanced traveler (or traffic) information systems.[62] How do drivers perceive these kinds of device and their service attributes? What is their response to the traffic information provided? In order to answer these and other questions, the Ph.D. candidate from Greece made use of two complementary datasets, one involving SmarTraveler, the first ATIS market product tested in the greater Boston area, and a second case study employing information gathered at the Golden Gate Bridge Corridor in San Francisco. To predict the traveler's behavior, Polydoropoulou combined RP and SP data[63] within a general modeling framework bearing the stamp of her supervisor

Ben-Akiva as the effects of ATISs on the user's reactions were assessed with the help of random-utility disaggregate models.

The third doctoral student whose research work would mark the development of the ITS project led by Ben-Akiva at MIT in the 1990s was Qi Yang. Awarded a bachelor of science degree in geography from Beijing University in 1983, Qi Yang obtained both his master's degree and his Ph.D. from MIT.[64] Though complementary to the research studies carried out by Ashok and Polydoropoulou, Yang's work displayed several singular features, starting with the term "laboratory" in its title: a *virtual* laboratory, with neither walls nor material equipment, to be more precise. Indeed, Yang's laboratory, initially called SIMLAB and renamed MIcroscopic Traffic SIMulation Laboratory (MITSIMLab) soon afterward, is a computer-based modeling system aiming to test and evaluate ATMS and ATIS designs. The modeling system consists of two central components: the microscopic traffic simulator[65] MITSIM (MIcroscopic Traffic SIMulator) coupled with a Traffic Management Simulator called TMS. MITSIM, a future winner of the *Discover Magazine* Award for Technological Innovation,[66] represents road networks in detail and simulates traffic flows at the vehicle level. A probabilistic model is used to capture the driver's route choice decisions in the presence of real-time traffic information—remember Polydoropoulou—provided by the virtual route guidance system. Within MITSIM, individual vehicles in the network are moved from their starting point to their destination and respond to the various traffic controls and guidance while interacting with each other. The vehicle movements are recorded by a surveillance module, representing traffic sensors and probe vehicles. TMS, the other central component of SIMLAB, represents and models the candidate ATMS/ATIS system under evaluation. Using the traffic data obtained by the surveillance module, TMS generates traffic management schemes that in turn are fed into MITSIM. It is worth noting that these management schemes are based on *anticipatory* traffic conditions—remember Ashok—predicted with the help of a specific mesoscopic traffic simulator[67] within TMS. (Virtual) drivers in the network respond therefore to the control measures and routing guidance generated by the TSM, and MITSIM simulates the resulting state of the network in detail. The surveillance system then informs TMS about the new state of traffic on the network, and control measures and routing strategies are generated in response. But why build such a virtual, computer-based laboratory? For the same reason that we simulate nuclear weapons that we want to test instead of building them and producing real explosions outdoors.[68] After all, a simulated system is much more flexible and convenient while costing less money to create than a real one. Of course, there is always the risk that the real system will actually behave differently from its digitized surrogate. But at least sometimes, something is better than nothing. Simulation-based evaluations for traffic management systems and strategies, such as ramp metering for freeway traffic control and urban traffic signal controls, had already been employed when Yang started working on SIMLAB.[69] Yet, a modeling system capable of both realistically simulating the traffic flow in the network *and* its dynamic interrelationships with the traffic management strategies generated by ATMS and ATIS under evaluation was undoubtedly an original undertaking. The author of the thesis could therefore proudly state that the "modelling system developed in this research is the first of its kind that has been implemented and whose usage has been demonstrated through a case study using a real network,"[70] namely the Amsterdam beltway, the same site that Ashok had used for testing prediction techniques for time-dependent O&D traffic matrices.

Judging from its reception both within the research community and by nonacademic actors, the ITS-related modeling research program carried out at MIT was a success. Among the stakeholders outside academia that showed a strong interest in the accomplishments of Ben-Akiva's MIT team was the R&D Office of the Federal Highway Administration. Just one month after it held a pre-proposal workshop in Chantilly, Virginia, in May 1994 to receive inputs from the professional transportation community, the FHWA initiated a DTA research project to address complex traffic control and other management issues in the information-based dynamic ITS environment. Following in the footsteps of the TRANSIMS project (chapter 5), the federal administration wanted to benefit from the end of the Cold War, and to place the skills and knowledge acquired by military-oriented research organizations over time in the service of civilian objectives. Oak Ridge National Laboratory (ORNL), set up in 1943,[71] was therefore designated as the manager for the development of a deployable, real-time DTA system designed to provide faithful and coherent real-time, trip guidance/information, including routing mode and departure time suggestions, for use by travelers, ATIS, and ATMS. More precisely, ORNL was charged with managing contracts for the development of dynamic traffic assignment procedures as well as testing and evaluating them.[72] Ajay Kumar Rathi, the author of a thesis on traffic flow simulation, was a key person in the management process.[73] Among the candidate DTA systems that were granted a contract by the National Laboratory in October 1995 was DynaMIT (DYnamic Network Assignment for the Management of Information to Travelers), proposed by the MIT team.[74]

However, though they were seduced by DynaMIT, the managers of the ORNL found another proposal, submitted by Mahmassani and his collaborators at the University of Texas at Austin, equally appealing. In October 1995, the Texan group learned that their project would also be backed with federal money, thus putting them into a competition with the proposal emanating from the team led by Ben-Akiva in Cambridge, Massachusetts.[75] As a matter of fact, since his early work conducted with Herman, Mahmassani had never stopped working on DTA issues. Like many of his colleagues,[76] Mahmassani would benefit from the Intermodal Surface Transportation Efficiency Act of 1991 and that year he succeeded in obtaining $559,016 in funding for "Traffic Modelling to Support Advanced Traveler Information Systems." Two years later, in November 1993, DYNASMART (DYnamic Network Assignment-Simulation Model for Advanced Road Telematics) was the subject matter of a voluminous research report coauthored by several scholars.[77] DYNASMART was, in fact, a highly collective undertaking in which many of Mahmassani's collaborators took part in the 1980s, including R. Jayakrishnan, an engineer of Indian origin whose doctoral research work would prove decisive for the project.[78] In addition to the contributions of an initial cohort of teammates, Mahmassani could rely for the further development of DYNASMART on a cosmopolitan group of Ph.D. supervisees at the University of Texas, including the Indian Srinivas Peeta, the Greek Athanasios Ziliaskopoulos, and the Taiwanese Ta-Yin Hu.[79]

In its revised 1994 version,[80] DYNASMART was composed of three main modules. The "simulation" module simulated traffic flows on the network with the help of a macroscopic model. Drawing inspiration from plasma physics, this component moves vehicles driven by motorists who belong to "multiple user classes with different vehicle performance characteristics." The "user behavior" module models how drivers respond to information and the various guidance strategies provided by the ITS. Unsurprisingly, drivers modeled by DYNASMART are the types of individuals found in Herbert Simon's books, since their route choice behavior exhibits a "bounded rationality." Seeking "satisficing" instead of optimal solutions, the drivers decide therefore to change their behavior only if the resulting gains exceed a certain threshold. "Path processing," the third module of DYNASMART, determines a series of network attributes, including travel times, for use in the user behavior module. Though "revised," the 1994 version of DYNASMART still remained "primarily a descriptive analysis tool for the evaluation of information supply strategies, traffic control measures and route assignment rules at the network level." However, its creators were hoping that their product would soon be able to evolve "towards a model that may be executed *on-line in quasi real-time* to support the functions of the system controller in the ATIS/ATMS."[81] Implemented on the University of Texas supercomputer CRAY Y-MP, while also able to run on other computer platforms with only minor modifications, DYNASMART was subject to a series of numerical experiments performed on two small transportation systems: first, a hypothetical test network consisting of 50 nodes and 168 links; and second, the much larger Austin core network composed of 676 nodes and 1,882 links.

Starting from October 1995, the intellectual arm wrestling between DynaMIT and DYNASMART under the federal oversight of ORNL would mobilize an army of master's and doctoral students in Cambridge, Texas, and later in Maryland where Mahmassani taught between 2002 and 2007 before heading for Northwestern University. Kazi Iftekhar Ahmed, John Alan Bottom, Constantinos Antoniou, and Ramachandran Balakrishna were part of Ben-Akiva's team,[82] while Yaser Hawas, Yi-Chang Chiu, Xuesong Zhou, Nhan H. Huynh, Akmal Saad Abdelfatah, Khaled F. Abddelghany, and Ahmed F. Abddelghany were involved in the crafting of DYNASMART.[83] For several years, the research reports multiplied as did the successive versions of DynaMIT and DYNASMART. It is worth noting that each team eventually produced two variants of their respective general modeling frameworks. DynaMIT and DYNASMART-X, designed for *real-time* management operations focused on route guidance, were delivered to ORNL for assessment in October 1998. DynaMIT-P and DYNASMART-P were the *offline* variants of the two modeling suites; geared toward *short-term* planning applications[84] and combining DTA with other travel demand modeling procedures, they would be subject to laboratory evaluation from October 1998 to June 2000.[85]

After a series of field tests[86] that completed the evaluation work carried out at ORNL by summer 2000, and after receiving $1,692,410 of federal money from January 1995 to May 2002, DYNASMART was declared the final winner of the contest. In summer 2004, the federal administration proudly announced that the planning version of DYNASMART would soon be available, and that two free training workshops would be offered by the FHWA and the University of Maryland in the Washington metropolitan area in fall 2004.[87] Following additional developments and tests, DYNASMART-P 1.3.0, which could run on the Windows operating system, was released by the FHWA through McTrans Center in February 2007. The released package came with a GIS data import and a network editing tool at the very modest price of $1,750 for its fully supported version, the latter including complete technical assistance and response to specific issues, while a limited-support version was available for $1,000.[88] In the early 2010s, after a series of applications and a promotional campaign in the United States, DYNASMART even crossed the Atlantic to land in the Netherlands.[89]

## From Academia to Practice

DYNASMART, and the DTA procedures it contains, seemed to effectively respond to real demands from the UTDM social world of the day. According to two Texan practitioners, the tool enables the engineer and the planner to "understand how traffic is going to react and how the network is going to be impacted based on any kind of disruption: Work zone, congestion, special events. Anything that's going to happen within a city that affects traffic, this simulation software can analyze it and depict what really will happen and let you anticipate future construction projects, too." As for the public and decision-makers, they can now be shown "what will work, what won't work, with a visual display, rather than boring them with statistics."[90] However, the most important and enduring impact of DYNASMART on the current American urban traffic forecasting landscape has probably been exerted through two spinoffs of the project. DynusT (DYNamic Urban Systems for Transportation) is a simulation-based mesoscopic DTA model created by the Taiwanese Yi-Chang Chiu, a former Ph.D. supervisee of Mahmassani.[91] While assistant professor at the University of Texas at El Paso from 2002 to 2006, Chiu spent "nights and weekends" on DynusT, and "all [his] passion went to this project,"[92] the development of which continued unabated after the author transferred to the University of Arizona. From January 2011 on, and for several years, DynusT evolved as a community-based open-source project overseen by Chiu himself, with the last version of the modeling suite released in May 2015. In 2012, the DynusT community featured among its active members several scholars from the universities of Arizona, Nevada, and Utah and the Texas Transportation Institute as well as a series of practitioners from large consulting transportation firms, including Parsons Brinckerhoff, AECOM, and PTV America. However, starting in 2017, the Arizona Board of Regents rescinded the open-source license of DynusT, making it a proprietary software licensed to Metropia. This firm was set up by Chiu himself in collaboration with the University of Arizona in 2012 to develop and deploy active demand management concepts and technology (the DynusT Lab at the University of Arizona has remained the lead research and development unit for DynusT).[93] DTALite, a lightweight dynamic traffic assignment package, is another outgrowth of DYNASMART. Also an open-source DTA software package, DTALite is presented as a tool able to model large networks with a minimal set of data and computing resources.[94] Its principal developer, Xuesong Zhou, a Chinese colleague of Chiu in Arizona for years, defended his Ph.D. thesis under the supervision of Mahmassani in 2004.[95] Along with DTALite, Zhou has also been the main system architect and developer for NEXTA (Network EXplorer for Traffic Analysis), a user-friendly open-source graphical interface for DTA datasets. NEXTA is also related to the DYNASMART project through the latter's Graphical Input Editor 1.0, initially developed by ITT industries in 2004. It is worth noting that NEXTA had also been used as the visualization program for TRANSIMS and DynusT for some time, though for about a half dozen years the latter has been coupled with DynuStudio, a graphical and data management system developed and commercialized by RST International founded in 1991. When in 2017 DynusT became a proprietary software package commercialized by Metropia, RST partnered up with Metropia to jointly market the DynusT/DynuStudio suite.

Though eventually DynaMIT lost to DYNASMART in the DTA competition organized by the federal administration, it has not disappeared from the modeling scene. In the mid-2010s, the software was in its 2.0 version,[96] and there was also another form of the MIT undertaking, called DynaMIT-MPI, using distributed memory architecture.[97] But DynaMIT would have a life outside MIT as well, since some talented members of the team gathered around Ben-Akiva as part of the MIT project—including Ramachandran Balakrishna according to whom DynaMIT "introduced generations of grad students to data-driven debugging, identifying bugs based on unexpected model results"[98]—have ended up at Caliper, the well-known proprietary software company. Looking for a tool for its TransCAD users (chapter 4), in the mid-1990s Caliper commenced development of a DTA project for travel planning purposes, which was based on the work already carried out by an expert in traffic assignment techniques, the academic Bruce Janson. This started with XuJun Eberlein, a 1995 MIT Ph.D. holder,[99] and was continued several years later when, late in 2004, Song Gao, a supervisee of Ben-Akiva,[100] joined the firm to work on this topic with Rabinowicz and Jonathan Brandon, holder of a master's degree in operations research from the Tel Aviv University in 1989. The other flagship software package from Caliper, TransModeler, a traffic simulation software with DTA capabilities, has also been closely associated with two other MIT "defectors," Qi Yang, the main creator of MITSIMLab,[101] and Srinivasan Sundaram, who had worked on the planning version of DynaMIT while a master's degree student.[102] A "talented and spirited co-worker" (Slavin's words), Qi Yang joined Caliper full-time in 1998, after having been a summer intern four years or so prior to that, and started working on the TransModeler project along with a few key software engineers from the firm, aided by Rabinowicz and Slavin as advisors. Probably the first attempt to develop a GIS suitable for traffic data and traffic microsimulation, from the end of 2006 on TransModeler has been granted four patents, enabling Caliper to differentiate it from MITSIMLab among others. First released in December 2005, TransModeler is now (as of June 2021) in its sixth edition.[103]

A similar process involving both academia and the private sector would occur several times in the field of dynamic traffic assignment in the 2000s, starting with DynaMIT's rival, DYNASMART. Working with S. Lee, while building on the work carried out by Carlos Daganzo on the "discretization" of the classical hydrodynamic traffic model,[104] Athanasios Ziliaskopoulos, a former Ph.D. student and collaborator of Mahmassani, developed, in the mid-1990s, an original traffic simulator called RouteSim, with a view to replacing the corresponding DYNASMART module.[105] At the same time, Ziliaskopoulos and some of his colleagues at Northwestern University were introducing a series of other changes into DYNASMART in a bid to strengthen its capacity to manage large databases. The final outcome of their efforts was an "Internet-based GIS that aims to integrate spatio-temporal data and models for a wide range of transport applications: Planning, engineering and operational."[106] Christened VISTA (Visual Interactive

System for Transportation Algorithms) by its designers, this original product for which the customer no longer needs to install software on their personal computer, and which operated under a license allowing academics access to the source code,[107] would lend his name to a firm: VISTA Transport Group was created in April 2004 in Evanston, Illinois, near Northwestern University where the product was launched, by several academics including, in addition to Ziliaskopoulos, Kyriacos Mouskos, a 1991 Ph.D. holder from the University of Texas at Austin, and S. Travis Waller, who defended his thesis at Northwestern University in 2000.

INRO, the Canadian firm set up in 1976, and now part of Bentley Systems, has built its reputation on a solid and continuously renewed expertise in sophisticated static assignment techniques (chapter 4). It was thus somehow predestined to eventually join the club of DTA software providers. INRO's current DTA model suite is called Dynameq (for DYNAMic EQuilbrium).[108] Like Emme, the firm's flagship product, Dynameq owes a great deal to the university system. In the second half of the 1990s, INRO's president Michael Florian launched, through the Department of Computer Science and Operational Research at the University of Montreal, several master's theses and doctoral dissertations on the various aspects of DTA.[109] Michael Mahut's doctoral research study proved most pertinent.[110] In his dissertation, the young scholar proposed a flow model for dynamic network loading derived from a simplified car-following relationship, which shares several similarities with cellular automaton (CA) traffic models (chapter 5). However, in Mahut's model time is continuous, making it considerably easier to calibrate against empirical data. At the 2004 annual meeting of the Transportation Research Board, an INRO cadre, including Mahut and Florian along with staffers from Calgary, Canada, delivered a lecture presenting the calibration of a simulation-based DTA model on a portion of the city of Calgary road network.[111] Dynameq was first released in March 2006,[112] while its 4.1 version became available to practitioners in July 2018.

Dynameq was not the only DTA Canadian proprietary software on the American market in the 2000s. INTEGRATION also owes a great deal to the academic world, since its intellectual fathers were Samuel Yagar, already mentioned in this chapter, and his student Michel van Aerde. Van Aerde began working on INTEGRATION during his doctoral thesis, which he defended in 1985 at the University of Waterloo, Canada, and kept on developing it until his death in 1999. For the most recent version of INTEGRATION, Aerde had drawn inspiration upon the cutting-edge work carried out by Mahmassani, Ben-Akiva, and their teams as well as by the TRANSIMS group. Work on INTEGRATION was continued by Aerdre's coworker and former student Hesham A. Rakha (1993 Ph.D.), currently a professor at Virginia Tech.[113] Last but not least, like Caliper and INRO, Citilabs and PTV America have also provided practitioners with their own DTA-related products since the mid-2000s.[114]

Given this surge of interest in dynamic traffic assignment both within academia and by all significant transportation planning proprietary software firms, it comes as no surprise that such major actors of the UTDM social world as the Transportation Research Board and the various federal agencies of the DOT have initiated and backed, often through the TMIP, a series of actions geared to the further development of efficient DTA techniques and their spread to the community of practitioners. Thus, a group of specialists, including Chiu (DynusT), Mahut (Dynameq), Balakrishna (DynaMIT) and Waller (VISTA), was commissioned by TRB to draw up a DTA *Primer* that appeared in the early 2010s,[115] while the 2012 edition of the TRB-sponsored manual on urban travel demand modeling included for the first time several passages on DTA.[116] Still in 2012 a *Guidebook* devoted to DTA saw the light of day under the aegis of the Federal Highway Administration, while a few years later the same federal agency released, through the TMIP, another DTA *Practical Guide*.[117] From the end of the 2000s on, several webinars dedicated to DTA issues have been held with the help of experts from private firms under the aegis of the TMIP,[118] while additional web-based resources include now the relevant page of the Travel Forecasting Resource website backed by the TRB.[119]

How have the main potential end clients of dynamic traffic assignment techniques, essentially the many MPOs and state departments of transportation, reacted to all these DTA tools—including the assignment module of TRANSIMS that does have DTA capabilities—progressively becoming available to the urban traffic forecasting community?[120] Unsurprisingly, they displayed an early interest in these advanced modeling techniques while also experiencing moments of disappointment. By 2008 the largest known DTA application had probably been carried out in Atlanta, Georgia, by Parsons Brinckerhoff, which utilized a VISTA package on behalf of the Georgia Department of Transportation. Approximately forty runs of the model were required to diagnose coding and software errors, while the execution time was approximately one week per run! Though the resulting model managed to satisfactorily replicate the observed conditions at the end of the field test, the amount of work needed to render it operational along with the run time required eventually prevented it from being used in studies as originally intended (figure 7.1).[121]

However, the difficulties that the first DTA models inevitably faced when confronted with real-life networks did not seem to dishearten their potential users. In fall 2009, the San Francisco County Transportation Authority (SFCTA) created, and successfully used, a subarea DTA model to examine the implications of closing some ramps on US 101 for the purposes of construction. In 2010, the authority received a grant from the FHWA to implement, with the help of PB, a dynamic traffic assignment model covering all of San Francisco, a project known as "DTA Anyway." For both projects Dynameq was used as the DTA platform, while some code was developed in an open-source environment.[122]

From 2010 on actions appeared to accelerate, and by fall 2012 no fewer than eighteen DTA projects of various kinds were reported across the country, with the oldest one dating back to 2005, either completed or under way, and using a variety of software.[123] A survey conducted in March 2015 and dealing with the twenty largest metropolitan planning organizations in the country and three smaller ones with a strong reputation for innovation in travel forecasting showed that DTA techniques were about to become a serious alternative to the static assignment procedures that had dominated urban traffic forecasting till then.[124] Not content with resorting to DTA techniques for specific projects, several planning agencies across the country decided to integrate DTA procedures as a permanent and versatile tool into their entire urban travel forecasting modeling suite. Of the twenty-three MPOs that responded to the March 2015 survey, after a two-year development period Portland and Phoenix could boast of having production-use DTA models. Seven other agencies reported having DTA models under development, while the Los Angeles model was still in a "predevelopment" stage.[125] In 2015, both Portland and Phoenix were using DynusT software,[126] four other MPOs intended to follow suit, while one each of the remaining four agencies was considering using DYNASMART, DTALite, TransModeler, and Aimsun, a software originating in Spain.

**TABLE 1 Dynamic User-Equilibrium Summary Statistics After First Signal Retime**

| vol7Range | Volume Range | Link Counts | 6–7 a.m. Count | 6–7 a.m. Flow | Rel. Error | % RMSE |
|---|---|---|---|---|---|---|
| 0 | <500 | 91 | 28,544 | 43,276 | 51.6% | 126.5 |
| 1 | 500–999 | 46 | 32,877 | 32,797 | −0.2% | 79.4 |
| 2 | 1,000–1,999 | 34 | 47,934 | 39,065 | −18.5% | 62.7 |
| 3 | 2,000–4,999 | 21 | 72,760 | 75,987 | 4.4% | 43.6 |
| 4 | 5,000+ | 16 | 107,936 | 86,434 | −19.9% | 45.6 |
| | Total | 208 | 290,051 | 277,559 | −4.3% | 77.6 |

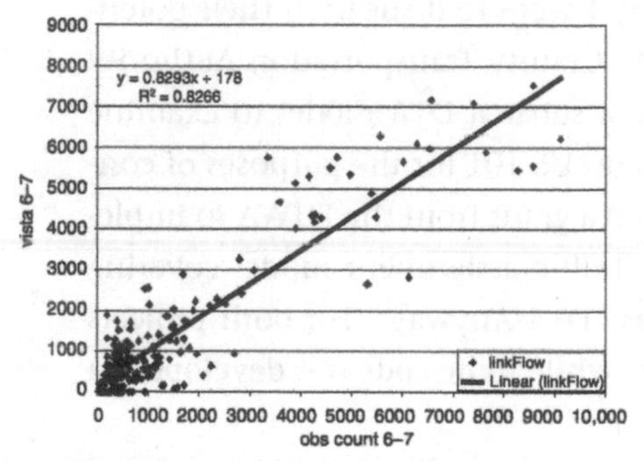


FIGURE 2 Dynamic user–equilibrium solution after first signal retime.

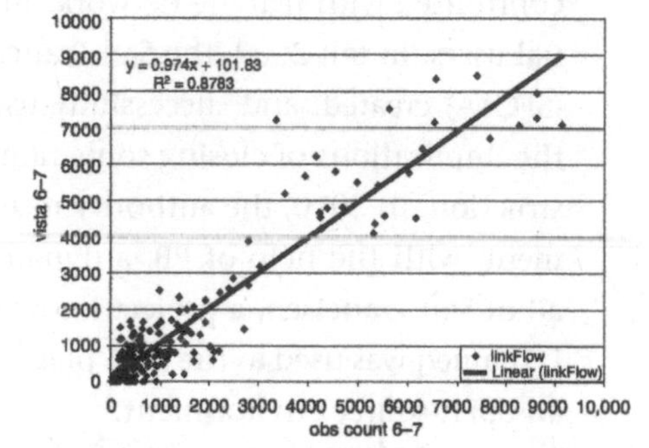


FIGURE 3 Dynamic user–equilibrium solution after fourth signal retime.

**TABLE 2 Dynamic User–Equilibrium Summary Statistics After Fourth Signal Retime**

| vol7Range | Volume Range | Link Counts | 6–7 a.m. Count | 6–7 a.m. Flow | Rel. Error | % RMSE |
|---|---|---|---|---|---|---|
| 0 | <500 | 91 | 28,544 | 42,952 | 50.5% | 119.0 |
| 1 | 500–999 | 46 | 32,877 | 33,295 | 1.3% | 72.0 |
| 2 | 1,000–1,999 | 34 | 47,934 | 40,493 | −15.5% | 62.4 |
| 3 | 2,000–4,999 | 21 | 72,760 | 83,557 | 14.8% | 39.2 |
| 4 | 5,000+ | 16 | 107,936 | 103,381 | −4.2% | 32.7 |
| | Total | 208 | 290,051 | 303,678 | 4.7% | 63.0 |

resolving the equilibrium settings, and so forth. The flows seemed to be converging toward the observed counts, and the traffic control settings seemed to be converging to a stable set of parameters. This outcome is exactly what would be desired in practice, yet nothing in the theory indicates that this will happen. There is no model specification for the problem of simultaneously computing traffic control settings and DTA solutions. It is a bi-level optimization problem that has no particularly useful formulation—at least none that would predict a convergent solution. Yet the experience indicates that the solution was moving toward convergence.

While these results look promising, they do not tell the whole story. These results were for a fairly low level of demand (6:00 to 7:00 a.m.); results were not shown for subsequent hours (i.e., 7:00 to 8:00 a.m. or 8:00 to 9:00 a.m.). (They have been calculated but are not sufficiently converged or calibrated at this point.) In fact, flows are generally a little more than half of what their counts are in these later periods. The challenge is to identify the reasons for these poor DTA results and develop a strategy once the causes are understood. The potential causes are many: ill-defined demand or temporal distribution of demand, network coding problems, model calibration parameters, or even incorrectly defined counts. One must try to identify causes of the underestimation of flows by building reports and analysis procedures that will help inform other DTA models and not just try to find some settings to which the DTA is particularly sensitive and modify those to calibrate this one model.

## CONCLUSIONS

This paper describes the experience of using DTA to calculate regionwide time-dependent flows for the purpose of specifying time-dependent origin–destination flows through a focused area for which detailed traffic

## Integrated Models, and the (Probable) End of an Era

It is worth noting that of the nine MPOs that reported having dynamic traffic assignment models in production work or in development by 2015, all but one either had an operational activity-based modeling suite or one that was under development. This must not come as surprise. While ABM represents the "demand" side of urban traffic forecasting, meaning the travel of individuals and households, and DTA "the supply" side of the equation, the effect of travel on the transportation system, ABM and DTA share much from a conceptual point of view. Both are highly disaggregated systems since they operate at the level of "individuals," whether persons or vehicles, while having a fine-grained time and space resolution (remember that most of traditional forecasting procedures within the FSM are aggregate and zone-based, and tend to use large time periods, typically morning peak, midday, afternoon peak, and night). By the very end of the 2010s existing operational activity-based models mostly had been coupled with aggregate static assignment models and DTA techniques usually had performed in tandem with aggregate travel demand models.[127] Nonetheless, the fact that microsimulation has been brought to a certain level of maturity on both the demand and supply sides of urban travel demand modeling increasingly has made the perspective of ABM-DTA *integration*[128] one of the most promising avenues, at least for those parts of the UTDM social world in the vanguard of the search for advanced modeling techniques. No wonder that a series of projects aiming to integrate ABM and DTA for maximum advantage of the disaggregate nature of both modeling

**Figure 7.1**
Tables and figures extracted from a publication concerned with the largest known dynamic traffic assignment application by 2008; the study was carried out in Atlanta, Georgia, by the consulting firm Parsons Brinckerhoff, which utilized a VISTA software package on behalf of the Georgia Department of Transportation. *Source:* James Hicks, "Dynamic Traffic Assignment Model Breakdown," in *Innovations in Travel Demand Modeling* (Conference Proceedings 42), *Volume 2: Papers*, ed. Transportation Research Board/National Research Council (Washington, DC: Transportation Research Board, 2008), 101–108, tables 1–2 and figures 2–3 , 107. Copyright © National Academy of Sciences. Reproduced with permission of the Transportation Research Board.

systems have been conducted from the end of the 2000s across the United States. Soon these kinds of "experimental" studies ceased to be the preserve of academics involved in cutting-edge travel modeling practices,[129] and started to spread to the practice-oriented segment of the UTDM social world, especially thanks to a series of steps taken by the federal administration and the Transportation Research Board. Thus, in January 2009 Cambridge Systematics was granted $2,649,999 to develop an integrated advanced travel demand modeling suite for the Sacramento area, joining together DaySim, as the ABM model, and DynusT and FAST-TrIPs (Flexible Assignment and Simulation Tool for Transit and Intermodal Passengers),[130] as the dynamic assignment models (for highways and transit systems).[131] A few months later, in August 2009, another ABM expert—the Resource Systems Group—received the amount of $1.9 million to integrate DaySim and the DTA component of TRANSIMS for Jacksonville, Florida. A smaller demonstration modeling suite would also be created for Burlington, Vermont, while the work done was later transferred to Tampa, Florida.[132] Both pilot cases emphasized open-source software in order to facilitate broader transfer of research experience.

As soon as these first two studies had been brought to completion, on February 4–5, 2014, twenty-four representatives from the TRB, the American Association of State Highway and Transportation Officials, several federal agencies including the FHWA, state DOTs, MPOs, academics, and the private sector participated in an Implementation Planning Workshop (IPW). This resulted in the definition of four research programs collectively known as the Advanced Travel Analysis Tool. Among these "capacity products," in the parlance of the IPW, was the C10 program, which aimed to integrate activity-based models with DTA for both highways and transit systems. Participants even agreed that the majority of available funding should go to the integration issue. In July 2014 a panel of experts selected four C10 pilot projects out of the twenty applications.[133] The studies were initiated in late 2014, and the bulk of the activity took place in 2015 and 2016.[134] In addition to a series of national outreach activities,[135] project teams stayed in touch with each other via quarterly conference calls, while further coordination occurred between the Ohio and Atlanta projects as they were using similar underlying models and the same consulting team, namely Parsons Brinckerhoff. Important as it was for familiarizing the UTDM social world with the integration issue, the C10 program has not been the only federal initiative in this domain. Thus, during the period October 2017 to April 2018, the consulting firm Resource Systems Group allied with the proprietary software company Caliper to carry out an FHWA-sponsored "experiment" on the impact of connected vehicle (CV) and autonomous vehicle (AV) technologies on regional transportation systems with the help of an integrated model, coupling DaySim with TransModeler, for Jacksonville, Florida.[136] And like DTA and ABM, the integration issue currently benefits from the web-based resources developed by the TRB Special Committee ADB45.[137]

In 1973 the *Journal of the American Institute of Planners* published an article intended "to evaluate, in some detail, the fundamental flaws in attempts to construct and use large models and to examine the planning context in which the models, like dinosaurs, collapsed rather than evolved."[138] Though the main target of the article's criticism was the large land use models of the day, many of the flaws by which these systems were supposed to be plagued could also be seen in the various urban traffic forecasting procedures at that time. The imminent demise of large-scale planning models affirmed by the author of the article never occurred, and this false prediction should remind us that talking about the future is always a tricky affair. It is probably still early and too bold to suggest that the urban travel demand modeling community is about to definitively turn its back on the four-step model. However, it is reasonable to contend that after dominating the UTDM social world for so many years, the FSM has already ceased to be the only game in town, and sooner rather than later it will be consigned to the past for good.

# Conclusion

The first few, sporadic attempts to forecast urban traffic in U.S. cities date back to the interwar period. A century later, and especially from the 1960s on following the advent of the Interstate Highway System, a fair chunk of which would run through metropolitan areas, urban travel demand modeling has grown into a commonplace and widespread activity performed by a host of individuals and organizations on a daily basis throughout the United States. After starting out as a largely civil engineering undertaking, essentially served by practitioners, as time went on UTDM has grown into a thriving academic field and has even become a striking example of the integration of several disparate disciplines, including social sciences, within a single practice-oriented modeling field. Now that the preceding chapters have introduced to the reader the essential details and episodes in American urban travel demand modeling, it may be useful to step back and take a more comprehensive look at this specific modeling domain, filled as it is with arcane technicalities, yet also tightly woven into the fabric of millions of Americans' everyday life. For better and for worse, urban travel forecasting has helped shape the landscape of large cities across the nation since the 1950s.

I have three goals in the following discussion: First, to take a high-altitude flight over American urban travel demand modeling territory to identify a number of core elements and patterns of its development *in the long run*. Second, to offer explanations of these core elements and patterns. Third, to use UTDM as the lens through which to observe larger phenomena currently under investigation by historians and other social scientists, ranging from the creation of practice-oriented bodies of knowledge requiring insights and skills from diverse fields to the history of the American

State, and from the increasing commercialization of academic knowledge to the development of knowledge-intensive firms.

## A Panoramic View of American Urban Travel Demand Modeling

### Intellectual Shifts

Although the history of American urban travel demand modeling is told here by a single historian, it is the collective creation of an entire "social world." UTDM has been, in fact, the complex and largely unpredictable outcome of the combined efforts of traffic forecasters and their many collaborators, ranging from travel survey designers to computer code writers, and from R&D funders to marketing specialists and even judges. Unsurprisingly, a highly heterogeneous and evolving historical entity populated by many and variegated elements, humans as well as nonhumans,[1] urban travel demand modeling has experienced a series of, and sometimes sweeping, changes in all its major components, intellectual and material, institutional and organizational, professional and cultural, and financial and social.

Regarding the intellectual component of UTDM—the *general principles* underlying the modeling knowledge and practices produced within it in particular—these changes tend to move in *three* main directions. Think one last time of the four-step model, the first comprehensive output of the UTDM social world when FSM took center stage at the end of the 1960s. By predicting *steady* flows, mostly related to work trips, over a representative time framework, usually peak hours, the FSM is *static* in nature. It is also zone-based and strongly *aggregate*, as it directly estimates the total number of trips undertaken by one or a few homogeneous groups of travelers moving from one urban zone to another, with the predicted amount of travel assigned to the network also represented as an *aggregate* continuous flow. In addition to being static and aggregate, the FSM in its incipient phase was also *analytical* from a mathematical point of view, since the underlying mathematical process could be completed in a finite number of operations. Now consider the urban traffic forecasting landscape depicted in the third part of the book, through the examples of activity-based modeling and dynamic traffic assignment, which, taken together, strove to be the FSM's definitive successors. Both ABM and DTA are *dynamic*, *disaggregate*, and *(micro)simulation-based*.[2] Thus, the ABM attempts to capture the various *temporal* aspects of people's travel behavior, including the choice of departure time and the reaction of trip-makers to real-time information, while the DTA procedures take account of *time variations* in network traffic flows and conditions. ABM and DTA are *disaggregate* in nature as well, since the former focuses on *individual* travelers and households, and the latter on *separate* vehicles, or clusters of vehicles. Both came to turn away from analytical mathematical formulae through the intensive use of *simulation* techniques.[3]

These developments concerning the *content* of urban travel forecasting knowledge and practices have taken place within an environment that has also undergone dramatic changes. To begin with, the mainframes of the 1960s pale into insignificance when compared with the computers and the available computational resources of today. However, advances in information technology were neither the sole nor even the most material driving force behind the shifts UTDM has experienced so far. Urban travel demand models have evolved and significantly changed in the long run because the problems they have had to cope with and the requirements to be met also transmuted over time. While they were originally designed to properly size and assess heavy transportation infrastructures, essentially urban highways, traffic forecasting models increasingly came to be used to help planners and transportation engineers confronted with new challenges—such as original forms of suburban congestion and air pollution—to make the most of existing capacity through a series of policy instruments, including public transport fares and real-time information systems. These challenges as well as a growing interest in social justice and concerns about discrimination issues, often enshrined into federal acts and other state laws, pushed the UTDM social world to develop more *individual-centered* modeling frameworks, able to predict the responses of the trip-makers and households to the various transportation policy "inputs" envisioned, assess their impact on various specific groups, and evaluate them in terms of both effectiveness and equity.[4] Speaking of the "environment" in which urban travel demand modeling has been embedded, a number of developments in the conceptual and mathematical structures of traffic forecasting models can be regarded as more or less felicitous attempts by the various segments of the UTDM social world to keep up with the economic, social, and demographic changes taking place in American society at large. Rising individual differences both in and among groups of people, growing ethnic diversification in metropolitan areas, shifts in home and work roles as well as changes in gender conditions[5] have brought about dramatic changes in household and individual travel behavior and mobility patterns, which newer traffic forecasting models have sought to grasp and account for. In order to deal with this general societal trend toward diversified lifestyles underpinned by an individualist view and a person-centered experience of social relations, the UTDM social world has increasingly focused on *individuals* as well as their differences and interactions, instead of on groups and zones, and progressively injected a great deal of economic and social sciences into the inner structures of models in order to augment their behavioral realism and, hopefully, their predictive power.

## An Increasingly Rich Social World

A problem-solving device, urban travel demand modeling has been constantly rooted in engineering and planning since its very beginnings in the early post–World War II period. Moreover, for several years practice and practical needs appeared to be the main driving forces behind the development of urban traffic forecasting, as federal, state, and metropolitan agencies, on the one hand, engineering consulting firms, on the other, proved a fertile ground for the development of original modeling work. However, as time went on a new kind of actor stepped onto the modeling stage never to exit it since: the *American university*. This growing *academicization* of the domain had a twofold consequence. First, in steadily supplying the traffic forecasting labor market with significant numbers of young people who benefited from a formal training, the university helped make the urban travel demand modeler into a recognizable and distinct professional figure within the overall American transportation community. Second, the academicization of the field deepened the division of labor within the UTDM social world as well, since a large portion, but not the entirety (more on this follows), of innovative modeling work has increasingly become the province of people engaged in research and development as their main occupation.[6]

After having been transformed from the 1960s on into an academic object within the university, mostly in civil engineering departments, UTDM has seen the range of actors serving it widen even further. Thus, the 1970s and 1980s witnessed the advent of two new types of stakeholders: the transportation consulting firm closely related to the academic world;

and the company specialized in proprietary urban traffic forecasting-related software. A decade later, a number of U.S. national laboratories, including Los Alamos National Laboratory, decided to recycle the competences they had honed during the Cold War years for the purposes of transportation modeling, while from the mid-2000s on a host of more or less informal groups, usually composed of both practitioners and academics and gravitating toward open-source software have also made their presence increasingly felt. Not surprisingly, both the multiplication and diversification of types of actors inhabiting the UTDM social world have been accompanied by the proliferation of the disciplinary backgrounds from which the urban traffic forecasting community is drawn from,[7] a process also reflecting the increasing sophistication of practices and techniques developed within this specific modeling domain over time.

This growing academicization of urban travel forecasting quickly, and heavily, impacted on the more practice-oriented segment of the UTDM social world as well. Indeed, from the early 1970s on, all major consulting firms and proprietary software companies active in this modeling field have become home to Ph.D. and master's degree holders, forged strong bonds with the academic world, and actively sought to incorporate a good deal of theoretical innovations, produced mostly within the university system but also by themselves, into their standard consulting activities and commercial products. As a result, American urban travel demand modeling saw the balance of expertise between the public sector—essentially represented by federal agencies, state transportation departments, and local transportation planning organizations—and the private sector—comprised of engineering consulting firms, proprietary software companies, and more recently, firms specialized in the collection and processing of data that travel demand models are fed with—progressively tilt toward the latter, especially from the 1970s on, when the (relative) withdrawal of the federal government[8] made it easier for market logic to gain strength and spread across urban traffic forecasting as a whole. This now long-standing supremacy of the market-oriented segment of the UTDM social world over its public counterpart does not appear to have been seriously challenged by the more recent attempt by the federal administration to return as a major actor in the field, especially through the rather controversial TRANSIMS project in the 1990s, or by the more recent development of open-source software packages reaping significant benefits from public funding.

### The Specialization Process

Judging from its trajectory in the long run, one can rightly conclude that urban travel demand modeling has been capable of dealing with a series of evolving historical settings through a *growing sophistication* of both its modeling content and the structure of the "social world" involved. This sophistication has been, to some degrees at least, the result of a *twofold specialization*. While issues about traffic and its modeling made their appearance in the interwar years, and were pursued even until the early 1960s, as part of the *larger* "urban question," as the UTDM social world increasingly gained a foothold both in the university and the transportation community, its members, at least many of them, would growingly cut off their concerns and work from *broader* settings, thus distinguishing their activities from those of *urban planners* and *other related professions*—we'll call this process *first specialization*. In addition, they would progressively show little interest in matters other than those in which they had *specialized within* the field of urban traffic forecasting and therefore deployed the bulk of their efforts to come to terms with the *specific*, narrow, and increasingly technical problems they were encountering *within* their modeling subfields (*second specialization*). Thus, when at the turn of the 1930s specialists working for the Bureau of Public Roads started seeing themselves as public servants uncommitted to specific policy, whose mission solely consisted in *satisfying* the mobility needs and desires of American urbanites, the *urban highway* emerged as the normal way to transport people in American cities (chapter 2). In the view of BPR staffers, people's individual choices and personal decisions were all that was needed to produce the city, and no further urban planning was therefore required. Traffic experts simply had to accommodate people's "demand" through *building* transportation systems, especially urban highways, the design and sizing of which came to be merely a "technical" question that engineers were able to answer accurately by capturing the drivers' desires through proper survey techniques and effective travel forecasts. Not surprisingly, the highly political debates over the "good city" and the impact of highways on its shape and daily functioning that took place in the interwar period[9] progressively faded away after the 1940s. "Specialization" forces were also at work in the travel demand modeler "decision" to assume a future land pattern *as given* and *exogenous*, thus turning a blind eye to the fact that the planned transportation system would probably affect both residential and employment patterns in the surrounding urban area, while the resulting changes in land use were very likely to influence in turn the number of trips, their destination, and modes.[10] But, there is more. The very four-step model that was granted paradigmatic status for the UTDM social world in the 1960s and dominated the modeling scene for decades since then, can be read, even if it does not solely boil down to this, as another vehicle for the aforementioned "specialization process." As a matter of fact, by adopting FSM, which breaks down a rather seamless phenomenon such as the decision to make a trip into distinct entities (four sequential subdecisions) that can be treated *separately* with the help of *specific* modeling practices, both academics and practitioners involved in the field channeled most of their efforts into solving the many "puzzles" they individually encountered *within* each one of the four steps.

I have just given some examples illustrating the intellectual grounds in which the twofold specialization process was rooted. In addition, one can mention a series of institutional factors as well. As the cases of the Chicago Area Transportation Study and Traffic Research Corporation (chapter 2) clearly show, in the 1950s and the early 1960s traffic forecasting studies were collectively produced by people coming from several disciplinary backgrounds and collaborating within multidisciplinary teams, while frequently publishing in journals dedicated to comprehensive urban issues, such as the *Journal of the American Institute of Planners*. It comes as no surprise that both CATS and TRC as well as similar organizations were able to produce both sophisticated *land use* and *transportation models*.[11] However, as time went by people involved in urban traffic forecasting began to receive specific modeling-focused training within engineering and planning departments, do modeling work in university centers dedicated to transportation research, work for specialized transportation consultancies, publish in (new) specific journals geared to transportation issues, such as *Transportation Research* and *Transportation Science*, and design and make increasing use of specialized software, hence progressively cutting themselves off from broader urban issues.[12]

### Routine versus Novelty

In the preceding paragraphs, emphasis was placed on the changes American urban traffic forecasting has undergone over time. However, in its long life it has also experienced significant time spans marked by more or less *routine* practices. Since urban travel demand modeling is a highly complex undertaking involving a large set of heterogeneous elements—federal agencies and local governments, consulting firms, algorithms and codes, travel surveys, university training and Ph.D. programs, journals and webinars, and so on—the introduction of significant changes within it is, more often than not, a cumbersome and costly affair in many respects. Indeed, substituting new practices for the prevailing ones is very likely to entail producing original kinds of data to be fed into the fresh models—and thus devising and implementing new, and often unfamiliar forms of surveys—writing original codes to embed the new models in computers, training people in original data gathering and modeling practices, and so on.[13] And practices are not the only items changed in the process: People must learn to work with the new tools as well. If for some members of the UTDM social world—think of the academics concerned—the production of novelty can be regarded as a professional duty, for many others it can require great effort to learn and master new ways of doing things, and challenge their interest, both cognitive and material. According to an eminent member of the UTDM social world, it can even frighten them, especially government officials who balk at the prospect of new methods that are likely to produce forecasts quite different from earlier ones, and which may thereby discredit their agency.[14]

And there is more. Sticking to the "old" and "familiar" while resisting the "new" can be a rational choice. The reader surely remembers that it took ABM several decades before it eventually reached operational status in the 2000s (chapter 6). During all this time, though cognizant of its inherent flaws, the UTDM social world largely persisted in backing and resorting to the FSM while incrementally improving in a rather piecemeal way either its main constituents—by grafting disaggregate mode choice models on an otherwise heavily aggregate modeling approach (chapter 3)—or the entire procedure, by introducing feedback between its various components, for example (chapter 5). But even after operational activity-based modeling had eventually gotten off the ground, people still had good reasons not to *discontinue* old practices altogether and overnight. As FSM and ABM are multipurpose, versatile entities, both endowed with great inner complexity, it would be highly unlikely that *one* of them *systematically* outperforms its competitor in all cases, under all conditions and circumstances, and in respect of all evaluation criteria the different actors inhabiting the UTDM social world may use. Thus, on the one hand, activity-based modeling systems can rightly boast that they represent both individual and household travel behavior and decision-making more realistically than the aggregate four-step approach does, and are hence more sensitive to a series of emerging transportation policies as well as to welfare and equity issues.[15] However, on the other hand, in several modeling settings the predictive capacities of advanced modeling techniques have neither always lived up to expectations, nor have their performances proved systematically superior to those displayed by older approaches.[16] Moreover, added sensitivity usually comes with supplementary costs. Compared with more traditional forecasting techniques, activity-based models are usually much more computationally demanding and time-consuming, and experience more "computational friction"[17] since the number of their parameters and submodels is usually much greater.[18] Advanced modeling practices are also very likely to suffer from higher *costs* associated with model development and implementation but also maintenance, be this in terms of time, money, or other institutional resources such as qualified and trained staff.[19] Hence, in the absence of a *single overarching* criterion by which the pros and cons of alternative and competing sets of modeling practices can be weighed up once and for all, one should not expect that practitioners, notwithstanding their ability to locate and specify the various shortcomings of established practices,[20] will swiftly abandon their usual ways of doing things, and easily convert en masse to new approaches.

Another factor that accounts for the (more or less) *stabilization* of urban traffic forecasting practices for significant periods of time can be traced to the deliberate efforts and decisions made by several influential federal agencies, such as the Bureau of Public Roads and the Urban Mass Transportation Administration in the 1960s and the 1970s, to pick what they then regarded as the "best practices" in the field, and promulgate them nationwide with the help of manuals, training courses, and, especially, free computer program suites among practitioners. In more recent periods, the rather *oligopolistic* structure of UTDM consultancy and software markets is also likely to have acted as a supplementary stabilization force.[21]

### Discontinuity (in Content) versus Continuity (in the System of Actors)

I have already referred to the emergence of new kinds of collective actors inhabiting the UTDM social world as time went by, such as university departments and research centers, consulting firms closely linked to academia, proprietary software companies, national laboratories, and groups gravitating toward open-source software for the more recent periods. However, shifting the *scale* of analysis[22] in order to focus on *individuals* rather than groups and institutions reveals, alongside shifts and changes, some interesting *continuities*. Indeed, though we are lacking enough statistical evidence for the moment to draw solid conclusions, the career paths followed by several individuals featured in this book clearly show that it is not uncommon for new organizations to be created or largely staffed by the very same individuals.[23] Accordingly, it would be interesting to differentiate between the intellectual aspects of UTDM, on the one hand, its institutional and "individual" features, on the other, as far as the issue of continuity is concerned. Compared with the number of sharp discontinuities that have characterized the intellectual growth and development of urban traffic forecasting over time, it is rather continuities or smoother shifts that seem to prevail at the institutional and individual levels.[24] Indeed, new and original practices have often been introduced and gained traction within the travel demand modeling community through the action of the *same* individuals and the *same* organizations at different junctures in their respective trajectories. Again, the fact that the UTDM social world tends to change more gradually and more continuously than its various outcomes should not come as a surprise. Only changes that succeed in mobilizing enough people willing to cooperate on a regular basis in order to sustain and further the initial idea, to capture existing resources, be they financial, material, or human, and then to harness them toward a shared goal are likely to survive, develop, and gain currency within a complex social setting.[25] And yet sheer quantity may not suffice to bring about significant changes. Novelties must be given the blessing of people whose voices are vested with great authority in their community. In point of fact, ABM grew increasing successful when several heavyweights in the UTDM social world were witnessed being won over to the new approach in the 1990s.[26]

## The Driving Forces behind the Vitality of American UTDM

### UTDM and the American University System

The number and intensity of changes urban travel demand modeling has undergone since its coming into being in the early post–World War II years bear witness to American urban traffic forecasting's capability to periodically rejuvenate itself. And though as time went by American supremacy in this modeling field progressively came to be less unequivocally established than in the past, the American UTDM social world proved able enough to display significant innovation capacity, albeit at varying pace, throughout the period stretching from the beginnings of the field up to the present. Two factors stand out when it comes to explaining the dynamism displayed by the American UTDM in the long run:[27] the American university system and, more broadly, the research and development system of the country, as well as the development of a series of knowledge-intensive firms that developed strong bonds with academia.[28]

As a matter of fact, several well-established characteristics of American R&D, in which universities have pride of place, proved particularly conducive to the growth of urban travel demand modeling in the United States. From the 1950s on, American universities increasingly regarded themselves as research and knowledge "factories," the end products of which should be *useful* to American society.[29] As a practice-oriented field, urban traffic forecasting has been therefore warmly welcome by a wide range of university departments. Urban traffic forecasting also benefited from the postwar general pattern of *science and government partnering* in the United States, according to which the federal administration would grant contracts to universities as well as to the private sector to do research and development rather than enlist the services of individual scientists and engineers in state-run research organizations[30] (although the TRANSIMS and the more recent POLARIS projects heavily involved national laboratories, all in all they have not called into question the aforementioned pattern as far as urban travel demand modeling is concerned).

Useful and knowledge-friendly, the post–World War II American research university system has also consisted in a highly devolved entity, formed by a large number of higher education institutions enjoying a considerable degree of autonomy. In this field, crisscrossed by intense intellectual rivalry, original knowledge and new subject areas as well as the relevant skills rarely remain confined to a specific locale. Instead, they are reproduced and further developed in several places, in large part through intense academic mobility, another prominent feature of the system. The resulting competition and emulation can therefore foster further developments and innovations.[31] These general characteristics of the American R&D system obviously worked to the advantage of travel demand modeling and its development in the long run. Indeed, as has been amply seen, the various state governments and, especially, the federal government massively funded both theoretical and more practice-oriented research on urban traffic forecasting within the university system. Once universities became home to urban traffic forecasting, their *decentralized* and *competitive* character has worked for the further development and diffusion of UTDM. And, as occurred in other academic disciplines, the diaspora of Ph.D. holders played an important part in the dissemination of this modeling field across the American university system. With regard to postgraduate degrees, one should also mention here the international openness and the attractiveness of the American higher education system after the end of World War II, which proved greatly beneficial to urban traffic forecasting, especially in the more recent periods. As a result, the share of people today who were born outside the United States and even attended foreign schools at the undergraduate level account for a significant part of the academic segment of the American UTDM social world. In shifting from its academic to its practice-oriented segment, one can also find the same openness and attractiveness characterizing the national university system, since a growing number of foreign-born and foreign-educated members of the UTDM social world have decided to settle and work in the United States, while non-American firms involved in the field have also been increasingly active on American soil.

Contrary to other national systems, such as that of France,[32] the American system of higher education shuns excessive *functional differentiation*. It follows that most American engineering departments are part and parcel of the university world. As a result, at the institutional level, the research and teaching of the sciences as well as a professional education in engineering have been often performed within the same establishment in which many "pure" and "applied" disciplines and research programs coexist and interact with each other. At the individual level, people tasked with producing new knowledge and those charged with training practitioners who will be able to put that knowledge to practical use are usually the same. These two characteristics of the American university, which facilitate the communication between academia and the professional realm, have proved to be a positive factor in the development of innovative theories and practices within urban traffic forecasting. The emergence and growth of the disaggregate approach in the 1970s and the 1980s is an excellent case in point. Let us recall that disaggregate techniques mainly resulted in applying a specific set of major, even mainstream, *economic theories*—the random utility theory, in particular—to a *practical problem* that transportation engineers had been struggling with for a long time, that is, predicting the behavior of travelers, especially with regard to mode choice. No wonder several seasoned economists within academia, among them the Nobel Prize laureate Daniel McFadden, played a prominent role in the birth and successive refinements of disaggregate techniques, which were rather quickly adopted, further developed, and intensively taught by scholars in civil engineering departments, such as Moshe Ben-Akiva and his colleagues at MIT from the mid-1970s on. This process, whereby persons in "Role A" (for example, McFadden and other academic economists) make use of the usual means and resources related to this role in order to achieve the aims of "Role B" (urban traffic forecasting researchers and practitioners),[33] similarly applied to the TRANSIMS project, in which the intellectual and material resources at the disposal of post–World War II national laboratories as well as specific know-how and skills acquired by Cold War scientists were channeled into civil transportation modeling.

## The Academic and the Practitioner

I have already discussed the growing differentiation of the role of the researcher and that of the practitioner as American urban traffic forecasting evolved. However, even though from time to time scholars involved in the field wished for more interactions and links between academics and groups pursuing more practice-oriented objectives,[34] all in all there have been few barriers to communication between the academic and the professional segments of the UTDM social world. In fact, both have developed shared long-term goals and recruited from overlapping labor markets, while benefiting from, and even frequently sharing, common (and usually federal) funding arrangements. There are several reasons why American urban travel demand modeling has proved able, on the whole, to establish permanent and strong links between the research and its more practice-oriented realms. First of all, serving a field rooted in professional practices from its incipient stages, many researchers in UTDM have rarely yearned for academic reputation only,[35] while, despite the increasing academicization of the field, demand for practical relevance never became extinct. At the same time, following in the footsteps of many other American industrial firms, most if not all large consultancies and proprietary software companies involved in urban traffic forecasting from the 1970s on have provided their "knowledge workers"[36] with a campus-like work atmosphere, enabling their contributions to new knowledge and research-based operational outcomes. In fact, some of the most talented and energetic scholars involved in urban traffic forecasting pursued, simultaneously or alternately, a double career as academics and practitioners,[37] being thus able to transfer research findings to the development of original operational practices. The most important firms in the field have been home to a great deal of Ph.D. and master's degree holders and have encouraged them to follow research activities along with their more practice-oriented tasks. Moreover, practitioners and full-time researchers have often been members of the same professional organizations,[38] published in the same journals[39] and read the same authoritative manuals and paradigmatic texts,[40] attended the same meetings, and interacted with each other at the same forums. A series of "boundary objects," such as computer packages, and "boundary organizations"[41] have rather successfully prevented a strict cleavage between the activities and outcomes of the academics and those of the more practice-oriented members of the UTDM social world.

## Placing American UTDM within Broader Contexts

### Modeling Complex Systems

Though models and modeling have captured the attention of scholars (other than modelers) from many walks of academic life for several years now, the rapidly proliferating literature on the subject still remains uneven in several respects.[42] In fact, most studies have been rather restricted to models within mature, well-established scientific fields, including physics, climate science, biology, and economics, while from a methodological point of view it is the epistemological perspectives and issues that have had pride of place.[43] In offering an in-depth historical and sociological account of an engineering modeling field that seeks to address societal and environmental challenges, this book contributes therefore to filling a gap in the existing body of research, especially through casting additional (and, hopefully, original) light on both the structure and history of knowledge-based systems that address vexing practical problems[44] and are firmly anchored in decision-making.[45]

To start with, in contrast with modeling developed within mature scientific fields, urban travel forecasting presents itself as a body of knowledge and practice that takes no note of established disciplinary boundaries, including those between engineering and human sciences, while, in addition, it hinges on a particularly large array of actors coming from varied backgrounds and belonging to more than one discipline.[46] Exchanging structure for history, UTDM also seems to illustrate another general trend in modeling complex systems. While in the past engineers and other like-minded people resorted to modeling techniques that make massive use of mere statistical relationships between aggregated data, and often based on analogies between natural and social phenomena—in the manner of the gravity model studied herein—they now increasingly seem to favor much more *disaggregate* approaches. These pay close attention to the elementary entities—often called *agents*—actions, and relations of which the system under investigation is made.[47] Usually called agent-based modeling (also abbreviated ABM), this new line of modeling seeks to identify and articulate the underlying micro-mechanisms that generate the macro-phenomenon under study and to explain how the combined actions performed by the elementary entities of the system, the agents, produce together various intended as well as unintended global outcomes. In seeking to grasp the inner structure and intimate functioning of the system under investigation, the modeler hopes therefore to both increase his or her understanding of it and to envision a much larger range of policies than before thanks to (hopefully) reliable "*what-if* inferences,"[48] which disaggregate techniques make possible or easier to implement.[49] Focused on specific transportation agents—be they individual travelers and households (this is the case with activity-based modeling) or individual (or group of) vehicles (dynamic traffic assignment)—and their action as well as the outcomes of their interactions, recent urban travel forecasting can be regarded as a member of the larger agent-based modeling family. Having the peculiarity of heavily amalgamating engineering science-based models with modeling blocks built on theoretical frameworks provided by the social sciences, both activity-based and DTA modeling can therefore enrich and widen our understanding of the universe of agent-based techniques and their use for both theoretical (explanation) and practical purposes (policies).[50]

Urban travel demand modeling can also legitimately be considered an illustration, and rather an early one, of a series of other general trends in modeling complex systems. Remember that although they are predicated on increasingly sophisticated theoretical frameworks drawing on a variety of engineering and social sciences, urban travel demand models have been crafted with the explicit view to having practical implications. As a result, unlike modeling systems developed within physics and other basic sciences, they have not primarily been designed to *test* theories but instead to *apply* them to practical settings in order to produce reliable predictions to serve public policy goals, the latter necessarily involving some, albeit not always explicitly stated, value judgments.[51] In this respect, urban traffic forecasting models share a series of commonalities with other recent problem-solving modeling systems resorting to approximations and characterized by high modularity, which makes it difficult to discern and localize the various sources of success and failures, and to attribute them to a specific set of particular components of the modeling suite. These features make UTDM a good candidate, besides climate models for example, for the further investigation of a series of contemporary issues regarding modeling and simulation. The various factors, be they intellectual or social, at work in the sanctioning and validation of a model, the sources of its credibility and the trust accorded to its outputs, the way value judgments operate in the building of the model and in its assessment, the sinuous path from diverse theoretical frameworks to applications, the strategies to which modelers resort in order to increase confidence in their results, the role of "fictions" (such as the proverbial rational trip-maker optimizing his or her travel behavior, posited by the utility-based disaggregate approach), and even the role of inconsistencies[52] in the functioning of a model, are some of these issues.

Another general tendency of which the trajectory, especially in the most recent periods, of urban traffic forecasting has also been indicative is the trend, observed in climate modeling too,[53] toward increasingly feeding models with *calculated* data rather than *actual* observations and other real-world elements. As a matter of fact, the *synthetic population generators* widely used by activity-based modeling systems are intellectual devices that produce, on the basis of empirical evidence and with the help of statistical techniques, artificial outputs, aptly named synthetic population, which serve as *surrogate* observation data in place of actual information. The same kind of substitution of computer-generated information for real-world datasets also applies to the most recent travel surveys. As has been seen in chapter

6, in order to obtain more reliable and accurate data, from the early 2000s on travel surveys have increasingly made use of global positioning system techniques instead of being grounded, as they were in the past, first in face-to-face encounters between the interviewer and the traveler in the latter's dwelling, and, later, in computer-assisted telephone interviews.[54] However, although GPS techniques, either using customized devices or smartphones, are very effective at recording time and positional characteristics of travel, they remain totally silent on trip purpose and travel mode choice, both of which had been important attributes in the traditional travel surveys as well as essential pieces of information for travel modeling purposes. Hence the necessity of GPS *data processing* able to convert raw spatial information, such as position in space and travel times, into artificial data that can be used by urban traffic forecasting models, including information on the mode used and the motives lying behind the decision to undertake a specific travel (both not directly observed by GPS technology).[55]

In addition, like their colleagues in climate modeling,[56] and even preceding them, modelers serving urban traffic forecasting have also put into execution, since the middle of the 1960s at least, the notion of *model intercomparison*, by regularly running different models under common specifications, while comparing their respective performances with the help of *standardized set* of data.[57] As a result, a series of specific sites, including the metropolitan areas of Washington, DC, as well as the road networks of the cities of Sioux Falls, Chicago, and Winnipeg—and the list is growing[58] —have emerged, and distinguished themselves as outdoor laboratories of sorts.

It is no exaggeration to say that in the absence of information technology, urban traffic forecasting, a data-hungry modeling field, would in all likelihood be a nonstarter. As a matter of fact, the interwar tabulating machines and their coding techniques, the mainframes of the 1950s and 1960s with their programming languages, the PC revolution of the 1980s and 1990s, and the more recent high-speed computations performed on new generations of parallel computers have all been part and parcel of American urban travel demand modeling. Many information technology (IT) major actors, such as IBM and CDC, but also General Electrics and the Los Alamos National Laboratory indeed helped to shape urban traffic forecasting over time. In this respect, UTDM and its development constitute a field of inquiry that should be of interest to IT historians.[59] But there is more. As both the TRANSIMS enterprise in the 1990s and the POLARIS project more recently suggest, urban traffic forecasting is also indicative of the more general trend toward the increasing role played by *computing*—hardware, but also software and all elements of computational modeling as well—in both the physical sciences[60] as well as engineering and other more practice-oriented disciplines.[61] And as the same undertakings clearly show, this rise to prominence of computational techniques is likely to alter not only the *intellectual* components of the field but its social and professional dimensions as well. As we have seen through the examples of TRANSIMS and POLARIS, computing techniques specialists, capable of conceiving and putting to use generic tools geared to responding to a great variety of problems, came to engage in a dialogue, interact and collaborate, but also compete with people whose expertise in contrast is based upon years of acquaintance with the specific phenomenon, system, or problem subjected to computation.

As a number of illustrations contained in this book amply suggest, urban traffic forecasters have proved very keen on images, and repeatedly made extensive use of many sorts of visual representations, be they displayed on printed paper or, more and more frequently as time went by, on a computer screen.[62] These visual displays are addressed in the first place to the members of the UTDM social world in order to help them in analyzing and exploring the massive datasets produced in the various stages of the modeling process. However, since models are part of the decision-making, they must also be understandable by or at least appeal to politicians and other decision-makers as well. It is not by chance that the more recent visual displays, marked by interactive user interfaces and an increasing realism based on 3D and animation techniques, have also allowed travel forecasters to plot and animate their results in ways that make them more salient, understandable, informative, and appealing, but also luring,[63] to nonexperts.

## UTDM and Contemporary Historical Issues

A specter has been hovering over the American historiography of science and technology for several decades: the Cold War.[64] Undoubtedly, the post–World War II contest between the United States and the now defunct Soviet Union as well as the aftermath of this rivalry were successfully smuggled into urban travel demand modeling in several ways. To begin with, urban travel forecasting would draw upon several disciplinary bodies of knowledge that benefitted from the Cold War competition, especially operations research to name just one. It is highly significant that one of the most utilized procedures within UTDM, the "Frank-Wolfe algorithm" used for the traffic assignment step, was the outcome of a research contract between Princeton University and the Office of Naval Research, one of the largest funders of science and technology research on behalf of the U.S. Department of Defense.[65] And as the TRANSIMS project as well as the more recent POLARIS undertaking also amply testify to, a number of federal agencies heavily involved in nuclear weapons design, such as the Los Alamos and Argonne Laboratories, actively sought nonmilitary uses for their supercomputer facilities and modeling expertise in the post-Cold War age. Yet, although it reaped benefits from spin-offs from military research and development, and occasionally one can come across U.S. Army-connected institutions, including the RAND Corporation, all in all, UTDM proved to be a rather *pure civilian* project. By bringing the gamut of nonmilitary actors to the fore, and by shifting the focus to the outcomes of their efforts regarding R&D and university training, this study therefore casts light on less explored areas in post–World War II American science and technology and, in doing so, it helps redress imbalances in historiographical attention.[66]

As a mostly civilian project, American urban traffic forecasting did not seem to suffer the secrecy that surrounded military-related research during the Cold War. However, secrecy did penetrate UTDM through market forces, which cast a certain veil over urban traffic forecasting knowledge and practices, and progressively made a significant part of them a calculated mix of openness, through publications, and deliberate secrecy, through proprietary software, for example. Though much less well publicized than other fields subject to market forces from the 1970s on, such as biotechnology,[67] UTDM also forms an early example of the now famous "triple helix of entrepreneurial science," in which forces from academia, the for-profit sector, and government have become intertwined in the pursuit of research and development agendas, and the transformation of their outcomes into market products and services.[68] In this respect, the establishment of Cambridge Systematics in 1972 by four MIT professors would be an early manifestation of a more general pattern within the UTDM social world, which persisted into the present: That of the research-driven transportation firm, be it a consultancy or a proprietary software company, which has distinguished itself as a specific, and effective, mechanism linking the university system to the market. Heavily populated by academics as well as Ph.D. and master's degree holders, cultivating strong links with academia on a systematic basis, these travel demand modeling-related firms have gradually supplanted government departments and agencies as the chief repository of urban forecasting traffic expertise in the U.S. In this respect, they represent a more general tendency affecting many spheres of expertise, including law, for contemporary states to resort to the consulting services of private firms.[69]

Making society into a legible entity in order to (re)shape it: According to the political scientist and anthropologist James C. Scott, this has been a central goal pursued by the modern state,[70] sometimes also dubbed the "Information State,"[71] since the systematic gathering of information thanks to which individuals are "caught in powerful force fields of public and private knowledge"[72] has proved heavily influential in this process. The various travel surveys featured in this book undoubtedly proceeded from the will of the American State to identify the travel patterns and to decipher the desires of American households regarding mobility, in order to develop infrastructures able to accommodate current and, especially, future traffic. They form, in this respect, an additional twentieth-century chapter, unexplored by the existing literature so far, of the already lengthy book relating the intimate relationships the United States has established with information from the colonial era to the present.[73]

This study can also enter into a dialogue with other research projects concerned with the building of the modern bureaucratic state in the United States, and the ways the latter has exercised its power over time. It has been widely held that American bureaucracy is hobbled by staff lacking in skills, competences, and good training, especially when compared to their European counterparts; it has also been argued that "the horizontal separation of powers and vertical federalism diminished the centralization of governing and institutional resources that are necessary for strong national administrative state capacity."[74] Following extensive research studies of specific administrations,[75] as well as comparative work,[76] this "weak state" view has come recently under severe attack.[77] An increasing number of scholars has produced a finely shaded account of the American State's form and development, especially by stretching the notion of the state beyond the federal administrative apparatus, and by stressing the capacity of the central government in extending its authority by enlisting the state and other local governments, and even private organizations that are not conventionally considered to be part of the state at all, to achieve public aims.[78] The history of urban traffic forecasting provides supplementary evidence against the thesis of the "weak" American State. The Bureau of Public Roads—and the same applies to several state and local transportation agencies, including the Chicago Area Transportation Study in the 1950s and 1960s—proved, in fact, capable of producing high-quality and autonomous bureaucracy. From the 1920s on, the BPR started hiring large numbers of people with up-to-date training in a great variety of disciplines. In addition, it created a centralized system of internal training and socialization for young recruits, while a significant number of its chiefs enjoyed, in the manner of Thomas H. MacDonald, long tenure and their opinions carried significant weight in federal policy. Let us also recall that, strong and effective during its lifespan, the BPR orchestrated a series of federal-state-local relationships as well, while its action in urban traffic forecasting (and beyond) greatly benefited from volunteer efforts on the part of private actors, such as the Automotive Safety Foundation.[79] And it is also worth noting that even after the disbanding of the BPR, the federal administration has displayed effective capacity concerning transportation modeling without a "strong" bureaucracy, through a series of partnerships with consulting firms, proprietary software companies, and university research centers.

Shifting from the public realm to the for-profit sector, the history of urban traffic forecasting brings to the foreground a particular type of industrial organization that has received little attention so far from business historians and other scholars studying organization structures and which are more or less focused on the large-scale companies:[80] the knowledge-intensive firm, which has forged close links with the university system, more akin to research-related institutions than to a traditional profit-making company. As a matter of fact, this kind of organization, which can be a rather small-scale autonomous firm or an organizational unit within a much larger corporation,[81] is characterized by weak bureaucratic structures, loose hierarchical controls, and greater levels of autonomy conferred on its highly trained, university-based knowledge workers, who display work practices and professional codes that are often governed by academic norms—think of the professional norm of publication, for example. Business historians could therefore explore in much more detail than was done in this book the workings—both structures and processes—of the firms featured in this study,[82] which, alongside their bonds with academic environments, have also developed a series of original devices and strategies—including patronizing user groups, organizing user forums, and even bestowing prizes on innovative customers—in a bid to retain their clients, but also to enlist their collective intelligence and experience in the improvement of the company's outputs.[83] Scholars interested in the various models of software licensing could also gain insights from the history of UTDM, which has intensely experienced, especially in more recent periods, both commercial and open-source software approaches.[84]

There are narratives with a preordained end, and others with a suspenseful one. The story related in this book falls into the latter category, as it ends in the present time, while urban traffic forecasting continues. Neither the historical actors nor the historian studying them know how the story ends. What does the future have in store for urban travel demand modeling? Modelers, like prophets, allow themselves to proffer predictions. Historians refrain from attempting endeavors such as these, preferring to "leave the future to the future,"[85] and to the people who are making it.

# Notes

## Introduction

1. Philip Roth, *American Pastoral* (London: Vintage Books, 1998), 40–41.

2. Robert Barkley Dial, "Probabilistic Assignment: A Multipath Traffic Assignment Model Which Obviates Path Enumeration" (Ph.D. diss., University of Washington, 1970), biographical note, 155.

3. Total enrollment figures in the institutions of higher education soared from 2,444,900 in 1949–1950 to 8,004,660 in 1969–1970; Thomas D. Snyder, ed., *120 Years of American Education: A Statistical Portrait* (Washington, DC: U.S. Department of Education, 1993), 75.

4. See chapter 2.

5. Information contained in "Dial_hudhistory_8.15.07.doc": this file was sent to me by the author during my research stay in the United States (2010–2012). Unless otherwise specified, information about Dial's life and career is drawn from this document, and, secondarily, from the short biographical note inserted by the author in his Ph.D. dissertation, "Probabilistic Assignment."

6. Robert Barkley Dial, "Automated Street Address to Grid Coordinate Conversion System" (master's thesis, University of Washington, 1965).

7. See chapter 2.

8. See, for example, Alan M. Voorhees & Associates Inc., *Urban Mass Transit Planning Project—Factors Influencing Transit Planning*, Technical Report #1 for Department of Housing and Urban Development (McLean, VA: n.p., October 1966); Alan M. Voorhees & Associates, *Urban Mass Transit Planning Project Computer Program Specifications*, Technical Report #2 for Department of Housing and Urban Development (McLean, VA: n.p., October 1966).

9. G. H. Johnson and R. B. Dial, *Transportation Planning Programs: Road System* (Downsview in Toronto: D.H.O., January 1969). On the career of TRIPS, see especially chapter 4.

10. See chapter 2.

11. Vergil G. Stover and J. T. Brudeseth, *Texas Traffic Assignment Practice* (College Station: Texas Transportation Institute, July 1967; revised October 1967).

12. "Dial_hudhistory_8.15.07.doc."

13. Dial, "Probabilistic Assignment."

14. Robert B. Dial, "A Probabilistic Multipath Traffic Assignment Model Which Obviates Path Enumeration," *Transportation Research* 5, no. 2 (June 1971): 83–111. On the reception of Dial's article up to 2016, see Nikunja M. Modak, José M. Merigó, Richard Weber, Felipe Manzor, and Juan de Dios Ortúzar, "Fifty Years of Transportation Research Journals: A Bibliometric Overview," *Transportation Research Part A* 120 (February 2019): 188–223 (on 195 and 197).

15. See chapter 2.

16. See chapter 3.

17. See especially Anonymous, *New Systems Requirements Analysis Program. A Procedure for Long Range Transportation (Sketch) Planning* (n.p.: UMTA, 1973).

18. Robert B. Dial, "Urban Transportation Planning System: Philosophy and Function," *Transportation Research Record*, no. 619 (1976): 43–48 (on 48). On three major contracts regarding UTPS (in the range of $1.1–$1.5 million) that were passed between UMTA and a number of consultancies from March 1972, see UMTA, *Innovation in Public Transportation, Fiscal Year 1975* (Washington, DC: n.p., n.d.), 44–46. More on UTPS in chapters 3 and 4.

19. UTPS would be widely accepted by the transportation planning community, with the number of users increasing from about fifty in July 1973 to more than 275 in July 1975. In addition, more than eight hundred people have attended the various training sessions that have been offered (UMTA, *Innovation in Public Transportation, Fiscal Year 1975*, 43).

20. See chapter 4.

21. Robert B. Dial, "A Path-Based User-Equilibrium Traffic Assignment Algorithm That Obviates Path Storage and Enumeration," *Transportation Research Part B* 40, no. 10 (December 2006): 917–936 (on 936). On Dial's latest work, see Ehsan Jafari, "Network Modeling and Design: A Distributed Problem Solving Approach" (Ph.D. diss., University of Texas at Austin, 2017). It is worth noting that the so-called Algorithm B devised by Dial has been incorporated into the proprietary computer suite developed by Caliper (email from Howard Slavin to the author dated January 22, 2021).

22. See, for example, letter from Michael Florian to Marvin Manheim, dated January 6, 1976; letter from Robert Dial to Moshe Ben-Akiva, dated April 4, 1977 (both in Massachusetts Institute of Technology, Libraries, Department of Distinctive

Collections, Marvin L. Manheim Papers (hereinafter Manheim's Personal Records), MC-0330, Box 8, Folder 16; Box 6, Folder 17, https://archivesspace.mit.edu/repositories/2/resources/831.

23. In September 2019, the (international) LinkedIn group "Transport & Traffic Modelling" numbered around twelve thousand members; https://www.linkedin.com/groups/2168238/profile, accessed September 16, 2019.

24. One can find many job advertisements on the Internet, under the rubric of "transportation modeler."

25. See, for instance, Zuoyue Wang, "Transnational Science during the Cold War: The Case of Chinese/American Scientists," *Isis* 101, no. 2 (June 2010): 367–377.

26. On the evolving forms of the American city, see, for example, Sam Bass Warner and Andrew Whittemore, *American Urban Form: A Representative History* (Cambridge, MA: MIT Press, 2012).

27. On the relationships between the intra-urban transportation system and the spatial form and organization of the American metropolis, see especially Peter O. Muller, "Transportation and Urban Form: Stages in the Spatial Evolution of the American Metropolis," in *The Geography of Urban Transportation*, 4th ed., ed. Genevieve Giuliano and Susan Hanson (New York: The Guilford Press, 2017), 57–85.

28. Practice has been a fashionable notion in the history, sociology, and even philosophy of science for several years. I find particularly useful MacIntyre's conception of practice as a "complex, collaborative, socially organized, goal-oriented, sustained activity" (Daniel J. Hicks and Thomas A. Stapleford, "The Virtues of Scientific Practice: MacIntyre, Virtue Ethics, and the Historiography of Science," *Isis* 107, no. 3 [September 2016]: 449–472).

29. Yet, a handful of urban transportation "mega-projects" have moved forward in recent years as well. See Alan Altshuler and David Luberoff, *Mega-Projects. The Changing Politics of Urban Public Investment* (Washington, DC: The Brookings Institution, 2003).

30. A number of references will be given in the conclusion of the book as well as in other appropriate places.

31. Louis Menand, *The Metaphysical Club: A Story of Ideas in America* (New York: Farrar, Straus and Giroux, 2001); Donald MacKenzie, *An Engine, Not a Camera: How Financial Models Shape Markets* (Cambridge, MA: MIT Press, 2006); Mieke Boon and Tarja Knuuttila, "Breaking Up with the Epochal Break: The Case of Engineering Sciences," in *Science Transformed? Debating Claims of an Epochal Break*, ed. Alfred Nordman, Hans Radder, and Gregor Schiemann (Pittsburgh: University of Pittsburgh, 2011), 66–79.

32. Several historians and sociologists of science and technology including Robert E. Kohler, David E. Rowe, Karin Knorr Cetina, (the late) Ann Johnson, and Paul N.

Edwards, to name some, have resorted to the organizing metaphor of "production." However, the details of the analytical framework built on the idea of "production" that I make use of in this book are mine.

33. For the idea of social world, see especially Howard S. Becker, *Art Worlds* (Berkeley: University of California Press, 1982); Howard S. Becker, *Telling About Society* (Chicago: University of Chicago Press, 2007). Joan H. Fujimura also made use of this notion in his book *Crafting Science: A Sociohistory of the Quest for the Genetics of Cancer* (Cambridge, MA: Harvard University Press, 1996), and mentions some other research studies building on it. It is also worth noting that the notion of social world as it is used in this book often parallels the notion of "instrumental community" coined by Cyrus C. M. Mody, *Instrumental Community: Probe Microscopy and the Path to Nanotechnology* (Cambridge, MA: MIT Press, 2011), chapter 1.

34. I have preferred to resort to the notion of social world instead of other possible candidates, including the concept of *field* put forward by Pierre Bourdieu for example, because of Becker's emphasis on production and cooperation (instead of competition), which is in line with my own conception of modeling as a production process. As a matter of fact, the "social world" is the (collective) actor that operates the "production process" with the help of "means of production" and "raw materials." To avoid any ambiguities, I would like to stress here that the terms "urban travel demand modeling" and "UTDM social world," though closely related to each other, denote different orders of reality. UTDM refers to a heterogeneous and evolving historical entity populated both by humans and nonhumans; a social world is composed instead of human actors only, with individuals and groups inhabiting organizations and institutions. Thus, the gravity model first articulated by Voorhees and enthusiastically embraced by many other traffic forecasters in the 1950s, or the codes written by Dial in the 1960s are parts of UTDM but neither belongs to the UTDM social world; they are instead specific historical outcomes of the latter. This doesn't mean that nonhuman entities do not matter; they do interact with people, do affect them (see also note 40), and can play a significant role in historical outcomes. This is why they are so pervasive throughout the book, in the shape of "raw materials" and "means of production," for example. But I am an old-fashioned historian who is convinced that nonhuman actors cannot talk without the help of human beings, including the historian, claiming to speak on their behalf.

35. Peter L. Galison, "Trading Zone: Coordinating Action and Belief," in *The Science Studies Reader*, ed. Mario Biagioli (New York: Routledge, 1999), 137–160.

36. On what precedes, see especially David Kaiser, Kenji Ito, and Karl Hall, "Feynman Diagrams in the USA, Japan, and the Soviet Union," *Social Studies of Science* 34, no. 6 (December 2004): 879–922; Timothy Lenoir, *Instituting Science: The Cultural Production of Scientific Disciplines* (Stanford: Stanford University Press, 1997).

37. Susan Leigh Star and James R. Griesemer, "Institutional Ecology, 'Translations' and Boundary Objects: Amateurs and Professionals in Berkeley's Museum of Vertebrate Zoology, 1907–39," *Social Studies of Science* 19, no. 3 (August 1989): 387–420.

38. John Parker, "On Being All Things to All People: Boundary Organizations and the Contemporary Research University," *Social Studies of Science* 42, no. 2 (April 2012): 262–289.

39. Simon Schaffer, Lissa Roberts, Kapil Raj, and James Delbourgo, eds., *The Brokered World: Go-Betweens and Global Intelligence, 1770–1820* (Sagamore Beach, MA: Watson Publishing International LLC, 2009).

40. On the general issue of "co-shaping" of people (scientists and engineers) and the outcomes of their action, see especially Pierre Teissier, Cyrus C. M. Mody, and Brigitte Van Tiggelen, eds., "From Bench to Brand and Back: The Co-shaping of Materials and Chemists in the Twentieth Century," special issue, *Cahiers François Viète*, Series III—no. 2 (2017).

41. Robert E. Kohler, "Moral Economy, Material Culture, and Community in Drosophila Genetics," in Biagioli, *The Science Studies Reader*, 243–257; Janet Atkinson-Grosjean and Cory Fairley, "Moral Economies in Science: From Ideal to Pragmatic," *Minerva* 47, no. 2 (June 2009): 147–170.

42. For example, should results reached by a for-profit entity thanks to government money be considered public, and thus freely available to anyone within the urban traffic forecasting community?

43. Lenoir, *Instituting Science*; David Kaiser, ed., *Pedagogy and the Practice of Science: Historical and Contemporary Perspectives* (Cambridge, MA: MIT Press, 2005).

44. James A. Evans, "Industry Collaboration, Scientific Sharing, and the Dissemination of Knowledge," *Social Studies of Science* 40, no. 5 (2010): 757–791.

45. Joseph Rouse, "Power/Knowledge," in *The Cambridge Companion to Foucault*, ed. Garry Gutting (Cambridge: Cambridge University Press, 1994), 72–114. See also Fritz Ringer, "The Intellectual Field, Intellectual History, and the Sociology of Knowledge," *Theory and Society* 19, no. 3 (June 1990): 269–294.

46. For a vigorous plea against novelty-centered approaches in the field of technological studies, see, in the first place, David Edgerton, "Innovation, Technology, or History: What Is the Historiography of Technology About?," *Technology and Culture* 51, no. 3 (July 2010): 680–697.

47. For these two terms, see Galison, "Trading Zone."

48. Be they successful practitioners in the private sector or well-known academics. For an approach focusing on the existence of various segments within a professional

world, see the now classic article by Rue Bucher and Anselm Strauss, "Profession in Process," *American Journal of Sociology* 66, no. 4 (January 1961): 325–334.

49. David Hackett Fischer, *Paul Revere's Ride* (Oxford: Oxford University Press, 1994), xv, 374.

50. John W. Servos, *Physical Chemistry from Ostwald to Pauling: The Making of a Science in America* (Princeton, NJ: Princeton University Press, 1990), 75.

51. "Commemorative Practices in Science: Historical Perspectives on the Politics of Collective Memory," annual issue, *Osiris* 14 (1999); Michelle L. Meade, Celia B. Harris, Penny Van Bergen, John Sutton, and Amanda J. Barnier, eds., *Collaborative Remembering: Theories, Research, and Applications* (Oxford: Oxford University Press, 2018).

52. Konstantinos Chatzis, "De l'Importation de Savoirs Américains à la Création d'une Expertise Nationale: La Modélisation des Déplacements Urbains en France, 1950–1975," in *De l'Histoire des Transports à l'Histoire de la Mobilité?*, ed. Mathieu Flonneau and Vincent Guigueno (Rennes: PUR, 2009), 159–169; Konstantinos Chatzis, "La Modélisation des Déplacements Urbains en France depuis les Années 1980, ou la Domination Progressive du Champ par le Secteur Privé," *Flux*, nos. 85–86 (July–December 2011): 22–40.

53. Konstantinos Chatzis, "La Modélisation des Déplacements Urbains aux États-Unis et en France, 1945 à nos Jours: Administrations, Universités et Bureaux d'Études," vol. III (HDR diss., Université Paris-Est, 2013). "Habilitation" is the highest academic qualification a scholar with a Ph.D. can achieve in France. In the humanities, the Habilitation process requires, among other things, a book project based on original research. A PDF version of my Habilitation thesis has been available since 2013 at http://temis.documentation.developpement-durable.gouv.fr/docs/Temis/0079/Temis-0079070/20902.pdf, accessed December 15, 2021.

54. Daniel Lee Kleinman and Steven P. Vallas, "Science, Capitalism, and the Rise of the 'Knowledge Worker': The Changing Structure of Knowledge Production in the United States," *Theory and Society* 30, no. 4 (August 2001): 451–492.

55. Be that as it may, there are few academic works dealing with specific aspects of the book's subject matter from a social and human sciences perspective. See, for example, P. Anthony Brinkman, "The Ethical Challenges and Professional Responses of Travel Demand Forecasters" (Ph.D. diss., University of California at Berkeley, 2003); Cheryl Deutsch, "The Origins of Metropolitan Transportation Planning in Travel Demand Forecasting, 1944–1962" (master's thesis, University of California at Los Angeles, 2013). The chair for both works was Martin Wachs, an academic heavily involved in urban transportation planning (Martin Wachs, "Becoming a Reflective Planning Educator," *Journal of the American Planning Association* 82, no. 4 [Autumn 2016]: 363–370). See also Harry T. Dimitriou, *Urban Transport Planning: A Developmental Approach* (New York: Routledge, 1992), part I.

56. John R. G. Turner, "The History of Science and the Working Scientist," in *Companion to the History of Modern Science*, ed. R. C. Olby, G. N. Cantor, J. R. R. Christie, and M. J. S. Hodge (London: Routledge, 1990), 23–31.

57. See, for example, Michael Patriksson, *The Traffic Assignment Problem—Models and Methods* (Utrecht: VSP Publishers, 1994; new edition, Mineola, NY: Dover Publications, 2015), a work that features more than a thousand references. Another character of the story told in this book, Peter R. Stopher, has recently proposed his version of UTDM's trajectory in Peter R. Stopher, "History and Theory of Urban Transport Planning," in *Handbook on Transport and Urban Planning in the Developed World*, ed. Michiel C. J. Bliemer, Corrine Mulley, and Claudine J. Moutou (Cheltenham, UK: Edward Elgar, 2016), 36–52. An older historical survey of the field was produced by Daniel Brand in the early 1970s: see Daniel Brand, "Travel Demand Forecasting: Some Foundations and a Review," in *Urban Travel Demand Forecasting*, ed. Daniel Brand and Marvin L. Manheim (Washington, DC: Highway Research Board, 1973), 239–282. Other relevant references will be provided later at appropriate places.

58. David Boyce and Huw Williams, *Forecasting Urban Travel: Past, Present and Future* (Cheltenham, UK: Edward Elgar Publishing, 2015).

59. David Edward Boyce, "The Effect of Direction and Length of Person Trips on Urban Travel Patterns: An Examination of an Assumption of Urban Trip Distribution Models" (Ph.D. diss., University of Pennsylvania, 1965). Before becoming involved in transportation modeling, Huw Williams had graduated in physics and received his doctorate in theoretical physics from Oxford University.

60. The authors had already published a series of articles and chapters in edited books in the period 2005–2012.

61. In order not to unnecessarily multiply the references to Boyce and Williams, I quote the latter only when I have been unable to locate the information drawn from their work in other primary/secondary sources.

62. Boyce and Williams, *Forecasting Urban Travel*, 12. On a series of differences between historical work done by specialists involved in the subject matter of their study and work performed by scholars specializing in history and social sciences, see Thomas Haigh, "The History of Information Technology," *Annual Review of Information Science and Technology* 45, no. 1 (2011): 431–487. It goes without saying that good, as well as bad, historical work is not the preserve of professionally trained historians.

63. To use Neil J. Smelser's terms, the analytical framework used for this study consists of "empty theoretical boxes" that must be filled with empirical evidence; see Neil J. Smelser, *Social Change in the Industrial Revolution: An Application of Theory to the British Cotton Industry* (Abingdon: Routledge, [1959] 2006), chapter 2. Their only function is to provide useful guidelines for a systematic description of the evolution

of urban travel demand modeling over time while indicating the empirical points of investigation. It is evident that, as an analytical tool, the framework allows neither empirical conclusions nor predictions to be drawn as to the essential characteristics and the trajectory over time of the object under investigation.

64. For the relevance of the "Kuhnian" picture of scientific development to technological knowledge and practice-related issues, see, for example, Rachel Laudan, ed., *The Nature of Technological Knowledge: Are Models of Scientific Change Relevant?* (Dordrecht, Holland: D. Reidel, 1984).

## Chapter 1

1. U.S. Department of Transportation/Federal Highway Administration, *Highway Statistics: Summary to 1985* (n.p.: FHWA, n.d.), 25, 39. As far as the story told in this book is concerned there is no need to delve into the reasons why Americans eventually proved to be so fond of private cars. For the reader who wants to familiarize themselves with the motorization of the American nation, the recent book by Christopher W. Wells, *Car Country: An Environmental History* (Seattle: University of Washington Press, 2012) is an excellent starting point. The work carried out by Michael L. Berger, *The Automobile in American History and Culture: A Reference Guide* (Westport: Greenwood Press, 2001) remains the best evaluative survey of relevant bibliographic material published before 2000.

2. See references in Wells, *Car Country*, 339–340.

3. Clay McShane, "The Origins and Globalization of Traffic Control Signals," *Journal of Urban History* 25, no. 3 (March 1999): 379–404, especially 380–381.

4. Quoted by Wells, *Car Country*, 87–88. Information on traffic congestion in various cities, including Washington, DC, Chicago, and Boston, can be found in McShane, "The Origins and Globalization of Traffic Control Signals"; Paul Barrett, *The Automobile and Urban Transit: The Formation of Public Policy in Chicago, 1900–1930* (Philadelphia: Temple University Press, 1983); Asha Elizabeth Weinstein, "The Congestion Evil: Perceptions of Traffic Congestion in Boston in the 1890s and 1920s" (Ph.D. diss., University of California at Berkeley, 2002).

5. Wells, *Car Country*, 88.

6. Executive Committee of the National Conference on Street and Highway Safety, *Guides to Traffic Safety . . . approved by the Fourth National Conference, May 23–25, 1934* (Washington, DC: n.p., May 1934), 7 and 9.

7. Executive Committee of the National Conference on Street and Highway Safety, *Guides to Traffic Safety*, foreword.

8. Kerry Segrave, *Parking Cars in America, 1910–1945: A History* (Jefferson, NC: McFarland & Company, 2012).

9. Robert M. Fogelson, *Downtown: Its Rise and Fall, 1880–1950* (New Haven: Yale University Press, 2001); Alison Isenberg, *Downtown America: A History of the Place and the People Who Made It* (Chicago: University of Chicago Press, 2004).

10. Harlan Paul Douglass, *The Suburban Trend* (New York: The Century Co., 1925); Kenneth T. Jackson, *Crabgrass Frontier: The Suburbanization of the United States* (New York: Oxford University Press, 1985).

11. This trend can easily be seen in Los Angeles, but it also played out across the country. A series of relevant studies are nicely summarized by Wells, *Car Country*.

12. Blaine A. Brownell, "Urban Planning, the Planning Profession, and the Motor Vehicle in Early Twentieth-Century America," in *Shaping an Urban World*, ed. Gordon E. Cherry (New York: St. Martin's Press, 1980), 59–77 (on 69–72); Jeffrey Brown, "From Traffic Regulation to Limited Ways: The Effort to Build a Science of Transportation Planning," *Journal of Planning History* 5, no. 1 (February 2006): 3–34 (on 11).

13. Peter D. Norton, *Fighting Traffic: The Dawn of the Motor Age in the American City* (Cambridge, MA: MIT Press, 2008), 153.

14. H. J. Loman, "Automobile Insurance," *Annals of the American Academy of Political and Social Science* 161 (May 1932): 85–90; Richard O. Bennett, "Effect of Accident and Cost Trends on Automobile Insurance Premiums," *Traffic Quarterly* 3, no. 4 (October 1949): 334–347.

15. For an application of this notion originating in STS studies to car-related issues, see Norton, *Fighting Traffic*.

16. For a contemporary view, see, among others, "The Automobile: Its Province and Its Problems," special issue, *Annals of the American Academy of Political and Social Science* 116 (November 1924); "Planning for City Traffic," special issue, *Annals of the American Academy of Political and Social Science* 133 (September 1927); the publications resulting from the various National Conferences on Street and Highway Safety, which took place in 1924, 1926, 1930, and 1934, respectively.

17. Logan Waller Page, "Salutatory," *Public Roads* 1, no. 1 (May 1918): 3.

18. Maureen Flanagan, *America Reformed: Progressives and Progressivisms, 1890s–1920s* (New York: Oxford University Press, 2006); Michael McGerr, *A Fierce Discontent: The Rise and Fall of the Progressive Movement in America, 1870–1920* (Oxford: Oxford University Press, 2005). In addition to scientific management methods, the Progressive Era also saw the emergence and consolidation of the City Manager Movement, designed to combat corruption in local government by promoting effective management, and the City Functional Movement (also known as City Practical Movement), aiming to improve the workings of the urban machinery as a whole. On the City Manager Movement, see Bradley Robert Rice, *Progressive Cities: The Commission*

*Government Movement in America, 1901–1920* (Austin: University of Texas Press, 1977). On the City Functional Movement, see, among others, the recent study by Daniel Baldwin Hess, "Transportation Beautiful: Did the City Beautiful Movement Improve Urban Transportation?," *Journal of Urban History* 32, no. 4 (May 2006): 511–545, which contains a rich bibliography.

19. Nelson P. Lewis, "Planning of Streets and Street Systems," in *Highway Engineers' Handbook*, ed. Arthur H. Blanchard (New York: John Wiley & Sons; London: Chapman & Hall, 1919), 363–418 (on 399).

20. See, for example, the two recent survey articles: Ann Johnson, "Engineering"; Sarah K. A. Pfatteicher, "Ethics and Professionalism in Engineering," both in *The Oxford Encyclopedia of the History of American Science, Medicine, and Technology*, vol. 1, ed. Hugh Richard Slotten (New York: Oxford University Press, 2014), 302–318 and 343–354, respectively.

21. Information on these interwar traffic surveys can be found in The Erskine Bureau for Street Traffic Research, *Street Traffic Bibliography. A Selected and Annotated Bibliography of the Literature of Street Traffic Control and Related Subjects 1920–1933* (Cambridge, MA: n.p., 1933); Robert Emmanuel Barkley, "An Annotated Bibliography on Traffic Volume Studies and Origin-Destination Surveys" (master's thesis, Purdue University, 1950); Bureau of Public Roads, *A Bibliography of Highway Planning Reports* (Washington, DC: U.S. Government Printing Office, 1950).

22. Both through individual corporations, such as the Metropolitan Life Insurance Company (1868), and its various collective bodies, especially the Accident Prevention Department of the Association of Casualty and Surety Companies.

23. Donald A. Krueckeberg, ed., *The American Planner: Biographies and Recollections* (New Brunswick, NJ: Center for Urban Policy Research, 1994); William Fulton, "Portrait of a Consultant in Hard Times: John Nolen, 1928–1937," *Journal of Planning History* 16, no. 1 (February 2017): 24–49.

24. Joseph Heathcott, "The Whole City Is Our Laboratory: Harland Bartholomew and the Production of Urban Knowledge," *Journal of Planning History* 4, no. 4 (November 2005): 322–355.

25. The first name of the firm was Kelker, De Leuw & Company; see Institute of Transportation Engineers, *Pioneers of Transportation* (Washington, DC: Institute of Transportation Engineers, 2011), 20.

26. "Brennan Heads Water Bureau; Marsh Retained," *Pittsburgh Press*, December 31, 1924, 1–2 (on 2).

27. Burton W. Marsh, "Traffic Engineering—Whence and Whither," *Traffic Quarterly* 25, no. 4 (October 1971): 505–520 (especially 509–510).

28. "Marsh Considers Philadelphia Post," *Pittsburgh Press*, March 30, 1930, 1. A short note on Marsh can be found in Institute of Transportation Engineers, *Pioneers of Transportation*, 39.

29. McShane, "The Origins and Globalization of Traffic Control Signals," 393.

30. Henry K. Evans, ed., *Traffic Engineering Handbook* (New Haven: Institute of Traffic Engineers, 1950), xiii.

31. The BPR would change its name as well as the department under the jurisdiction of which it operated several times during its institutional life, which came to an end in 1970 (see chapter 3), https://www.archives.gov/research/guide-fed-records/groups/030.html, accessed January 14, 2021. For the sake of simplicity, this federal agency will be designated throughout the entire book by the single term Bureau of Public Roads. On BPR and its history, see Bruce E. Seely, *Building the American Highway System: Engineers as Policy Makers* (Philadelphia: Temple University Press, 1987). Other references will be given at appropriate places.

32. Federal Highway Administration, *America's Highways, 1776–1976: A History of the Federal-Aid Program* (Washington, DC: U.S. Government Printing Office, 1977); Kimberley S. Johnson, *Governing the American State: Congress and the New Federalism, 1877–1929* (Princeton, NJ: Princeton University Press, 2007), chapter 5.

33. Federal Highway Administration, *America's Highways*, 214.

34. There are several monographs dealing with the history of individual state highway departments, including North and South Carolina, Texas, Kansas, Michigan, Alabama, Wisconsin, and Arkansas. See https://www.fhwa.dot.gov/infrastructure/bibstate.cfm, accessed January 14, 2021.

35. Now the American Association of State Highway and Transportation Officials or AASHTO.

36. American Association of State Highway Officials, *A Story of the Beginning, Purposes, Growth, Activities and Achievements of AASHO, 1914–1964* (Washington, DC: AASHO, 1965); Chris Becker, *AASHTO 1914–2014: A Century of Achievement for a Better Tomorrow* (Washington, DC: AASHTO, 2014).

37. See especially Frederick W. Cron, "Highway Design for Motor Vehicles—A Historical Review," eight parts, *Public Roads* 38, no. 3 (December 1974): 85–95; 38, no. 4 (March 1975): 163–174; 39, no. 2 (September 1975): 68–79; 39, no. 3 (December 1975): 96–108; 39, no. 4 (March 1976): 163–171; 40, no. 1 (June 1976): 9–18; 40, no. 2 (September 1976): 78–86; 40, no. 3 (December 1976): 93–100; Bruce E. Seely, "The Scientific Mystique in Engineering: Highway Research at the Bureau of Public Roads, 1918–1940," *Technology and Culture* 25, no. 4 (October 1984): 798–831; Bruce E. Seely, "Inventing the American Road: Innovations Shaping the American Freeway,"

in *From Rail to Road and Back Again? A Century of Transport Competition and Interdependency*, ed. Ralf Roth and Colin Divall (Burlington: Ashgate, 2015), 229–247.

38. Page, "Salutatory," 3.

39. Harry H. Harrison, "Why Do States Need Traffic Engineers and How Are Their Departments Organized and Administered," in *1937 ITE Proceedings*, ed. Institute of Traffic Engineers (New York: Office of the Secretary, n.d.), 4–7.

40. My reckoning, based on a list of names (with related information) in Maxwell Halsey, *Traffic Accidents and Congestion* (New York: John Wiley & Sons; London: Chapman & Hall, 1941), 362–390.

41. For a recent overview, see Roger L. Geiger, *The History of American Higher Education: Learning and Culture from the Founding to World War II* (Princeton, NJ: Princeton University Press, 2015).

42. Roger L. Morrison, "Traffic Engineering Training," in *1941 ITE Proceedings*, ed. Institute of Traffic Engineers (New York: Office of the Secretary, n.d.), 161–164 (on 161–162); Halsey, *Traffic Accidents and Congestion*, 107–110.

43. Currently known as the Eno Transportation Foundation. See John A. Montgomery, *Eno—the Man and the Foundation: A Chronicle of Transportation* (Westport: Eno Foundation for Transportation, 1988).

44. On the birth of the Bureau for Street Traffic Research, see, among others, Norton, *Fighting Traffic*, 163–166.

45. Miller McClintock, "The Street Traffic Problem" (Ph.D. diss., Harvard University, 1924). McClintock's thesis, produced as part of the university's Public Administration doctoral program, was published as a book in 1925: Miller McClintock, *Street Traffic Control* (New York: McGraw-Hill Book Company, 1925). On McClintock's career, see especially Norton, *Fighting Traffic*, 163–169, 202–205, 234–235.

46. Maxwell Halsey, *Training Traffic Engineers, Origins and Functions of the Bureau for Street Traffic Research* (n.p.: Yale University, n.d.; reprinted from *Yale Scientific Management*, Winter Issue, 1939); Bureau for Street Traffic Research, Yale University, "Graduate Training Program Outline of Courses—1939–1940," typescript (New Haven: Yale University, 1939). Both documents are held at Frances Loeb Library at Harvard Graduate School of Design.

47. Halsey, *Traffic Accidents and Congestion*; Theodore M. Matson, Wilbur S. Smith, and Frederick W. Hurd, *Traffic Engineering* (New York: McGraw-Hill Book Company, 1955).

48. Woodrow W. Rankin, *Bureau of Highway Traffic: History, 1925–1982* (Cheshire, CT: Bureau of Highway Traffic Alumni Association, 1997). The Records of Bureau of Highway Traffic for the entire period 1926–1979 are now held at the

Special Collections Library of Pennsylvania State University, Collection Number 442, https://aspace.libraries.psu.edu/repositories/3/resources/2384, accessed October 7, 2022. Many Bureau of Highway Traffic-related documents can also be found in the Frances Loeb Library at Harvard Graduate School of Design.

49. On MacDonald, see especially Earl Swift, *The Big Roads: The Untold Stories of the Engineers, Visionaries, and Trailblazers Who Created the American Superhighways* (New York: Houghton Mifflin Harcourt Publishing Company, 2011); Tom Lewis, *Divided Highways: Building the Interstate Highways, Transforming American Life* (Ithaca, NY: Cornell University Press, 2013).

50. The organization was born as the Advisory Board on Highway Research in 1920 and renamed the Highway Research Board in 1924–1925. Two books relate its history: Highway Research Board, *Ideas & Actions: A History of the Highway Research Board* (Washington, DC: Highway Research Board, ca. 1971); Sarah Jo Peterson, *The Transportation Research Board, 1920–2020: Everyone Interested Is Invited* (Washington, DC: National Academies Press, 2020).

51. Highway Research Board, *Ideas & Actions*, 29.

52. Anonymous, "New Committees for Highway Research," *Engineering News–Record* 88, no. 6 (February 9, 1922): 237.

53. A list of HRB's Committees and publications from 1920–1970 can be found in Highway Research Board, *Ideas & Actions*, 201 and 203–205. On highway research done in the United States from 1920 to 1940, see especially A. R. Rankin, ed., *Highway Research 1920–1940* (Washington, DC: Highway Research Board and American Association of State Highway Officials, 1940).

54. On professions and professionalization, see Rolf Torstendahl and Michael Burrage, eds., *The Formation of Professions: Knowledge, State and Strategy* (London: Sage Publications, 1990); Rolf Torstendahl and Michael Burrage, eds., *Professions in Theory and History. Rethinking the Study of the Professions* (London: Sage Publications, 1990). For a recent survey of the issue by a specialist, see Magali Sarfatti Larson, "Professions Today: Self-criticism and Reflections for the Future," *Sociologica, Problemas e Praticas* 88 (2018): 27–42.

55. Including *Public Roads*, the in-house research journal of the Bureau of Public Roads (set up in 1918); *American Highways*, AASHO's journal starting in 1922; and *Highway Research Board Proceedings*.

56. Such as *The American City*.

57. Including the *Annals of the American Academy of Political and Social Science*.

58. On the birth and the early years of ITE, see Anonymous, "The Early Years: Establishing an Identity," *ITE Journal* (August 1980): 10–15; Marsh, "Traffic Engineering—Whence and Whither," 512–515.

59. Wilbur S. Smith, "Report of Secretary-Treasurer, October 1, 1944—September 30, 1945," in *1945 ITE Proceedings*, ed. Institute of Traffic Engineers (New York: Office of the Secretary, n.d.), 146–149 (on 146).

60. W. L. Carson, "Findings of ITE Membership Survey," in *1959 ITE Proceedings*, ed. Institute of Traffic Engineers (Washington, DC: ITE, n.d.), 117–141 (on 117, 119). Useful information on the profession in the first post–World War II decades is provided by Fred W. Hurd, "The Future Role of Traffic Engineering," *Engineering Bulletin of Purdue University* 48, no. 4 (July 1964): 51–58.

61. Especially *Traffic Engineering*, the ancestor of the current journal of the institute, and the annual *ITE Proceedings*. Information about ITE publications can be found in Anonymous, "ITE Publications," *ITE Journal* (August 1980): 32–35; Anonymous, "Postwar to the 70's: Decades of Change," *ITE Journal* (August 1980): 17–26.

62. Harold F. Hammond and Leslie J. Sorenson, eds., *Traffic Engineering Handbook* (New York: Institute of Traffic Engineers and National Conservation Bureau, 1941); Evans, *Traffic Engineering Handbook*.

63. Hammond and Sorenson, *Traffic Engineering Handbook*, preface.

64. On competition and interaction between professions, see Andrew Abbott, "Linked Ecologies: States and Universities as Environments for Professions," *Sociological Theory* 23, no. 3 (September 2005): 245–274.

65. Halsey, *Traffic Accidents and Congestion.*

66. Miller McClintock, "The Traffic Survey," *Annals of the American Academy of Political and Social Science* 133 (September 1927): 8–18 (quotation on 8); emphasis mine. See also Lewis, "Planning of Streets and Street Systems"; Jacob Viner, "Urban Aspects of Highway Finance," Part I: *Public Roads* 6, no. 11 (January 1926): 233–235, 240, and Part II: *Public Roads* 6, no. 12 (February 1926): 260–268. Supplementary references can be found in Mark Foster, "City Planners and Urban Transportation: The American Response, 1900–1940," *Journal of Urban History* 5, no. 3 (May 1979): 365–396.

67. Wells, *Car Country*, 137.

68. For a recent survey of these measures and actions, see Wells, *Car Country*, 137–139. Supplementary information can be found in McShane, "The Origins and Globalization of Traffic Control Signals"; Norton, *Fighting Traffic*; for thorough case studies, see Barrett, *The Automobile and Urban Transit*; Weinstein, "The Congestion Evil"; Scott Bottles, *Los Angeles and the Automobile: The Making of the Modern City* (Berkeley: University of California Press, 1987). Two older publications are still very useful: Harold M. Lewis in consultation with Ernest P. Goodrich, *Highway Traffic Including a Program, by Nelson P. Lewis, for a Study of All Communication Facilities within the Region of New York and Its Environs; Regional Survey, Volume III* (New York: Regional Plan of New York and Its Environs, 1927), especially chapter 9; William

Phelps Eno, *The Story of Highway Traffic Control, 1899–1939* (Saugatuck, CT: The Eno Foundation for Highway Traffic Control, 1939).

69. On Major Traffic Street Plans, see especially Brown, "From Traffic Regulation to Limited Ways"; Jeffrey R. Brown, Eric A. Morris, and Brian D. Taylor, "Planning for Cars in Cities: Planners, Engineers, and Freeways in the 20th Century," *Journal of the American Planning Association* 75, no. 2 (Spring 2009): 161–177.

70. On techno-politics and the techno-political object, see, for example, Tony Bennett and Patrick Joyce, eds., *Material Powers: Cultural Studies, History, and the Material Turn* (New York: Routledge, 2010).

71. See, among others, Halsey, *Traffic Accidents and Congestion*, 24.

72. Among them are New York, New Jersey, Detroit, and especially Los Angeles with its Arroyo Seco Parkway, renamed the Pasadena Freeway.

73. Joseph F. C. DiMento and Cliff Ellis, *Changing Lanes: Visions and Histories of Urban Freeways* (Cambridge, MA: MIT Press, 2013), especially chapters 1, 2, and 3; Timothy Davis, "The Rise and Decline of the American Parkway," in *The World beyond the Windshield*, ed. Christof Mauch and Thomas Zeller (Athens: Ohio University Press, 2008), 35–58; Brown, Morris, and Taylor, "Planning for Cars in Cities"; Wells, *Car Country*, 139–142, 338–339.

74. McClintock, "The Traffic Survey," 8.

75. Ian Hacking, "Biopower and the Avalanche of Printed Numbers," *Humanities in Society* 5, nos. 3–4 (Summer/Fall 1982): 279–295.

76. George M. Smerk, *The Federal Role in Urban Mass Transportation* (Bloomington and Indianapolis: Indiana University Press, 1991), especially chapters 3 and 4; the author provides an extensive bibliography on 313–320; The Federal Transit Administration, *Historic Context Report for Transit Rail System Development Nationwide*, (Washington, DC: U.S. Department of Transportation/Federal Transit Administration, June 2017), especially for its bibliography.

77. For a first acquaintance with the subject, see Paul Barrett and Mark H. Rose, "Street Smarts: The Politics of Transportation Statistics in the American City, 1900–1990," *Journal of Urban History* 25, no. 3 (March 1999): 405–433 (407–413 in particular). Clay McShane, *Down the Asphalt Path: The Automobile and the American City* (New York: Columbia University Press, 1994) also gives several interesting pieces of information.

78. Henry W. Blake and Walter Jackson, *Electric Railway Transportation* (New York: McGraw-Hill Book Company, 1917), 45. On this traffic survey, see especially A. Merritt Taylor, "The Solution of a City's Transit Problem," *The Electric Journal* 11, no. 10 (October 1914): 514–542; Anonymous, "Traffic Count in Philadelphia," *Electric Railway Journal* 40, no. 17 (November 1912): 949–952.

79. On the use of mechanical means of calculation in traffic surveys in the 1910s, see R. H. Horton, "Methods of Observing and Analyzing Passenger Traffic: The Merits of Several Methods of Obtaining Data Are Pointed Out, Together with Helpful Advice on How to Use Traffic Study Data," *Electric Railway Journal* 54, no. 2 (July 1919): 75–78. On the various uses of punched cards and tabulating machines in the early twentieth century, see Lars Heide, *Punched-Card Systems and the Early Information Explosion, 1880–1945* (Baltimore: Johns Hopkins University Press, 2009).

80. Blake and Jackson, *Electric Railway Transportation*, 48–52 (on Boston); Barrett and Rose, "Street Smarts," 410–411, 429 (on Detroit, Chicago, and Pittsburg).

81. Anonymous, "A Traffic Census That Recorded Vehicle Movements in Business District of Chicago," *Engineering and Contracting* 49, no. 23 (June 5, 1918): 564–566; Sidney J. Williams and Peter J. Stupka (prepared under the direction of), "Engineering Manual for Traffic Surveys," typescript (Washington, DC: Federal Emergency Relief Administration, March 1934).

82. Winters Haydock, "How the Pittsburgh Traffic Count Was Made," *Engineering and Contracting* 58, no. 14 (October 1922): 325–326.

83. Harold M. Lewis, "The New York City Motor Traffic Problem," *Annals of the American Academy of Political and Social Science* 116 (November 1924): 214–223 (on 215).

84. For a list, see The Erskine Bureau for Street Traffic Research, *A Selected and Annotated Bibliography*, chapter 10; Barkley, "An Annotated Bibliography"; Mark S. Foster, "The Model-T, the Hard Sell, and Los Angeles's Urban Growth: The Decentralization of Los Angeles during the 1920s," *Pacific Historical Review* 44, no. 4 (1975): 459–484 (on 466); Fogelson, *Downtown*, chapter 6.

85. Harold M. Lewis, "Routing Through Traffic," *Annals of the American Academy of Political and Social Science* 133 (September 1927): 19–27 (on 21–22); emphasis mine.

86. Be that as it may, cordon counts have not disappeared from the traffic scene; they have remained valuable in determining enforcement and local control measures, evaluating the effects of minor improvements, and showing traffic trends.

87. By the early 1940s, there were three such "indirect methods": license plate; return postal cards; tag of car. For a brief presentation of each method, see Hammond and Sorenson, *Traffic Engineering Handbook*; and, especially, Barkley, "An Annotated Bibliography."

88. Michigan State Highway Department et al., *Street Traffic: City of Detroit, 1936–1937* (Lansing: Franklin De Kleine Company, 1937), chapter 4. A more sophisticated version of the license plate recording method was used on a large scale in Chicago on September 9, 1941. See Cook County Highway Department, *A Traffic Survey of the Chicago District* (Chicago: The Inland Press, April 1943).

89. In the 1920s, several urban origin and destination surveys involving active interaction between the "observer" and the driver, through roadside interviews for example, were conducted to determine new bridge locations or the feasibility of bypass routes. One such O&D study was carried out by the Traffic Division of the New York Police Department in June 1925 (Lewis in consultation with Goodrich, *Highway Traffic*, appendix E, 155–158. Appendix E of the report also expounds a number of other O&D studies of New York City's traffic in the second half of the 1920s). An extensive urban roadside interview type of O&D survey was also made in Boston as early as in 1927–1928 (Barkley, "An Annotated Bibliography," 46–47, 259–260).

90. For these first surveys, see Herbert S. Fairbank, "Highway Transport Research," typescript, ca. 1920, https://planeandtrainwrecks.com/Search, accessed January 17, 2021. For a history of the traffic census in the United States before 1920, see A. N. Johnson, "The Traffic Census" (with discussion), *Public Roads* 3, no. 27 (July 1920): 16–28.

91. Anonymous, "New Traffic Survey Starts in Connecticut: State and Federal Government Join in Most Comprehensive Program Ever Attempted," *Engineering News-Record* 89, no. 14 (October 1922): 579.

92. On McKay, see Robert J. Lampman, *Economists at Wisconsin: 1892–1992* (Madison: Department of Economics/University of Wisconsin, 1992), 305–307.

93. Bureau of Public Roads, *A Bibliography of Highway Planning Reports*; Barkley, "An Annotated Bibliography."

94. Such as Oswald Milton Elvehjem (1894–1930), who received a Ph.D. in economics in 1925; Leroy Elden Peabody (1894–1956), who completed a master's degree in mathematics in 1916; and Charles W. Vickery (1906–1982), who defended his doctoral thesis in mathematics in 1932, to name a few. More information is provided in Chatzis, "La Modélisation des Déplacements Urbains," vol. III, 59–61.

95. Anonymous, "A Study of Highway Traffic in the Cleveland Regional Area," *Public Roads* 9, no. 7 (September 1928): 129–138, 152.

96. See the now classic volume by Nancy E. Rose, *Put to Work: The WPA and Public Employment in the Great Depression*, 2nd ed. (New York: Monthly Review Press, 2009), see also the more recent work by Sandra Opdycke, *The WPA: Creating Jobs and Hope in the Great Depression* (New York: Routledge, 2016). For the views expressed by the BPR on the issue of employment at that time, see Herbert S. Fairbank, "Roads to Prosperity," typescript, 1931, https://planeandtrainwrecks.com/Search, accessed January 17, 2021.

97. Herbert S. Fairbank, "Rural Highway Planning Survey and Highway Transportation Study," typescript, 1935, 10, https://planeandtrainwrecks.com/Search, accessed January 17, 2021.

98. It should be recalled here that in order to resolve the long-standing problem of how to pay for the improvements to the nation's highway network, the states resorted to gasoline taxes levied on highway users. Oregon first introduced such a tax in 1919, and by 1929 all forty-eight states had followed suit. The federal government began to levy its own gas tax in 1932 (see Wells, *Car Country*, chapter 5).

99. Robert H. Paddock and Roe P. Rodgers (reported by), "Preliminary Results of Road-use Studies," *Public Roads* 20, no. 3 (May 1939) 45–54, 62–63 (quotation on 45); emphasis mine.

100. Herbert S. Fairbank, "Utilization of the Planning Survey," typescript, 1938, 11, https://planeandtrainwrecks.com/Search, accessed January 17, 2021; emphasis mine. On the debates about the "Highway Finance Problem," see also Herbert S. Fairbank, "Planning Our Highway System," typescript, ca. 1935–1936, https://planeandtrainwrecks.com/Search, accessed January 17, 2021; Fairbank, "Rural Highway Planning Survey"; and Viner, "Urban Aspects of Highway Finance."

101. The questions could be asked either through roadside interviews or by means of franked cards distributed to motorists, who filled them in and returned them afterward to the BPR. For the last method, see especially L. E. Peabody (reported by), "The Western States Traffic Survey," *Public Roads* 13, no. 1 (March 1932): 1–18 (especially 3).

102. T. M. C. Martin and Homer L. Baker, "The Application of Road-use Survey Methods in Traffic Origin and Destination Analysis," *Public Roads* 22, no. 3 (May 1941): 59–62, 66 (quotation on 59).

103. Robert H. Paddock, "Some Characteristics of Motor-Vehicle Travel," *Public Roads* 23, no. 2 (April 1942): 17–29 (on 18).

104. Barkley, "An Annotated Bibliography," 212.

105. Paddock and Rodgers, "Preliminary Results of Road-use Studies," 45.

106. For thorough presentations and analyses of the data gathered, see Herbert S. Fairbank, "Newly Discovered Facts About Our Highway System," typescript, 1938, https://planeandtrainwrecks.com/Search, accessed January 17, 2021; Fairbank; "Utilization of the Planning Survey"; Bureau of Public Roads, *Toll Roads and Free Roads* (Washington, DC: U.S. Government Printing Office, 1939), 4–13, and passim; Martin and Baker, "The Application of Road-use Survey Methods"; Paddock, "Some Characteristics of Motor-Vehicle Travel."

107. The quotations are drawn from the recollections of "Ted Holmes on Thomas MacDonald and Herbert Fairbank," https://www.fhwa.dot.gov/infrastructure/holmes.cfm, January 17, 2021. On Fairbank, see also David C. Oliver, "In the Shadow of a Giant: H. S. Fairbank and Development of the Highway Planning Process," *Transportation Quarterly* 45, no. 4 (October 1991): 611–630.

108. Herbert S. Fairbank, "Post-War Planning for Highway Construction," typescript, 1943, 9–10, https://planeandtrainwrecks.com/Search, accessed January 17, 2021; emphasis mine.

109. Herbert S. Fairbank, "Highway Needs of Expanding Urban Areas," typescript, 1948, 4, https://planeandtrainwrecks.com/Search, accessed January 17, 2021; emphasis mine.

110. Peter Wagner, *A Sociology of Modernity: Liberty and Discipline* (New York: Routledge, 1994), 28–29.

111. Some twenty years later, Francis C. Turner, appointed director of the BPR by President Johnson, would express the same view. See Highway Research Board, *Highway and Urban Transportation in the 1970s and 1980s* (Washington, DC: Highway Research Board, 1971), 4. Neither Fairbank nor Turner was alone in believing that "the free individual has been justified as his own master; the state as his servant" (Dwight D. Eisenhower, cited by Cotten Seiler, *Republic of Drivers. A Cultural History of Automobility in America* (Chicago: University of Chicago Press, 2008), 69. Regarding themselves as civil servants putting their expertise at the service of their fellow countrymen, they expressed a widely shared conception of the role of engineers and planners in twentieth-century America (see the relevant references provided by Konstantinos Chatzis, "Capter et Cartographier les Pratiques et Désirs des Américains en Matière de Mobilité Urbaine (de l'Entre-Deux-Guerres aux Années 1960)," *Flux*, nos. 111–112 (January–June 2018): 57–79 (on 61). BPR staffers seemed to combine here a "Jeffersonian" vision of a Republic of free and self-sufficient individuals with a "Hamiltonian" vision of a strong federal government actively engaged in the accomplishment of large-scale public works projects in order to meet individual needs. On the historical roots of these conceptions, see, for example, the recent book by Colin Woodard, *American Character: A History of the Epic Struggle between Individual Liberty and the Common Good* (New York: Penguin Books, 2017), especially chapters 3 and 4.

112. I follow here Theodore M. Porter, "Quantification and the Accounting Ideal in Science," in Biagioli, *The Science Studies Reader*, 394–406. Traffic engineers would emphasize themselves the importance of numbers as a bulwark against what they considered were arbitrary decisions made by politicians (Chatzis, "Capter et Cartographier les Pratiques," 69–70).

113. Anonymous, "The Connecticut Transportation Survey: Digest of the Report of a Survey of Transportation on the State Highway System of Connecticut," *Public Roads* 7, no. 6 (August 1926): 109–118 (on 118).

114. J. G. McKay and O. M. Elvehjem (reported by), "The Main Highway Transportation Survey. A Preliminary Report," *Public Roads* 6, no. 3 (May 1925): 45–58 and 67–68.

115. McKay and Elvehjem, "The Main Highway Transportation Survey," 55.

116. Herbert S. Fairbank, "Use of Forecasts in Highway Management," typescript, 1940, https://planeandtrainwrecks.com/Search, accessed January 17, 2021.

117. J. Gordon McKay (reported by), "Digest of Report of Ohio Highway Transportation Survey," *Public Roads* 8, no. 4 (June 1927): 61–71, 88; J. Gordon McKay (reported by), "Digest of Vermont Highway Transportation Survey," *Public Roads* 8, no. 10 (December 1927): 215–224, 229; L. E. Peabody, "Highway Traffic Analysis Methods and Results," *Public Roads* 10, no. 1 (March 1929): 1–10.

118. Thus, traffic observed in New Hampshire in 1931 proved to be "greater by 15.3 per cent than the amount forecast in 1926." See L. E. Peabody (reported by), "The New Hampshire Traffic Survey," *Public Roads* 13, no. 8 (October 1932): 131–136 (on 131).

119. Peabody, "The New Hampshire Traffic Survey," 132. An explanation for this better correlation is given by a famous interwar traffic engineer Nathan Cherniack, "Measuring the Potential Traffic of a Proposed Vehicular Crossing," *Transactions of the American Society of Civil Engineers* 106 (1941): 520–576.

120. Anonymous, "The Connecticut Transportation Survey," 118. Concerning railway traffic prediction in the early 1910s, see, for example, George Rathjens, "Method of Estimating the Probable Volume of Railway Traffic," *Association of Engineering Societies* 46, no. 3 (March 1911): 218–233.

121. On Goodrich, see Institute of Transportation Engineers, *Pioneers of Transportation*, 13.

122. In fairness to BPR traffic specialists, it should be noted here that although they did favor a forecasting approach based on statistical correlations, in their analyses of traffic data they were able to identify a series of (causal) factors accounting for the traffic patterns they were about to observe. Thus, comfort, rapidity, and even new scenery were regarded as factors affecting drivers' route choice (see Peabody, "The New Hampshire Traffic Survey," 131). Some of these factors would play an important role in the building of more behavioral-centered modeling approaches after World War II (see chapters 3 and 4).

123. David A. Johnson, *Planning the Great Metropolis: The 1929 Regional Plan of New York and Its Environs* (New York: Routledge, 2003).

124. Lewis in consultation with Goodrich, *Highway Traffic*, 104; emphasis mine. For a full presentation of the model see Lewis in consultation with Goodrich, *Highway Traffic*, appendix D, 150–151. Goodrich mused much on urban traffic behavior and forecasting in the 1920s. See, for example, Ernest P. Goodrich, "The Design of the Street System in Relation to Vehicular Traffic," *The American City* 27, no. 4 (October 1922): 311; Ernest P. Goodrich, "Zoning and Its Relation to Traffic Congestion," *Annals of the American Academy of Political and Social Science* 133 (September 1927): 222–233.

125. It is worth noting that Goodrich and Whitten would work together. See, for example, Ernest P. Goodrich and Robert H. Whitten, *City Plan of the City of White Plains, New York* (White Plains, NY: Long's White Plains Print Shop, 1928).

126. The metropolitan area was divided into fifty zones and interviews were taken at 178 stations. See Barkley, "An Annotated Bibliography," 46–47, 259–260.

127. Robert Herbert Whitten, "The Traffic Analysis and Forecast and Its Relation to Thoroughfare Planning" (with discussion), *Planning Problems of Town, City and Region*, the 21st National Conference on City Planning, 1929 (Philadelphia: W. F. Fell Co., 1929), 179–196.

128. See, for example, "the traffic between any two areas of registration is directly proportional to the product of the registration and inversely proportional to the square of the highway distance between the two areas of registration"; George Hartley, "Traffic Surveys for Toll Bridges," *Roads and Streets* 79, no. 6 (June 1936): 37–41 (on 38).

129. On the term and the idea of "preconditions," see Ian Hacking, *The Emergence of Probability: A Philosophical Study of Early Ideas about Probability, Induction, and Statistical Inference* (New York: Cambridge University Press, [1975] 2006).

## Chapter 2

1. David M. Kennedy, *Freedom from Fear: The American People in Depression and War, 1929–1945* (New York: University of Oxford Press, 1999).

2. U.S. Bureau of the Census, *Historical Statistics of the United States, Colonial Times to 1970, Bicentennial Edition, Part 2* (Washington, DC: U.S. Government Printing Office, 1975), 716.

3. James T. Patterson, *Grand Expectations: The United States, 1945–1974* (New York: University of Oxford Press, 1996); Price Fishback et al., *Government and the American Economy: A New History* (Chicago: University of Chicago Press, 2007).

4. See, especially, Special Committee on Post-War Economic Policy and Planning, *The Role of the Federal Government in Highway Development. An Analysis of Needs and Proposals for Post-War Action* (Washington, DC: U.S. Government Printing Office, 1944).

5. See chapter 1.

6. National Interregional Highway Committee, *Interregional Highways* (Washington, DC: U.S. Government Printing Office, 1944), 52.

7. Bureau of Public Roads, *Guide for Forecasting Traffic on Interstate System* (Washington, DC: U.S. Government Printing Office, October 1956).

8. On the 1956 Federal-Aid Highway Act and the resulting Interstate System, see, among others, the following publications: Federal Highway Administration, *America's Highways*; Mark H. Rose and Raymond A. Mohl, *Interstate: Highway Politics and Policy since 1939* (Knoxville: University of Tennessee Press, 2012); DiMento and Ellis, *Changing Lanes*; Lewis, *Divided Highways*.

9. Federal Highway Administration, *America's Highways*, 276–277.

10. For a detailed history of the Home Interview O&D traffic survey from the early 1940s to the 1960s as well as the relevant bibliography, see Chatzis, "Capter et Cartographier les Pratiques." What follows is a summary of this article. I have also added a few references that are not featured in that study.

11. Jean M. Converse, *Survey Research in the United States: Roots and Emergence, 1890–1960* (New York: Routledge, 2017).

12. Diane L. Oswald, *Fire Insurance Maps: Their History and Applications* (College Station, TX: Lacewing Press, 1997).

13. Self-coding was supposed to facilitate coding operations and to reduce interview time. On the design of forms and their history, see, for example, Paul Stiff, Paul Dobraszczyk, and Mike Esbester, "Designing and Gathering Information: Perspectives on Nineteenth-Century Forms," in *Information History in the Modern World. Histories of the Information Age*, ed. Toni Weller (Basingstoke: Palgrave Macmillan, 2011), 57–88; Alison Black, Paul Luna, Ole Lund, and Sue Walker, eds., *Information Design: Research and Practice* (New York: Routledge, 2017), especially part 1: "Historical Perspectives."

14. Work, business, medical/dental, school, social recreation, change travel mode, eat meal, shopping, and serve passenger. The following simple example can help us to understand how HI works. Let us suppose that a spouse drives her husband to work and then drives back home. The interviewer should report *three* trips: *two* for the wife (with the purpose "serve passenger") and *one* for the husband (with the purpose "work").

15. Nonmotorized modes were first included in urban travel demand models in the United States at the turn of the 1980s only (Patrick A. Singleton, Joseph C. Totten, Jaime P. Orrego-Oñate, Robert J. Schneider, and Kelly J. Clifton, "Making Strides: State of the Practice of Pedestrian Forecasting in Regional Travel Models," *Transportation Research Record*, no. 2672 [2018]: 58–68). On the current state of practice regarding bicycling and walking trip modeling, see RSG and The RAND Corporation, *Evaluation of Walk and Bicycle Demand Modeling Practice* (n.p.: AASHTO, 2019).

16. On this "division of labor," see, among others, the recent study by Katherine J. Parkin, *Women at the Wheel: A Century of Buying, Driving, and Fixing Cars* (Philadelphia: University of Pennsylvania Press, 2017). People involved in urban travel demand modeling have recently explicitly recognized the relations between the

ways of collecting data (about) and modeling people's mobility practices by traffic experts with the *zeitgeist* of the era under investigation: See, especially, Alec T. Shuldiner and Paul W. Shuldiner, "The Measure of All Things: Reflections on Changing Conceptions of the Individual in Travel Demand Modeling," *Transportation* 40, no. 6 (November 2013): 1117–1131.

17. For a series of results obtained from accuracy tests, see Walter Y. Oi and Paul W. Shuldiner, *An Analysis of Urban Travel Demands* (n.p.: Northwestern University Press, 1962), 43. It is worth noting that in all the studies referred to in this book, the "predicted counts" fall considerably below the actual counts.

18. Lizabeth Cohen, *A Consumers' Republic: The Politics of Mass Consumption in Postwar America* (New York: Vintage Books, 2004).

19. On computers as a tool of management in the workplace, see, among others, David Lyon, *The Electronic Eye: The Rise of Surveillance Society* (Cambridge: Polity Press, 1994), especially chapter 7: "The Transparent Worker."

20. On the "competitors" of the Home Interview O&D traffic study, see, for example, Herbert Martell Edwards, "The Unidirectional Roadside Interview Method for Origin-Destination Surveys" (master's thesis, Purdue University, 1954); James Herbert Kell, "Comparison of Two Methods of Internal Origin-Destination Surveys: Home-Interview and Controlled Post-Card (Logansport, Indiana Traffic Survey)," *Engineering Bulletin of Purdue University* 37, no. 1 (January 1953); Donald E. Cleveland, ed., *Manual of Traffic Engineering Studies* (Washington, DC: Published by the Institute of Traffic Engineers, 1964), chapter 4.

21. The idea of "desire line" and of "desire line map" seem to have originated in Detroit in the early 1940s; see Lloyd B. Reid, "Expressways for Detroit," in *1943 ITE Proceedings*, ed. Institute of Traffic Engineers (New York: Office of the Secretary, n.d.), 88–91, especially figure no. 1, on 89. Reid spoke even of "desirable routes" (on 89).

22. According to the oldest, and more popular, method the desire lines were directly traced on the map from the center of an origin zone to the center of a destination zone. Where there were multiple trips occurring between them, the bundles of trips were represented by bands the width of which was proportional to the number of trips.

23. While deputy director at Tri-State Transportation Committee, J. Douglas Carroll Jr. (more on him later) explicitly used the term "consumer" when speaking of a highway user. See J. Douglas Carroll, "Interaction of Traffic and Land Use in Urban Areas—How These Apply in Long Range Planning," typescript, 1964, Cornell University Library, Division of Rare and Manuscript Collections, J. Douglas Carroll, Jr. Papers (hereinafter Carroll's Personal Records), 4178, Box 1, Folder 43, https://rmc.library.cornell.edu/EAD/htmldocs/RMM04178.html. On the idea of the driver "as an analogue of the sovereign consumer" in post–World War II America, see especially Seiler, *Republic of Drivers*.

24. For the four "components" of traffic expected to use the projected facility, see Earl Campbell, "Foreword," *Highway Research Board Bulletin*, no. 61 (1952): iii.

25. Campbell, "Foreword"; emphasis in original.

26. Like all students at the Yale Bureau of Highway Traffic, Campbell had written a research thesis, which he submitted in May 1946 and was entitled "Toll Bridge Influence on Highway Traffic Operation" (n.p.: ENO Foundation, 1947).

27. Highway Research Board, *Ideas & Actions*, 61, 88.

28. Circular letter from Earl Campbell to Traffic and Planning Engineers, dated April 19, 1949 (reproduced in Highway Research Board, *Route Selection and Traffic Assignment. A Compendium of Correspondence Relating to a Suggested Technique* [n.p.: Highway Research Board Correlation Service, 1950], ii).

29. Earl Campbell to H. G. Van Riper, dated November 12, 1948, in Highway Research Board, *Route Selection and Traffic Assignment*, 5.

30. See, for example, Roy E. Jorgensen, Roy W. Crum, Fred Burggraf, and W. N. Carey Jr., "Influence of Expressways in Diverting Traffic from Alternate Routes and in Generating New Traffic," *Highway Research Board Proceedings* 27 (1948): 322–330; R. M. Brown, "Expressway Route Selection and Vehicular Usage," *Highway Research Board Bulletin*, no. 16 (1948): 12–21. See also the (six) articles published in *Highway Research Board Bulletin*, no. 61 (1952).

31. For some of their comments, see also Deutsch, "Origins of Metropolitan Transportation Planning," 45–49.

32. James S. Burch to Earl Campbell, dated April 26, 1949, in Highway Research Board, *Route Selection and Traffic Assignment*, 25.

33. Campbell, "Foreword," in Highway Research Board, *Route Selection and Traffic Assignment*, i.

34. Highway Research Board, *Route Selection and Traffic Assignment*, 1–5, 6–11, 71–72, 73–79, respectively. See also Earl Campbell, "Theoretical Assignment of Diverted Traffic," *Traffic Engineering* 20, no. 2 (November 1949): 88–89.

35. Letter from E. H. Holmes to Earl Campbell, dated July 22, 1949, in Highway Research Board, *Route Selection and Traffic Assignment*, 52.

36. Darel L. Trueblood, "The Effect of Travel Time and Distance on Freeway Usage," *Public Roads* 26, no. 12 (February 1952): 241–250 (the study was also published in *Highway Research Board Bulletin*, no. 61 (1952): 18–37).

37. A good summary of these undertakings can be found in Robert E. Schmidt and M. Earl Campbell, *Highway Traffic Estimation* (Saugatuck, CT: The ENO Foundation for Highway Traffic Control, 1956), 131–133. See also the six studies published in

*Highway Research Board Bulletin*, no. 61 (1952) (including the study conducted on the Shirley Highway).

38. Earl M. Campbell, *Diversion of Traffic from City Streets to Expressways as a Basis for Traffic Assignment* (Washington, DC: Highway Research Correlation Service, 1951) (three case studies); Schmidt and Campbell, *Highway Traffic Estimation*, 129–133 (eight case studies).

39. E. Wilson Campbell and Robert S. McCargar, "Objective and Subjective Correlates of Expressway Use," *Highway Research Board Bulletin*, no. 119 (1956): 17–38 (Detroit); Karl Moskowitz, "California Method of Assigning Diverted Traffic to Proposed Freeways," *Highway Research Board Bulletin*, no. 130 (1956): 1–26 (California). For a short but well-informed history of the various diversion curves conceived in the United States by the mid-1950s, see Schmidt and Campbell, *Highway Traffic Estimation*, chapter 4.

40. On the mainframes of that time, see Paul E. Ceruzzi, *A History of Modern Computing* (Cambridge, MA: MIT Press, 2012).

41. E. Wilson Campbell, "A Mechanical Method for Assigning Traffic to Expressways (Appendix: 'Step by Step Machine Procedures,' by Robert E. Vanderford)," *Highway Research Board Bulletin*, no. 130 (1956): 27–46 (on 27).

42. See, for example, the papers published in *Highway Research Board Bulletin*, no. 130 (1956).

43. Bureau of Public Roads, *Electronic Computer Program for Traffic Assignment* (BPR Program no. T-4, developed by Edwards and Kelcey. Engineers and Consultants) (Washington, DC: n.p., n.d.).

44. On Fratar, see Institute of Transportation Engineers, *Pioneers of Transportation*, 33.

45. Leonard K. Eaton, *Hardy Cross: American Engineer* (Champaign: University of Illinois Press, 2006).

46. Thomas J. Fratar, "Vehicular Trip Distribution by Successive Approximations," *Traffic Quarterly* 8, no. 1 (January 1954): 53–65 (on 64).

47. On Fratar's contributions, see Thomas J. Fratar, "Forecasting Distribution of Interzonal Vehicular Trips by Successive Approximations" (followed by a discussion), *Highway Research Board Proceedings* 33 (1954): 376–384; Fratar, "Vehicular Trip Distribution."

48. The ratio of future trips (originating in and destined to) expected in a particular zone to present trips is called the "growth factor" for that zone.

49. Deutsch, "Origins of Metropolitan Transportation Planning," 41.

50. The "uniform factor," "average factor," "Fratar" method, and the "Detroit" method. See Glenn E. Brokke and William L. Mertz, "Evaluating Trip Forecasting Methods with an Electronic Computer," *Public Roads* 30, no. 4 (October 1958): 77–87 (published also in *Highway Research Board Bulletin*, no. 203 [1958]: 52–75). On the uniform factor and average factor methods, see Schmidt and Campbell, *Highway Traffic Estimation*, 196–199. On the Detroit variant, see especially Howard W. Bevis, "Forecasting Zonal Traffic Volumes," *Traffic Quarterly* 10, no. 2 (April 1956): 207–222.

51. Bureau of Public Roads, *Opportunities for Young Engineers in the Bureau of Public Roads* (Washington, DC: U.S. Government Printing Office, n.d. [ca. early 1960s]), 6–8.

52. William L. Mertz and Lamelle B. Hamner, "A Study of Factors Related to Urban Travel," *Public Roads* 29, no. 7 (April 1957): 170–174; William L. Mertz, "A Study of Traffic Characteristics in Suburban Residential Areas," *Public Roads* 29, no. 9 (August 1957): 208–212.

53. Lee Mertz, "Memories of 499," https://www.fhwa.dot.gov/infrastructure/memories.cfm, accessed January 17, 2021. See also the short notice on Mertz's life and career by his colleague Kevin E. Heanue, a civil engineer from Tufts University who holds a master's degree from Georgia Tech; https://www.fhwa.dot.gov/infrastructure/mertz.cfm, accessed January 17, 2021. Unless otherwise stated, information about Mertz is drawn from these two electronic documents.

54. On the computer-development research at the NBS, see, among others, Atsushi Akera, *Calculating a Natural World: Scientists, Engineers, and Computers during the Rise of U.S. Cold War Research* (Cambridge, MA: MIT Press, 2007), chapter 4.

55. The BPR ordered its first computer in fiscal year 1957 (July 1, 1956–June 30, 1957); see Bureau of Public Roads, *Annual Report, Fiscal Year 1957* (Washington, DC: U.S. Government Printing Office, n.d.), 37.

56. Bureau of Public Roads, *Electronic Computer Program for Forecasting Interzonal Traffic Movements (Fratar Method)*, Program No. T-1 (Washington, DC: Bureau of Public Roads, n.d.), George Mason University Libraries. Special Collections Research Center, William L. Mertz Papers (hereinafter Mertz's Personal Records), C0050, Box 18, Folder 36. Information concerning computer costs is given by William L. Mertz, "Review and Evaluation of Electronic Computer Traffic Assignment Programs," *Highway Research Board Bulletin*, no. 297 (1961): 94–105 (especially 103–104), https://scrc.gmu.edu/finding_aids/mertz.html.

57. On Voorhees, see Thomas B. Deen, "Alan Manners Voorhees, 1922–2005," *Memorial Tributes: National Academy of Engineering* 12 (2008): 320–326; the many obituaries that were published in all major newspapers of the country.

58. Fratar, "Forecasting Distribution," 383 (in the discussion part of the Fratar's contribution); emphasis mine.

59. Alan M. Voorhees, "The Relationship of the Cost of Public Utilities to the Subdivision Pattern" (master's thesis, MIT, 1949).

60. Alan M. Voorhees, "Basic Characteristics of Work Trips," unpublished research thesis, Bureau of Highway Traffic, Yale University, 1952 (a brief presentation of Voorhees's thesis can be found in Schmidt and Campbell, *Highway Traffic Estimation*, 71–74).

61. Fratar, "Forecasting Distribution," 383 (in the discussion part of the Fratar's contribution).

62. Alan M. Voorhees, "A General Theory of Traffic Movement," in *1955 ITE Proceedings*, ed. Institute of Traffic Engineers (New Haven: n.p., n.d.), 46–56.

63. It is worth noting that Voorhees makes no reference to any of the (few) interwar attempts to predict traffic with the help of models inspired by Newton's law of universal gravitation (see chapter 1). However, the author mentions "Reilly's Law" announced by William J. Reilly (1899–1970) in his *Law of Retail Gravitation* in 1931.

64. Gordon B. Sharpe, "Travel to Commercial Centers of the Washington Metropolitan Area," *Highway Research Board Bulletin*, no. 79 (1953): 1–15; Alan M. Voorhees, Gordon B. Sharpe, and J. T. Stegmaier, *Shopping Habits and Travel Patterns* (Washington, DC: n.p., 1955).

65. On the "modalization" issue ("it *seems* that"), see the pioneering work by Bruno Latour and Steve Woolgar, *Laboratory Life: The Construction of Scientific Facts*, 2nd ed. (Princeton, NJ: Princeton University Press, 1986).

66. Voorhees, "A General Theory of Traffic Movement," 50–52; emphasis mine.

67. Alan M. Voorhees, "Forecasting Peak Hours of Travel," *Highway Research Board Bulletin*, no. 203 (1958): 37–46.

68. Deutsch, "Origins of Metropolitan Transportation Planning," 43–44; Brian V. Martin, Frederick W. Memmott III, and Alexander J. Bone, *Principles and Techniques of Predicting Future Demand for Urban Area Transportation* (Cambridge, MA: MIT Press, 1966), 142–143; Rex H. Wiant, "A Simplified Method for Forecasting Urban Traffic," *Highway Research Board Bulletin*, no. 297 (1961): 128–145 (especially 129). Both the Fratar and the gravity methods seem to have been used in California too by the end of 1950s; see Sam Osofsky, "The Multiple Regression Method of Forecasting Traffic Volumes," *Traffic Quarterly* 13, no. 3 (July 1959): 423–445 (on 428).

69. Martin, Memmott, and Bone, *Principles and Techniques*, 36–60, 146–149, 152, 167–170, B-1–B-8. More specific references will be given in the following notes.

70. On the creation and first years of CATS, see especially Andrew V. Plummer, "The Chicago Area Transportation Study: Creating the First Plan (1955–1962). A Narrative," http://www.surveyarchive.org/Chicago/cats_1954-62.pdf, accessed November 18, 2020.

71. Letter from Bartelsm Teyer to Carroll, dated September 27, 1955, followed by a photostatic copy of the "Agreement" (Carroll's Personal Records, 4178, Box 1, Folder 10).

72. Suzanne Mettler, *Soldiers to Citizens: The G.I. Bill and the Making of the Greatest Generation* (New York: Oxford University, 2005).

73. G. Holmes Perkins to Carroll, dated September 19, 1945 and dated November 21, 1945 (Carroll's Personal Records, 4178, Box 1, Folder 9).

74. J. Douglas Carroll Jr., "Home-Work Relationships of Industrial Employees: An Investigation of Relationships of Living and Working Places for Industrial Employees with Attention to Implications for Industrial Siting and City Planning" (Ph.D. diss., Harvard University, 1950).

75. J. Douglas Carroll Jr., "A Continuing University–Community Research Project," *Journal of the American Institute of Planners* 19, no. 2 (1953): 78–86; J. D. Carroll Jr., *Report on the Parking Survey of the Flint Downtown Business District* (n.p.: University of Michigan, 1950); J. D. Carroll Jr., *Urban Land Vacancy: A Study of Factors Affecting Residential Building on Improved Vacant Lots in Flint, Michigan* (n.p.: University of Michigan, 1952).

76. Melvin R. Levin and Norman A. Abend, *Bureaucrats in Collision: Case Studies in Area Transportation Planning* (Cambridge, MA: MIT Press, 1971).

77. J. Douglas Carroll Jr., "Urban Transportation Planning," typescript, 1956 (Carroll's Personal Records, 4178, Box 1, Folder 17).

78. Detroit Metropolitan Area Traffic Study, *Report on the Detroit Metropolitan Area Traffic Study*, Part I: *Data Summary and Interpretation* (Lansing, MI: Speaker-Hines and Thomas, Inc., 1955); Part II: *Future Traffic and a Long Range Expressway Plan* (Lansing, MI: Speaker-Hines and Thomas, 1956).

79. Schmidt and Campbell, *Highway Traffic Estimation*, 209–211.

80. Roger L. Creighton, *Urban Transportation Planning* (Urbana: University of Illinois Press, 1970), 131.

81. H. James Brown, J. Royce Ginn, Franklin J. James, John F. Kain, and Mahlon R. Straszheim, *Empirical Models of Urban Land Use: Suggestions on Research Objectives and Organization* (New York: NBER, 1972), 6–7.

82. Schmidt and Campbell, *Highway Traffic Estimation*, 176–180; E. H. Holmes, "What's Ahead in Traffic Volumes," in *1950 ITE Proceedings*, ed. Institute of Traffic Engineers (New Haven: n.p., n.d.), 62–71; and especially chapter 1.

83. Carroll, "Urban Transportation Planning." Another pioneering study that used the land use–transportation approach to forecast trips was that of San Juan, Puerto

Rico, which began in 1948 (Ramiro Ramirez Carril, "Traffic Forecast Based on Anticipated Land Use and Current Travel Habits," *Highway Research Board Proceedings* 31 [1952]: 386–410).

84. See especially John R. Hamburg, *Land Use and Traffic Generation in the Detroit Metropolitan Area* (Detroit: DMATS, 1955).

85. Robert B. Mitchell and Chester Rapkin, *Urban Traffic: A Function of Land Use* (New York: Columbia University Press, 1954).

86. According to an interview with Carroll in 1978, quoted in Dimitriou, *Urban Transport Planning*, 37; Mitchell and Rapknin's book was reviewed in *CATS Research News* 1, no. 7 (1957): 15–16.

87. On the "GI generation" see David Kaiser, *American Tragedy. Kennedy, Johnson, and the Origins of the Vietnam War* (Cambridge, MA: The Belknap Press of Harvard University Press, 2000) (quotations on 8).

88. The bulk of the information about the CATS staff comes from *CATS Meow*, an in-house newsletter published by the Chicago Area Transportation Study. I also draw heavily from published obituaries. More details about the CATS team as well as precise references for each of its members can be found in Chatzis, "La Modélisation des Déplacements Urbains aux États-Unis et en France." Additional references will be found in subsequent notes.

89. On Hoch's doctoral thesis, see Marc Nerlove, *Essays in Panel Data Econometrics* (New York: Cambridge University Press, 2002), 18–21.

90. Information provided by Boyce and Williams, *Forecasting Urban Travel*, 33.

91. On the creation of the systems analyst as a distinct occupational category within programming, see the pioneering study by Philip Kraft, "The Routinizing of Computer Programming," *Sociology of Work and Occupations* 6, no. 2 (May 1979): 139–155.

92. I borrow the term "knowledge worker" from Kleinman and Vallas, "Science, Capitalism, and the Rise of the 'Knowledge Worker.'"

93. Chicago Area Transportation Study, *Final Report*, vol. 1: *Survey Findings* (Chicago: Western Engraving & Embossing Co., December 1959), acknowledgments.

94. On the crafting of such a report, see Wade G. Fox, "Preparation of Pittsburgh Area Transportation Study Final Report," *PATS Research Letter* 4, no. 2 (1962): 14–29.

95. Christopher R. Henke and Thomas F. Gieryn, "Sites of Scientific Practice: The Enduring Importance of Place," in *The Handbook of Science and Technology Studies*, 3rd ed., ed. Edward J. Hackett, Olga Amsterdamska, Michael Lynch, and Judy Wajcman (Cambridge, MA: MIT Press, 2008), 353–376.

96. J. D. Carroll Jr., "Outline of Proposed Chicago Traffic Study," typescript, May 17, 1955 (Carroll's Personal Records, 4178, Box 1, Folder 10), especially 6; E. W. Campbell, "Organizing a Continuing Agency for a Metropolitan Area Transportation Study," *Highway Research Board Proceedings* 38 (1959): 1–8.

97. An entire subunit, headed by Lucy F. Sassaman, was devoted to the quality control of the data obtained from the various traffic surveys: Lucy F. Sassaman, "An Approach to Quality Control," *CATS Research News* 1, no. 3 (1957): 9–11; Lucy F. Sassaman, "An Approach to Quality Control. The Second of the Two Articles on the Control of Quality in the Home Interview," *CATS Research News* 1, no. 4 (1957): 8–11.

98. Steven Shapin, *The Scientific Life: A Moral History of a Late Modern Vocation* (Chicago: University of Chicago Press, 2008).

99. Roger Creighton, "Coordinating Research," *CATS Research News* 1, no. 2 (1957): 5–7.

100. Plummer, "The Chicago Area Transportation Study," 9–10, 20, and passim. Creighton dedicated one of his books "To J. Douglas Carroll, Jr. who pioneered in so much of this work" (Creighton, *Urban Transportation Planning*). See also the more recent recollections by Alan E. Pisarski, "J. Douglas Carroll, Jr., Pioneer of Urban Transportation Planning," *TR News*, no. 283 (November–December 2012): 34.

101. Alan Black, "The Chicago Area Transportation Study: A Case Study of Rational Planning," *Journal of Planning Education and Research* 10, no. 1 (October 1990): 27–37 (on 34). Creighton talks also of "human atmosphere" that was "exciting" (quoted by Plummer, "The Chicago Area Transportation Study," 13).

102. It seems that Carroll himself and several members of the team used to play baseball on the CATS premises. See Black, "The Chicago Area Transportation Study," 34–35.

103. Steven Shapin, *A Social History of Truth: Civility and Science in Seventeenth-Century England* (Chicago: University of Chicago Press, 1994).

104. It is not by chance that a high percentage of the papers that came out of the CATS group were multiauthored.

105. Robert H. Kargon and Scott G. Knowles, "Knowledge for Use: Science, Higher Learning, and America's New Industrial Heartland, 1880–1915," *Annals of Science* 59, no. 1 (January 2002): 1–20.

106. Armour Research Foundation of Illinois Institute of Technology, *The Development of an Apparatus for the Generation of Two-Dimensional Density Functions*, Part A: "Technical Presentation"; Part B: "Time and Cost Estimates" (Proposal No. 56-865E) (Chicago: Armour Research Foundation of Illinois Institute of Technology, August 1956). For an early description of the Cartographatron, see *CATS Research News* 2,

no. 6 (1958). See also Chatzis, "Capter et Cartographier les Pratiques," which provides more information on the device, and gives several relevant references.

107. Armour Research Foundation of Illinois Institute of Technology, *Investigation of Methods for the Assignment of Trip Demand to a Road Network*, Part A: "Technical Presentation"; Part B: "Time and Cost Estimates" (Proposal No. 57-608E) (Chicago: Armour Research Foundation of Illinois Institute of Technology, April 1957), 1, 24.

108. Armour Research Foundation of Illinois Institute of Technology, *Investigation of Methods for the Assignment of Trip Demand to a Road Network*, Phase Report No. 1: "Algorithm for Finding Shortest Paths Between Pairs of Points in a Network," ARF Project E084 (Chicago: Armour Research Foundation of Illinois Institute of Technology, June 1957), 1, 16–19.

109. On the history of "shortest path" algorithms, see Alexander Schrijver, "On the History of the Shortest Path Problem," *Documenta Mathematica*, Extra Volume ISMP (2012): 155–167.

110. Edward F. Moore, "The Shortest Path Through a Maze," *The Annals of the Computational Laboratory of Harvard University* 30 (1959): 285–292. The ARF's researchers quoted the paper presented by Moore at the conference.

111. Robert E. Bixby, "A Brief History of Linear and Mixed-Integer Programming Computation," *Documenta Mathematica*, Extra Volume ISMP (2012): 107–121. Armour's researchers explicitly referred to George B. Dantzig, *Notes on Linear Programming*, Part 35: "Discrete Variable Extremum Problems" (Santa Monica: The RAND Corporation, December 1956).

112. George B. Dantzig, "Discrete-Variable Extremum Problems," *Operations Research* 5, no. 2 (April 1957): 266–277 (quotation on 266).

113. Armour Research Foundation of Illinois Institute of Technology, *Investigation of Methods for the Assignment of Trip Demand to a Road Network*, Phase Report No. 1, 1.

114. J. Douglas Carroll Jr., "A Method of Traffic Assignment to an Urban Network," *Highway Research Board Bulletin*, no. 224 (1959): 64–71 (quotation on 65); Britton Harris, "First *Environment and Planning* Lecture: Synthetic Geography: The Nature of Our Understanding of Cities," *Environment and Planning A* 17, no. 4 (April 1985): 443–464 (on 449). On the history of computer time-sharing practices, see Martin Campbell-Kelly and Daniel D. Garcia-Swartz, "Economic Perspectives on the History of the Computer Time-Sharing Industry, 1965–1985," *IEEE Annals of the History of Computing* 30, no. 1 (January–March 2008): 16–36.

115. The assignment issue received much attention from CATS staff from the beginning. Members of the team met with experts of assignment techniques (from Toronto, for example), compiled bibliographies, and published excerpts from and commenting on relevant works produced across the world, including theoretical

contributions. See, for example, *CATS Research News* 1, no. 2 (1957): 15–16; 1, no. 10 (1957): 12–14; 1, no. 11 (1957): 16–20; 1, no. 15 (1957): 4–6; 1, no. 19 (1957): 12–16; 1, no. 22 (1957): 14–16; 2, no. 5 (1958): 10–13. On the work done by Schneider on the assignment issue, see Chicago Area Transportation Study, *Final Report*, vol. 2: *Data Projections* (Chicago: Western Engraving & Embossing Co., July 1960), 107–110.

116. Morton Schneider, "Gravity Models and Trip Distribution Theory," *Papers and Proceedings of the Regional Science Association* 5, no. 1 (January 1959): 51–56 (quotations on 51 and 52); emphasis mine.

117. J. Douglas Carroll Jr., "Spatial Interaction and the Urban-Metropolitan Regional Description," *Papers and Proceedings of the Regional Science Association* 1, no. 1 (January 1955): 59–73; Howard W. Bevis, "A Model for Predicting Urban Travel Patterns," *Journal of the American Institute of Planners* 25, no. 2 (1959): 87–89. For a good survey article on the gravitational principle, which was published in the mid-1950s, see Gerald A. P. Carrothers, "An Historical Review of the Gravity and Potential Concepts of Human Interaction," *Journal of the American Institute of Planners* 22, no. 2 (1956): 94–102. It is worth noting that this journal published many articles written by CATS staffers in the 1950s.

118. Samuel A. Stouffer, "Intervening Opportunities: A Theory Relating Mobility and Distance," *American Sociological Review* 5, no. 6 (December 1940): 845–867 (on 846).

119. Schneider, "Gravity Models and Trip Distribution Theory," 52. Schneider's approach is described in Chicago Area Transportation Study, *Final Report*, vol. 2: *Data Projections*, 81–92, 111.

120. John T. Lynch, Glenn E. Brokke, Alan M. Voorhees, and Morton Schneider, "Panel Discussion on Inter-Area Travel Formulas," *Highway Research Board Bulletin*, no. 253 (1960): 128–138 (the various quotations can be found on 132 and 135).

121. For an overview, see Martin, Memmott, and Bone, *Principles and Techniques*.

122. John F. McDonald, "The First *Chicago Area Transportation Study* Projections and Plans for Metropolitan Chicago in Retrospect," *Planning Perspectives* 3, no. 3 (1988): 245–268 (on 255); and especially Chicago Area Transportation Study, *Final Report*, vol. 1: *Survey Findings*, chapter 5: "Trip Generation"; Chicago Area Transportation Study, *Final Report*, vol. 2: *Data Projections*, chapters 4 and 5.

123. On CATS' in-house land use model, see among others John R. Hamburg and Roger L. Creighton, "Predicting Chicago's Land Use Pattern," *Journal of the American Institute of Planners* 25, no. 2 (1959): 67–72.

124. J. Douglas Carroll Jr. and Howard W. Bevis, "Predicting Local Travel in Urban Regions," *Papers and Proceedings of the Regional Science Association* 3, no. 1 (January 1957): 183–197. See also Chicago Area Transportation Study, *Final Report*, vol. 1: *Survey Findings*, 7, 9.

125. John J. Howe, "Modal Split of CBD Trips," *CATS Research News* 2, no. 12 (1958): 3–10; Chicago Area Transportation Study, *Final Report*, vol. 1: *Survey Findings*, 69–75, 86–89, and passim.

126. For a good (and early) discussion about the order of the steps, see Peter R. Stopher and Arnim H. Meyburg, *Urban Transportation Modeling and Planning* (Lexington, MA: Lexington Books, 1975).

127. Chicago Area Transportation Study, *Final Report,* vol. 3: *Transportation Plan* (Chicago: Printed by Western Engraving & Embossing Co., April 1962).

128. Renamed the Tri-State Regional Planning Commission in the early 1970s.

129. More information about the mobility of CATS staffers as well as relevant references can be found in Chatzis, "La Modélisation des Déplacements Urbains aux États-Unis et en France"; see also John M. Dutton and William H. Starbuck, "Diffusion of an Intellectual Technology," in *Communication and Control in Society*, ed. Klaus Krippendorff (New York: Gordon and Breach Science Publishers, 1979), 489–511.

130. Richard M. Zettel et al., *Summary Review of Major Metropolitan Area Transportation Studies in the United States* (Berkeley: University of California, 1962).

131. Brand and Manheim, *Urban Travel Demand Forecasting*.

132. Mark Solof, *History of Metropolitan Planning Organizations* (Newark, NJ: North Jersey Transportation Planning Authority, 1998); Katherine F. Turnbull (Rapporteur), *The Metropolitan Planning Organization, Present and Future* (Washington, DC: Transportation Research Board, 2007); Gian-Claudia Sciara and Susan Handy, "Regional Transportation Planning," in Giuliano and Hanson, *Geography of Urban Transportation*, 139–163.

133. Federal Highway Administration, *Urbanized Area Transportation Planning Programs: Directory* (Washington, DC: U.S. Government Printing House, 1970).

134. E. H. Holmes, "The State-of-the-Art in Urban Transportation Planning or How We Got Here," *Transportation* 1, no. 4 (March 1973): 379–401.

135. Henry D. Quinby, "Traffic Distribution Forecasts—Highway and Transit," *Traffic Engineering* 31, no. 5 (February 1961): 22–29, 54–55.

136. "Charles DeLeuw, Engineer, Was 79," https://www.nytimes.com/1970/10/31/archives/charles-deleuw-engineer-was-79-transportation-expert-dies-led.html, accessed January 17, 2021.

137. In 1985, the same Parsons acquired another interwar traffic and planning engineering firm, Harland Bartholomew and Associates (see chapter 1), http://www.fundinguniverse.com/company-histories/the-parsons-corporation-history/; https://www.parsons.com/about/acquisitions/, accessed January 20, 2021.

138. The Cornell University Library holds several publications by Frederick T. Aschman, covering the period 1949–1977, including studies, papers, and articles by Barton-Aschman Associates, Collection Number 3500, http://rmc.library.cornell.edu/EAD/htmldocs/RMM03500.html#d0e182, accessed January 20, 2021.

139. Barton-Aschman Associates, *Traffic Parking and Transportation Analysis: Central Business District Area Boston, Massachusetts* (n.p.: n.p., 1965).

140. "Our Expansion and Strategic Acquisition," Parsons, https://www.parsons.com/about/acquisitions/, accessed January, 20 2021.

141. On Roger Creighton Associates and John Hamburg & Associates, see George Mason University Libraries. Special Collections Research Center, John R. Hamburg transportation papers (hereinafter Hamburg's Personal Records), C0073, https://scrc.gmu.edu/finding_aids/hamburg.html.

142. John A. Montgomery, *History of Wilbur Smith and Associates, 1952–1984* (Columbia, SC: Wilbur Smith and Associates, 1985). See also Chatzis, "La Modélisation des Déplacements Urbains aux États-Unis et en France," for more information about the firm and a list of supplementary references.

143. F. Houston Wynn, "Intracity Traffic Movements" (with discussion), *Highway Research Board Bulletin*, no. 119 (1956): 53–68 (on 68).

144. Wynn, "Intracity Traffic Movements"; F. Houston Wynn, "Studies of Trip Generation in the Nation's Capital, 1956–58," *Highway Research Board Bulletin*, no. 230 (1959): 1–52; F. Houston Wynn and C. Eric Linder, "Tests of Interactance Formulas Derived from O-D Data," *Highway Research Board Bulletin*, no. 253 (1960): 62–85.

145. On the Greater London study, see Boyce and Williams, *Forecasting Urban Travel*, chapter 3.

146. London innovations included: (1) the "category analysis" (used in the trip generation step of the FSM); (2) transit network assignment programs; and (3) procedures to deal with capacity-constrained travel demand in highly congested urban areas ([2] and [3] are used in the traffic assignment step of the FSM).

147. See, for example, Transportation Planning Program Exchange Group, *T-PEG Newsletter* (Autumn 1967): 6.

148. Ernest E. Blanche, "The Use of UNIVAC in Processing and Analyzing Origin-Destination Data for the Washington, D.C. Metropolitan Area," typescript (found in Mertz's Personal Records, C0050, Box 26, Folder 109).

149. For the programs developed in the 1960s by WSA and its various collaborators, including the General Electric Computer Department, see especially *Newsletter of the Transportation Planning Computer Program Exchange Group* 1, no. 2 (1963): 3; 1, no. 4 (1964): 6, 10; 2, no. 3 (1965): 5; 2, no. 4 (1965): 2, 10–11.

150. On Kates's contributions to the design of UTEC, Canada's first computer, see John N. Vardalas, *The Computer Revolution in Canada: Building National Technological Competence* (Cambridge, MA: MIT Press, 2001).

151. KCS was named after its founders: Kates; Leonard P. Casciato (1925–2020), an "engineering whiz kid" according to his obituary; and Joe Shapiro. See especially *IEEE Toronto Section Newsletter* 1, no. 2 (2020): 9–15.

152. The main information source here is Traffic Research Corporation, *History, Services and Experience Record* (Boston, New York, San Francisco, Toronto: n.p., n.d. [ca. 1966]). More information about the firm and its staffers can be found in Chatzis, "La Modélisation des Déplacements Urbains aux États-Unis et en France."

153. The firm had access to a Burroughs B-5500 and a UNIVAC 1107 machine.

154. N. A. Irwin and H. G. von Cube, "Capacity Restraint in Multi-Travel Mode Assignment Programs," *Highway Research Board Bulletin*, no. 347 (1962): 258–89; *Newsletter of the Transportation Planning Computer Program Exchange Group* 1, no. 1 (1963): 9–10; 1, no. 2 (1963): 7; 1, no. 4 (1964): 1–4; 2, no. 1 (1964): 8–9; 2, no. 3 (1965): 3–4.

155. Stopher and Meyburg, *Urban Transportation Modeling and Planning*, 189.

156. These were: (1) the ratio of door-to-door travel time via public transit to the door-to-door-travel time via private automobile; (2) the ratio of out-of-pocket cost via public transit to the out-of-pocket cost via private automobile; (3) the ratio of excess travel time via public transit to excess travel time via private automobile; (4) the economic status of the trip maker; and (5) trip purpose.

157. This comes as no surprise: in fact, instead of comparing the performances of the projected freeway with existing alternative roads, the comparison now concerns a road network and a public transit system.

158. H. G. von Cube, R. J. Desjardins, and N. Dodd, "Assignment of Passengers to Transit Systems," *Traffic Engineering* 28, no. 11 (1958): 12–14; Thomas B. Deen, William L. Mertz, and Neil A. Irwin, "Application of a Modal Split Model to Travel Estimates for the Washington Area," *Highway Research Record*, no. 38 (1963): 97–123.

159. Peat, Marwick, Livingston & Co., *Proposal for Phase: Concept Formulation, Columbia Public Transit Technical Study Demonstration Program* (n.p.: n.p., 1968), 37–38.

160. Kell, "Comparison of Two Methods."

161. JHK & Associates, *Specializing in Transportation: Planning, Management, Engineering and Operations, Research, Training* (n.p.: n.p., n.d. [ca. 1980]).

162. The collaborator in question is Thomas Deen; he provided an affectionate portrayal of Voorhees in Robert B. Noland, ed., "Interview with Thomas Deen, January 2013, conducted by Nicholas K. Tulach," typescript, April 2015, 28–29, https://vtc

.rutgers.edu/wp-content/uploads/2014/12/Interview_Tom_Deen_2013.pdf, accessed January, 20, 2021.

163. Keith Gilbert, quoted by Deutsch, "Origins of Metropolitan Transportation Planning," 18.

164. Noland, "Interview with Thomas Deen," 11. On the early years of AMV, one can also consult the speech delivered by Deen on February 27, 2018, https://www.youtube.com/watch?v=8zaAV8abe00, accessed January 20, 2021.

165. Walter G. Hansen, "Accessibility and Residential Growth" (master's thesis, MIT, 1959).

166. On Schultz's (brilliant) career, see also chapter 6.

167. On Deen's career, see Deutsch, "Origins of Metropolitan Transportation Planning," 7, 10, 17, 24–25, 52, 55–56, and especially George Mason University Libraries, Special Collections Research Center, Thomas B. Deen Papers (hereinafter Deen's Personal Records), C0106, https://scrc.gmu.edu/finding_aids/deen.html.

168. Thomas B. Deen, "Acceleration Lane Lengths for Heavy Commercial Vehicles," unpublished research thesis, Bureau of Highway Traffic, Yale University, 1956.

169. Planning Research Corporation was founded by five systems analysts from RAND Corporation in 1954.

170. Thomas B. Deen, "RDSA/AMV: What and Who Are We? What Do We Want to Be?" (A White Paper Prepared to Stimulate an Intra-Company Dialogue on Our Company's Current Course and Future Directions; to Be Presented to the Steering Committee, November 19, 1977)," typescript, figure no. 6. Deen's Personal Records, C0106, Box 6, Folder 4.

171. AECOM stands for Architects, Engineers, Construction, Operation, and Maintenance. As far as I know, the only scholarly work done on this firm is Aaron Cayer, "Shaping an Urban Practice: AECOM and the Rise of Multinational Architecture Conglomerates," *Journal of Architectural Education* 73, no. 2 (2019): 178–192.

172. Donald Edward Cleveland, "A Study of Highway Traffic Assignment" (Ph.D. diss., Agricultural and Mechanical College of Texas, 1962).

173. Modak et al., "Fifty Years of Transportation Research Journals."

174. Peterson, *The Transportation Research Board, 1920–2020*, chapter 9.

175. Gerald S. Cohen, Frank McEvoy, and David T. Hartgen, "Who Reads the Transportation Planning Literature?," *Transportation Research Record*, no. 793 (1981): 33–40.

176. *Journal of Transport Economics and Policy* and *Socio-economic Planning Sciences* also started appearing in 1967; *Environment and Planning* began its editorial career in

1969, while the first issue of *Transportation* was published in May 1972. On the history of *Transportation*, see especially Martin Richards, "Putting the Blue Pencil Away and Taking Down the Name Board," *Transportation* 40, no. 6 (November 2013): 1087–1104. These journals also rapidly became home to urban traffic forecasting literature.

177. Martin Wohl and Brian V. Martin, *Traffic System Analysis for Engineers and Planners* (New York: McGraw-Hill Book Company, 1967), especially chapters 5 and 6.

178. *Transportation Science* 2, no. 3 (August 1968): 283–284; *Transportation Research* 3, no. 2 (July 1969): 283–284.

179. M. P. S. Santos and M. G. de C. Braga, "Research Trends in Urban Transport: Some Empirical Evidence from Academic Research," *Transportation Research Part A* 22, no. 1 (1988): 57–70; see especially figure 3.

180. See, for example, Oi and Shuldiner, *An Analysis of Urban Travel Demands*; Martin, Memmott, and Bone, *Principles and Techniques*, ix.

181. The term *academia* refers here to all knowledge production-related institutions, and not only to the university system sensu stricto.

182. See chapter 1.

183. D. S. Berry and Beverly Hickok, "Partial Listing of University Programs in Transportation and Traffic Engineering in the U.S.," typescript (Berkeley: Institute of Transportation and Traffic Engineering, 1955). According to the authors, thirty or so higher education establishments offered transportation and traffic engineering courses in the mid-1950s. Most of them were civil engineering departments.

184. Marvin L. Manheim (with Earl R. Ruiter and Kiran U. Bhatt), *Search and Choice in Transport Systems Planning, Summary Report, Volume I of a Series* (Cambridge, MA: Transportation Systems Division/Department of Civil Engineering/MIT, 1968), 27–29; emphasis mine. The authors provide a series of older references expressing this point of view on page 27.

185. They had even labeled this increase "induced traffic" (see note 24 in this chapter).

186. On the Ford Foundation, established in 1936, see among others Inderjeet Parmar, *Foundations of the American Century: The Ford, Carnegie, and Rockefeller Foundations in the Rise of American Power* (New York: Columbia University Press, 2012).

187. "Research ANd Development." On RAND, established by the Douglas Aircraft Company for the Army Air Force in 1945 and which became an independent nonprofit corporation in 1948, see David Jardini, *Thinking through the Cold War: RAND, National Security, and Domestic Policy, 1945–1975* (ebook, 2013).

188. J. R. Meyer, J. F. Kain, and M. Wohl, *The Urban Transportation Problem* (Cambridge, MA: Harvard University Press, 1965).

189. John R. Meyer and Mahlon R. Straszheim, *Pricing and Project Evaluation*, vol. 1: *Techniques of Transport Planning*, ed. John R. Meyer (Washington, DC: The Brookings Institution, 1971), vii–viii.

190. Including Dantzig, Albert William Tucker (1905–1995), Harold William Kuhn (1925–2014), and Oscar Morgenstern (1902–1977), https://princetoninfo.com/mathematica-where-data-drives-decisions/, accessed November 20, 2019; MATHEMATICA, *Decomposition Principles for Solving Large Structured Linear Programs* (Princeton, NJ: MATHEMATICA, n.d., [ca. 1963]).

191. Robert A. Nelson, "Origin, Purpose, and Status of Project," and Paul W. Shuldiner, "Northeast Corridor Transportation Project: Structure and Operation of Model System," both in *Transportation Engineering Journal of ASCE* 96, no. TE4 (November 1970): 439–444, and 445–454, respectively.

192. Including the mathematician and operations research specialist James Michael McLynn (1925–1986) and the firm Systems Analysis and Research Corporation.

193. On these models, see Meyer and Straszheim, *Pricing and Project Evaluation*, chapter 9: "Modeling Intercity Passenger Demand," 137–164 (the chapter is a "special contribution" authored by John F. Kain); Frank S. Koppelman, Geok-Koon Kuah, and Moshe Hirsh, "Review of Intercity Passenger Travel Demand Modelling: Mid-60s to the Mid-80s," typescript (Evanston, IL: The Transportation Center and Department of Civil Engineering/Northwestern University, 1984).

194. Richard E. Quandt, *The Collected Essays of Richard E. Quandt*, vol. 1 (Aldershot: Edward Elgar Publishing Limited, 1992), xx–xxi.

195. At the same time that MATHEMATICA economists were developing their Abstract Mode models, another economist was about to devise his now famous theory according to which a consumer does not choose a good but rather its attributes (see Kelvin J. Lancaster, "A New Approach to Consumer Theory," *Journal of Political Economy* 74, no. 2 [April 1966]: 132–157). For an early formulation of the "Abstract Mode" modeling approach, see Richard E. Quandt and William J. Baumol, "The Demand for Abstract Transport Modes: Theory and Measurement," *Journal of Regional Science* 6, no. 2 (December 1966): 13–26. See also several essays in Richard E. Quandt, ed., *The Demand for Travel: Theory and Measurement* (Lexington, MA: Heath Lexington Books, 1970). On the first application of this modeling approach to *urban* travel, see Rodney Paul Plourde, "Consumer Preferences and the Abstract Mode Model: Boston Metropolitan Area" (master's thesis, MIT, 1968). For an application of an improved version of this approach, based on data from East Los Angeles, see Peter Gordon, Peter M. Theobald, and C. S. Williams, "Intrametropolitan Travel Demand

Forecasting Using an Abstract Modes Approach," *Transportation Research Part A* 13, no. 1 (February 1979): 49–56.

196. Now known as SRI International; Donald Nielson, *A Heritage of Innovation: SRI's First Half Century* (Menlo Park, CA: SPI International, 2006).

197. Philip Mirowski, *Machine Dreams: Economics Becomes a Cyborg Science* (Cambridge: Cambridge University Press, 2002); Frederic H. Murphy, "Economics and Operations Research," in *Encyclopedia of Operations Research and Management Science*, ed. Saul I. Gass and Carl M. Harris (Norwell: Kluwer, 2001), 225–231.

198. Lester K. Ford Jr. and D. Ray Fulkerson, *Flows in Networks* (Princeton, NJ: Princeton University Press, 1962).

199. William Thomas, *Rational Action: The Sciences of Policy in Britain and America, 1940–1960* (Cambridge, MA: MIT Press, 2015); Saul I. Gass and Arjang A. Assad, "History of Operations Research," in *Tutorials in Operations Research*, ed. Joseph Geunes (Hanover: INFORMS, 2011), 1–14.

200. For early examples of such a seduction, see Mitchell and Rapkin, *Urban Traffic*, chapter 9; Robert H. Roy, "Operations Research: The Multi-Discipline Technique," in *Highway Research Board Special Report no. 28*, ed. Highway Research Board (Washington, DC: Highway Research Board, 1957), 55–56; D. Howland, "Applications of Operations Research to Highway Problems," *Highway Research Board Proceedings* 37 (1958): 72–80.

201. See, among others: Committee on Theory of Traffic Flow, "Bibliography on Theory of Traffic Flow and Related Subjects, October 1960," *Operations Research* 9, no. 4 (1961): 568–574; Frank A. Haight, "Annotated Bibliography of Scientific Research in Road Traffic and Safety," *Operations Research* 12, no. 6 (November–December 1964): 976–1039.

202. For an early appraisal of some of these algorithms ten years or so after they were first used in transportation modeling, see Stuart. E. Dreyfus, "An Appraisal of Some Shortest-Path Algorithms," *Operations Research* 17, no. 3 (May–June 1969): 395–412.

203. Estimating the various constants and parameters in the structure of the model using data obtained from surveys of actual travel patterns.

204. The capacity of the model to reproduce observed traffic.

205. See, for example, D. A. D'Esopo and B. Lefkowitz, "An Algorithm for Computing Interzonal Transfers Using the Gravity Model," *Operations Research* 11, no. 6 (November–December 1963): 901–907. D'Esopo and Lefkowitz, both Stanford Research Institute staffers, had also authored an early computer suit for the gravity

model (D. A. D'Esopo, B. Lefkowitz, and M. Kopri, *Manual for the Gravity Model Computer Programs* (Menlo Park, CA: Stanford Research Institute, 1962).

206. Frederick J. Wegmann and Edward A. Beimborn, "Transportation Centers and Other Mechanisms to Encourage Interdisciplinary Research and Training Efforts in Transportation," *Highway Research Record*, no. 462 (1973): 1–13 (on 4).

207. For a panoramic view, see L. Hoel, "Analysis of Transportation Planning Education," *Transportation Engineering Journal* 96, no. TE2 (1970): 123–134; *Highway Research Record*, no. 462 (1973) (the entire issue is devoted to "Education in Transportation Systems Planning").

208. It was renamed the Purdue Road School Transportation Conference and Expo in 2015. The annual Purdue Road School Proceedings from 1924 to the present can be consulted at https://docs.lib.purdue.edu/roadschool/, accessed November 20, 2020.

209. It was renamed the Joint Transportation Research Program (JTRP) in 1997.

210. Now known as the Joint Transportation Research Program (JTRP), https://docs.lib.purdue.edu/jtrprogram/, accessed January 20, 2021; Harold L. Michael, "50 Years of the Joint Highway Research Project (JHRP)," *Engineering Bulletin of Purdue University* (Engineering Extension Series No. 158) (n.d.): 25–37.

211. JTRP technical reports from the mid-1950s on are available at https://docs.lib.purdue.edu/jtrp/, accessed January 15, 2021.

212. In a report dated 1965, Harold L. Michael mentioned four (out of fifty or so) research projects active in the JHRP in 1964–1965 that were dealing with various aspects of FSM. See Harold L. Michael, *Current Highway Research at Purdue: The Joint Highway Research Project* (Lafayette, IN: Purdue University, 1965), 19–22. In the bibliographical list of his 1974 report, Joseph A. Ansah cites several other Ph.D. dissertations dealing with various aspects of the FSM; see Joseph A. Ansah, *Destination Choice Modeling and the Disaggregate Analysis of Urban Travel Behavior* (West Lafayette, IN: Purdue University, 1974).

213. "The Legacy of TTI," https://tti.tamu.edu/about/history/, accessed January 20, 2021; Texas A&M Transportation Institute, *Texas A&M Transportation Institute—Over 65 Years of Innovation* (n.p.: n.p., n.d.).

214. For example, in the mid-1960s a certain Charles Blumentritt, holder of a master's degree from Texas A&M University (1964), conducted a computer study for traffic assignment on an IBM 7094 machine with a grant of $22,000 from the BPR (Anonymous, "Computer Plotted Networks for Traffic Assignment Can Indicate Future Interregional Routes," *Texas Transportation Researcher* 1, no. 3 [1965]: 2).

215. Vergil G. Stover, *A Study of Remainder Parcels Resulting from the Acquisition of Highway Rights-of-way* (Lafayette, IN: Purdue University, 1963). On Stover and his

performances concerning assignment algorithms see also the introduction to this book.

216. Vergil G. Stover, "The Texas Large Systems Traffic Assignment Package," *Traffic Quarterly* 21, no. 3 (July 1967): 339–354; Vergil G. Stover and J. D. Benson, *Traffic Projection and Assignment* (College Station: Texas Transportation Institute, 1971); Texas Transportation Institute, *A Snapshot of Travel Modeling Activities: The State of Texas*, prepared for Federal Highway Administration (n.p.: n.p., 2011).

217. Regional Oral History Office/The Bancroft Library, *Harmer E. Davis: Founder of the Institute of Transportation and Traffic Engineering* (Berkeley: University of California at Berkeley, 1997); ITTE, *Report of Activities, 1967–1970, with a Review of Institute Development from the Time of Its Founding in 1948* (Berkeley and Los Angeles: University of California, 1970). Chatzis in "La Modélisation des Déplacements Urbains aux États-Unis et en France" provides a more detailed account than the one offered herein as well as supplementary bibliographical references.

218. Harmer E. Davis, "Education and Research at the Institute of Transportation and Traffic Engineering of the University of California," in *1951 ITE Proceedings*, ed. Institute of Traffic Engineers (New Haven: n.p., n.d.), 95–102 (on 97). On the first ITTE staffers, see *Institute of Transportation and Traffic Engineering Quarterly Bulletin* 1, no. 1 (1948), 118.

219. My reckoning is based on "Roster, Institute of Transportation and Traffic Engineering, Berkeley, December 1969," typescript; "Graduate Students, Transportation Engineering, Fall Quarter 1969," typescript (rosters held by the University of California Libraries under the general heading "Rosters of Students and Directories of Faculty at the Institute of Transportation and Traffic Engineering, 1954–1971").

220. Wolfgang S. Homburger, *Traffic Estimation Computer Programs on the IBM 1620 for Instructional Purposes* (Berkeley: University of California at Berkeley, 1966). Another edition appeared in 1970.

221. "Transportation Planning Laboratory" (CE 253 ML); "Traffic Flow on Transportation Networks" (CE 252); "Transportation Demand Analysis and Forecasting" (CE 254); "Transportation Policy and Planning (CE 250)" (ITTE, *Report of Activities, 1967–1970*, 10).

222. Just three examples: Walter W. Mosher Jr., *A Capacity Restraint Algorithm for Assigning Flow to a Transport Network* (Los Angeles: ITTE, Department of Engineering/University of California, 1963); Tony M. Ridley, *General Methods of Calculating Traffic Distribution and Assignment* (Berkeley: Institute of Transportation and Traffic Engineering/University of California, 1968); and a minor classic, coauthored by the ITTE staffer and operations research specialist Robert M. Oliver: Renfrey B. Potts and Robert M. Oliver, *Flows in Transportation Networks* (New York: Academic Press, 1972).

223. Some illustrative examples: Abdollah Mogharabi, "Theory and Practical Application of the 'Intervening Opportunities Model'" (Berkeley: ITTE/UC, 1970); David A. Merchant, "Interzonal Travel Patterns in the Sacramento Area Using the San Diego Friction Factor Method" (Berkeley: Division of Transportation Engineering/ University of California, 1959); Niels O. Jorgensen, "Some Aspects of the Urban Traffic Assignment Problem" (Berkeley: ITTE/University of California, 1963); Paul William Shuldiner, "Traffic Generating Characteristics of Urban Residences" (Ph.D. diss., University of California at Berkeley, 1961); Peter Schwarz Loubal, "A Mathematical Model for Traffic Forecasting" (Ph.D. diss., University of California at Berkeley, 1968).

224. Adib Kanafani, *Transportation Demand Analysis*, 2nd ed. (New York: McGraw-Hill, [1978] 1983).

225. *Northwestern University Transportation Center, 1954–1974* (Annual Report and Twenty-Year Perspective of) (n.p.: n.p., n.d.), 4.

226. The Transportation Center, *A Report, 1959–1960* (Evanston, IL: n.p., n.d.), introduction.

227. *A Report from the Transportation Center at Northwestern University* (in reference to the academic year 1956–1957) (n.p.: n.p., n.d.), 2, 6–7.

228. A. Charnes and W.W. Cooper, "Extremal Principles for Simulating Traffic Flow in a Network," *Proceedings of the National Academy of Sciences of the United States of America* 44, no. 2 (1958): 201–204.

229. Transportation Center at Northwestern University, *Research/Education, 1962–63* (n.p.: n.p., n.d.), brochure with no pagination. For an outline of the graduate program in the early 1970s, see The Transportation Center/Northwestern University, *Annual Report 1973* (n.p.: n.p., n.d.), 12–14.

230. *Northwestern University Transportation Center, 1954–1974*, 20.

231. Transportation Center at Northwestern University, *Annual Report 1969–70* (n.p.: n.p., n.d.), 16.

232. See, for example, for the period 1961–1962: "A Study of the Use of Auto and Transit Modes of Travel in Urban Areas" ($22,000); for the period 1964–1965: "Non-Residential Trip Generation Analysis" ($24,971) (*Northwestern University Transportation Center, 1954–1974*, 14).

233. Just some early examples: Stanley Leon Warner, *Stochastic Choice of Mode in Urban Travel: A Study in Binary Choice* (n.p.: Northwestern University Press, 1962) (more on this research study in chapter 3); Oi and Shuldiner, *An Analysis of Urban Travel Demands*; Leon N. Moses and Harold F. Williamson Jr., "Value of Time, Choice of Mode, and the Subsidy Issue in Urban Transportation," *Journal of Political*

*Economy* 71, no. 3 (June 1963): 247–264; Paul W. Shuldiner, "Trip Generation and the Home," *Highway Research Board Bulletin*, no. 347 (1962): 40–59.

234. *MIT Bulletin* 98, no. 2 (1962): 110.

235. "The General Catalogue Issue for the Centennial Year 1960–61," special issue, *MIT Bulletin*, 95 (1960): 96.

236. In 1967, the Civil Engineering Systems Laboratory (CESL), set up in 1960, had an IBM 1130 and an IBM 360/40 (*MIT Bulletin* 103, no. 3 [1967], 58). CESL staffers and students could also make use of the mainframes installed at the MIT Computation Center, which in the early 1960s had developed the first "Compatible Time-Sharing Service" in the world: David Walden and Tom Van Vleck, eds., *The Compatible Time-Sharing System (1961–1973), Fiftieth Anniversary, Commemorative Overview* (Washington, DC: IEEE Computer Society, 2011); Akera, *Calculating a Natural World*, chapter 9.

237. *MIT Bulletin* 92, no. 6 (1957): 183.

238. A civil engineer who graduated from MIT in 1956, Wohl received his master's degree from the same institution in 1960. Wohl's personal records are held by George Mason University Libraries, Special Collections Research Center, Martin Wohl Papers, C0174, https://scrc.gmu.edu/finding_aids/wohl.html.

239. "The General Catalogue Issue for the Centennial Year 1960–61," special issue, *MIT Bulletin* 95 (1960): 185.

240. Brian Vivian Martin and Frederick William Memmott III, "Principles and Techniques of Predicting Future Demand for Urban Area Transportation" (master's thesis, MIT, 1961). The report was reprinted in January 1962, January 1963, September 1965, and July 1966; see n. 68 in this chapter.

241. Unless otherwise mentioned, I draw upon Manheim's CV, dated January 1977 (Manheim's Personal Records, MC-0330, Box 1, Folder 16).

242. Eugenie L. Birch, "Making Urban Research Intellectually Respectable: Martin Meyerson and the Joint Center for Urban Studies of Massachusetts Institute of Technology and Harvard University 1959–1964," *Journal of Planning History* 10, no. 3 (August 2011): 219–238.

243. Marvin L. Manheim, "Highway Route Location as a Hierarchically-Structured Sequential Decision Process: An Experiment in the Use of Bayesian Decision Theory for Guiding an Engineering Process" (Ph.D. diss., MIT, 1964).

244. Herbert A. Simon, *The Sciences of the Artificial*, 3rd ed. (Cambridge, MA: MIT Press, 1996), 125.

245. On the early years of the Transportation Systems Division, see the brochure titled *Transportation Systems Division* (Manheim's Personal Records, MC-0330, Box 2,

Folder 26); *MIT Bulletin* 103, no. 3 (1967): 131–134. On the curriculum developed by Manheim, see Marvin L. Manheim, "Education in Transportation Systems Analysis," prepared for the Conference on Highway Transportation Engineering Education, June 13–15, 1967 (Manheim's Personal Records, MC-0330, Box 4, Folder 19); "Catalog Revision," Department of Civil Engineering, MIT, January 4, 1968 (Manheim's Personal Records, MC-0330, Box 1, Folder 6).

246. "Transportation Systems Division," without pagination (1) (Manheim's Personal Records); emphasis mine. See also Marvin L. Manheim and Earl R. Ruiter, "The Transportation Laboratory: Teaching Fundamental Concepts of Transportation Systems Analysis," *Highway Research Record*, no. 462 (1973): 32–40. On the systems approach, very popular in the United States at that time, see Agatha C. Hughes and Thomas Parke Hughes, eds., *Systems, Experts, and Computers: The Systems Approach in Management and Engineering, World War II and After* (Cambridge, MA: MIT Press, 2000). Richard de Neufville, a colleague of Manheim at MIT, was a specialist of this approach. See Richard de Neufville and Joseph H. Stafford, *Systems Analysis for Engineers and Managers* (New York: McGraw-Hill, 1971); Richard de Neufville, "Role of Systems Analysis in Transportation Curricula," *Highway Research Record*, no. 462 (1973): 18–26.

247. Earl R. Ruiter, "Toward a Better Understanding of the Intervening Opportunities Model," *Transportation Research* 1, no. 1 (May 1967): 47–56.

248. On the setting up of Urban Systems Laboratory, see *MIT Bulletin* 105, no. 3 (1969), 122, 424–425, 477–483, and passim.

249. *Annual Report Urban Systems Laboratory, 1973–1974* (Urban Systems Laboratory Records, 1968–1974), 1, MIT Archival Collection—AC 366, folder "Description—USL Annual Report, 1973–74."

250. Marvin L. Manheim, "Research Proposals—Urban Systems Laboratory and Curriculum Development," Memorandum to C. L. Miller, Department of Civil Engineering, MIT, May 16, 1968 (Manheim's Personal Records, MC-0330, Box 3, Folder 19).

251. Marvin L. Manheim and Earl R. Ruiter, "DODOTRANS I: A Decision-Oriented Computer Language for Analysis of Multimode Transportation Systems," *Highway Research Record*, no. 314 (1970): 135–163.

252. Massachusetts Institute of Technology/Center for Transportation Studies, "Development Plan," typescript (Cambridge, MA, December 1973), Manheim's Personal Records, MC-0330, Box 1, Folder 13.

253. Marvin. L. Manheim, *Fundamentals of Transportation Systems Analysis* (Cambridge, MA: MIT Press, 1979).

254. Detailed information on R&D studies funded by the BPR in the 1960s can be found in the Bureau of Public Roads, *Highway Research and Development Studies Using*

*Federal-Aid Research and Planning Funds* (Washington, DC: U.S. Government Printing Office, n.d.), various fiscal years; U.S. Department of Transportation, FHWA, NHSB, *R&D Highway & Safety Transportation System Studies 1970* (in Progress during Fiscal Year 1970, July 1, 1969–June 30, 1970) (Washington, DC: U.S. Government Printing Office, n.d.).

255. NCHRP was born in 1962; it was sponsored by AASHO and the Bureau of Public Roads while being administered by the Highway Research Board (D. Grant Mickle, "The National Cooperative Highway Research Program," *Traffic Quarterly* 20, no. 4 [October 1966]: 483–501).

256. Thus, to mention just a few examples: in 1964 alone the Armour Research Foundation was granted $298,033 (the equivalent of around $2,420,000 in 2018) to study the various factors that influenced the mode choice in urban areas; Louis E. Keefer, the former CATS staffer, was hired to work on the influence of land use on urban travel patterns ($129,568); Alan M. Voorhees & Associates was funded $150,980 to work on the factors and trends in trip lengths; while the Yale Bureau of Highway Traffic was commissioned to evaluate predicted traffic usage of a major highway facility versus actual use ($99,675; see also chapter 4). See National Cooperative Highway Research Program, *Summary of Progress through 1988, Special Edition* (n.p.: Transportation Research Board, 1988), 120, 114, 115, 121.

257. The 1960s also witnessed the (relative) withdrawal from the urban traffic forecasting scene of another old organization involved in the field, the Institute of Traffic Engineers. Yet, the latter would continue sponsoring a series of publications that proved popular in the UTDM social world—including the *Traffic Engineering Handbook* (see chapter 1) and, especially the Institute of Transportation Engineers, *Trip Generation Manual* (Washington, DC: Institute of Transportation Engineers, 2017), now in its tenth edition—while its *Journal* has remained open to UTDM-related publications.

258. On the concept of organizational reputation, and its use in another setting, see Daniel Carpenter, *Reputation and Power. Organizational Image and Pharmaceutical Regulation and the FDA* (Princeton, NJ: Princeton University Press, 2010).

259. "Statement of Functions, Office of Research—Highway Planning Division, 3-28-61" (Mertz's Personal Records, C0050, Box 26, Folder 72).

260. An easily exploitable source of information for the various R&D activities carried out by BPR staffers in urban traffic forecasting in the 1950s and 1960s is the *Annual Report* of the federal agency (renamed *Highway Progress* from 1959–1965, and *Highways & Human Values* in 1966).

261. Lynch et al., "Panel Discussion on Inter-Area Travel Formulas," 138.

262. Brokke and Mertz, "Evaluating Trip Forecasting Methods."

263. Anthony R. Tomazinis, "A New Method of Trip Distribution in an Urban Area," *Highway Research Board Bulletin*, no. 347 (1962): 77–99.

264. U.S. Department of Commerce, *Highway Progress 1961* (Washington, DC: U.S. Government Printing Office, November 1961), 47–48; U.S. Department of Commerce, *Highway Progress 1962* (Washington, DC: U.S. Government Printing Office, November 1962), 36; Walter G. Hansen, "Evaluation of Gravity Model Trip Distribution Procedures," *Highway Research Board Bulletin*, no. 347 (1962): 67–76; Constantine Ben, Richard J. Bouchard, and Clyde E. Sweet, "Simplified Procedures for Determining Travel Patterns Evaluated for a Small Urban Area," *Public Roads* 33, no. 6 (February 1965): 112–124; Richard J. Bouchard and Clyde E. Pyers, "Use of Gravity Model for Describing Urban Travel: An Analysis and Critique," *Highway Research Record*, no. 88 (1965): 1–43; Clyde E. Pyers, "Evaluation of Intervening Opportunities Trip Distribution Model" (with discussion), *Highway Research Record*, no. 114 (1966): 71–98; Kevin E. Heanue and Clyde E. Pyers, "A Comparative Evaluation of Trip Distribution Procedures" (with discussion)," *Highway Research Record*, no. 114 (1966): 20–50 (it was also published in *Public Roads* 34, no. 2 [June 1966]: 43–51); Frank E. Jarema, Clyde E. Pyers, and Harry A. Reed, "Evaluation of Trip Distribution and Calibration Procedures," *Highway Research Record*, no. 191 (1967): 106–129. Clyde E. Pyers was a civil engineer with graduate studies in city and regional planning at the University of California at Berkeley.

265. Concerning assignment procedures, see Mertz, "Review and Evaluation"; Thomas F. Humphrey, "A Report on the Accuracy of Traffic Assignment When Using Capacity Restraint," *Highway Research Record*, no. 191 (1967): 53–75. Regarding the mode choice step, see especially Arthur B. Sosslau, Kevin E. Heanue, and Arthur J. Balek, "Evaluation of a New Modal Split Procedure" (with discussion), *Highway Research Record*, no. 88, (1965): 44–68 (published also in *Public Roads* 33, no. 1 [April 1964]: 5–9 and 12–19).

266. Glenn E. Brokke, "Urban Transportation Planning Computer System," in *Proceedings of National Conference, St. Paul, Minnesota, May 23–24, 1967*, ed. AASHO Committee on Electronics (n.p.: U.S. Department of Transportation, n.d.), 195–219.

267. Stephen George Jr., Paul Bliss, and Harry Reed, "Travel & Activity Log for Washington, DC, Trip (Traffic Assignment Computer Run at National Bureau of Standards, Washington, DC, June 15–25, 1960)," Minnesota Department of Highways, typescript, n.d. See also the testimony of Arthur Sosslau (more on Sosslau in chapter 4) concerning the Missouri State Highway Department in *Newsletter of the Transportation Planning Computer Program Exchange Group* 1, no. 1 (1963): 2.

268. See, for example, Bureau of Public Roads, *Urban Transportation Planning Course* (Washington, DC: Bureau of Public Roads, May 1970), Mertz's Personal Records, C0050, Box 26, Folder 98. The course proved to be a great success and was also

taught in the following decade. See, for example, FHWA and UMTA, *Urban Travel Demand Forecasting Course, November 3–7, 1975* (Washington, DC: n.p., n.d.).

269. U.S. Department of Commerce/Bureau of Public Roads, *Highways & Human Values* (*The Bureau of Public Roads Report for Fiscal Year 1966*) (Washington, DC: U.S. Government Printing Office, n.d.), 24.

270. Bureau of Public Roads, *Calibrating and Testing a Gravity Model for Any Size Urban Area* (Washington, DC: U.S. Government Printing Office, July 1963), reprinted in October 1965, November 1968, and November 1983; Bureau of Public Roads, *Calibrating and Testing a Gravity Model with a Small Computer* (Washington, DC: U.S. Government Printing Office, October 1963), reprinted in September 1965; Bureau of Public Roads, *Traffic Assignment Manual for Application with a Large, High Speed Computer* (Washington, DC: U.S. Government Printing Office, June 1964); Bureau of Public Roads, *Traffic Assignment and Distribution for Small Urban Areas* (Washington, DC: U.S. Government Printing Office, September 1965); Bureau of Public Roads, *Guidelines for Trip Generation Analysis* (Washington, DC: U.S. Government Printing Office, June 1967), new (updated) edition in 1975; Martin J. Fertal, Edward Weiner, Arthur J. Balek, and Ali F. Sevin (prepared by), *Modal Split: Documentation of Nine Methods for Estimating Transit Usage* (Washington, DC: U.S. Government Printing Office, December 1966).

271. Bureau of Public Roads, *Urban Transportation Planning: General Information and Introduction to System 360* (Washington, DC: U.S. Government Printing Office, June 1970); Bureau of Public Roads, *Urban Transportation Planning System 360: Program Documentation* (Washington, DC: U.S. Government Printing Office, June 1970). See also Transportation Planning Exchange Group, *Minutes* (of the meeting held October 28, 1968) (n.d.), 3–5; Transportation Planning Exchange Group, *Minutes* (of the meeting held May 12, 1969) (n.d.), 2–3; Transportation Planning Exchange Group, *Minutes* (of the meeting held October 20, 1969) (n.d.), 11–16.

272. However, the practitioner could make use of a mass transit planning system for IBM 7090 and 360 machines, developed by Alan M. Voorhees & Associates under contract with the U.S. Department of Housing and Urban Development (see this book's introduction). See also Transportation Planning Exchange Group, *Minutes* (of the meeting held October 28, 1968) (n.d.), 6; Transportation Planning Exchange Group, *Minutes* (of the meeting held May 12, 1969) (n.d.), 8.

273. Transportation Planning Exchange Group, *Minutes* (of the meeting held May 17, 1971) (n.d.), 5.

274. To obtain the IBM 360 battery one could just send a magnetic tape to the BPR office in Washington, DC. There was no charge for state highway departments, local transportation study groups, or local government bodies. Everyone else had to pay $40 to reimburse the government for copying the tape.

275. Mikaela Sundberg "Organizing Simulation Code Collectives," *Science & Technology Studies* 23, no. 1 (2010): 37–57, which contains a rich bibliography.

276. After the BPR disbanded in 1970, the UTP would continue to grow in the 1970s under the aegis of the Federal Highway Administration (see chapter 3).

277. Such as Peat, Marwick, Livingston & Co./Peat, Marwick, Mitchell & Co., Alan Manner Voorhees & Associates, and De Leuw, Cather & Company.

278. Such as the Institute of Transportation and Traffic Engineering and the University of California at Los Angeles.

279. Such as the Service Bureau Corporation, a subsidiary of IBM set up in 1957.

280. An excellent source for the study of this collective endeavor is the *Newsletter* issued by the Transportation Planning Computer Program Exchange Group that was set up in the early 1960s under the aegis of the Bureau of Public Roads (the first issue of the *Newsletter* was released on July 15, 1963).

**Chapter 3**

1. Rose and Mohl, *Interstate*.

2. *Survey Methodology: A Journal Published by Statistics Canada* 21, no. 1 (June 1995), contains a special memorial section in honor of Stanley L. Warner, including a bibliography of the author's principal publications.

3. Warner, *Stochastic Choice of Mode in Urban Travel*, acknowledgments and 4.

4. "Report of the President-Elect's Task Force on Transportation," typescript, dated January 5, 1969 (the copy I have consulted can be found in Manheim's Personal Records, MC-0330, Box 5, Folder 32); emphasis mine.

5. Raymond A. Mohl, "Citizen Activism and Freeway Revolts in Memphis and Nashville. The Road to Litigation," *Journal of Urban History* 40, no. 5 (September 2014): 870–893; Rose and Mohl, *Interstate*, chapter 9. For a contemporary criticism, see Ben Kelley, *The Pavers and the Paved* (New York: Donald W. Brown, 1971). See also the correspondence and other papers from the period of Francis C. Turner (1908–1999), https://planeandtrainwrecks.com/Search, accessed November 25, 2020. Turner was director of the BPR from 1957 to 1969 and appointed Federal Highway Administrator at the Department of Transportation in February 1969.

6. Rose and Mohl, *Interstate*, chapter 10; Raymond A. Mohl, "The Interstates and the Cities: The U.S. Department of Transportation and the Freeway Revolt, 1966–1973," *Journal of Policy History* 20, no. 2 (April 2008): 193–226.

7. David W. Jones Jr., *Urban Transit Policy: An Economic and Political History* (Englewood Cliffs, NJ: Prentice-Hall, 1985); Smerk, *The Federal Role in Urban Mass*

*Transportation*; Michael E. Kraft, "U.S. Environmental Policy and Politics: From the 1960s to the 1990s," *Journal of Policy History* 12, no. 1 (January 2000): 17–42; Edward Weiner, *Urban Transportation Planning in the United States: History, Policy, and Practice* (New York: Springer, 2016).

8. Highways and Urban Development, *Report on the Second National Conference* (Williamsburg: n.p., 1965).

9. As for April 1990, some forty HOV facilities in twenty metropolitan areas were in operation across the country. See Katherine F. Turnbull (prepared by), *HOV Project Case Studies. History and Institutional Arrangements* (Washington, DC: U.S. DOT, 1990), 2.

10. For an early systematic criticism of the aggregate nature of the four-step model and a plea for an alternative way of modeling travel, see Charles River Associates, *A Model of Urban Passenger Travel Demand in the San Francisco Metropolitan Area* (Boston: Charles River Associates, December 1967); Gerald Kraft and Martin Wohl, "New Directions for Passenger Demand Analysis and Forecasting," *Transportation Research* 1, no. 3 (November 1967): 205–230. See also Peter R. Stopher and Thomas E. Lisco, "Modelling Travel Demand: A Disaggregate Behavioral Approach—Issues and Applications," *Transportation Research Forum Papers—Eleventh Annual Meeting* 11 (1970): 195–214. For an early criticism of aggregation procedures regarding trip generation in particular (the first stage within FSM), see Oi and Shuldiner, *An Analysis of Urban Travel Demands*.

11. It is worth noting that variability within zones can be much greater even than between zones. See Christopher R. Fleet and Sydney R. Robertson, "Trip Generation in the Transportation Planning Process," *Highway Research Record*, no. 240 (1968): 11–31; Harold Kassoff and Harold D. Deutschman, "Trip Generation: A Critical Appraisal," *Highway Research Record*, no. 297 (1969): 15–30.

12. Daniel L. McFadden, "The Theory and Practice of Disaggregate Demand Forecasting for Various Modes of Urban Transportation," in *Emerging Transportation Planning Methods*, ed. William F. Brown, Robert B. Dial, David S. Gendell, and Edward Weiner (Washington, DC: U.S. Department of Transportation, 1978), 1–27 (on 2); emphasis mine.

13. The same sample indicates that model development and testing accounted for 6 percent of the study budget, and forecasting for another 6 percent. See Thomas J. Hillegass, "Urban Transportation Planning—A Question of Emphasis," *Traffic Engineering* 39 (June 1969): 46–48. Thomas James Hillegass (1943–2015) was a civil engineer who graduated from Villanova University and had a master's degree from Purdue University. In the 1970s, Hillegass closely worked with Robert Dial at UMTA, and was heavily involved in the very popular course in urban travel demand modeling organized then by the DOT (see FHWA and UMTA, *Urban Travel Demand Forecasting Course, November 3–7, 1975*; "Dial_hudhistory_8.15.07.doc").

14. On the distinction (and its history) between "statistical" and "structural" (or "causal") modeling, a distinction familiar to economists in the 1960s, see, for example, Piotr Tarka, "An Overview of Structural Equation Modeling: Its Beginnings, Historical Development, Usefulness and Controversies in the Social Sciences," *Quality& Quantity* 52, no. 1 (January 2018): 313–354. For an early treatment of the issue, see Guy H. Orcutt, "Actions, Consequences, and Causal Relations," *Review of Economics and Statistics* 34, no. 4 (November 1952): 305–313. For an early application of "structural modeling" to the trip generation step, see the study coauthored by two academics: Edward J. Kannel and Kenneth W. Heathington, "Structural Model for Evaluating Urban Travel Relationships," *Transportation Research Record*, no. 526 (1974): 73–82.

15. W. S. Robinson, "Ecological Correlations and the Behavior of Individuals," *American Sociological Review* 15, no. 3 (June 1950): 351–357. On the problem of the "ecological fallacy" regarding especially the trip generation models, see the early remarks by Kassoff and Deutschman, "Trip Generation"; Gerald M. McCarthy, "Multiple-Regression Analysis of Household Trip Generation—A Critique," *Highway Research Record*, no. 297 (1969): 31–43. The book coauthored by de Neufville and Stafford, *Systems Analysis for Engineers and Managers*, seemed to have played an important part in sensitizing transportation modelers to the "ecological fallacy" issue.

16. For a telling example showing how a model tending to predict the mode share on the sole basis of the walk time while ignoring the family size (the "hidden" variable) can lead to inaccurate forecasts, see McFadden, "The Theory and Practice," 7.

17. The quotation is drawn from Moshe Emanuel Ben-Akiva, "Structure of Passenger Travel Demand Models" (Ph.D. diss., MIT, 1973), 57; Ben-Akiva draws the expression "self-fulfilling prophecies" from Meyer and Straszheim, *Pricing and Project Evaluation*, 134.

18. See Daniel McFadden, "Disaggregate Behavioral Travel Demand's RUM Side—A 30 Years Retrospective," in *Travel Behaviour Research: The Leading Edge*, ed. David Hensher (Amsterdam: Pergamon, 2001), 17–63 (on 17).

19. Alice O'Connor, *Poverty Knowledge: Social Science, Social Policy and the Poor in Twentieth Century U.S. History* (Princeton, NJ: Princeton University Press, 2002); Michael A. Berstein, *A Perilous Progress: Economists and Public Purpose in 20th Century America* (Princeton, NJ: Princeton University Press, 2001).

20. Be that as it may, the UTDM social world would continue being interested in the other steps of the FSM in the 1970s and 1980s. For distribution and assignment steps, see this chapter and chapter 4. Regarding trip generation in the 1970s and the early 1980s, one can consult for example Robert L. Smith Jr. and Donald E. Cleveland, "Time-Stability Analysis of Trip-Generation and Predistribution Modal-Choice Models," *Transportation Research Record*, no. 569 (1976): 76–86; Lawrence C.

Caldwell III and Michael J. Demetsky, "Transferability of Trip Generation Models," *Transportation Research Record*, no. 751 (1980): 56–62; "Advances in Trip Generation and Quantitative Methods," special issue, *Transportation Research Record*, no. 874 (1982).

21. Thomas Edward Lisco, "The Value of Commuters' Travel Time: A Study in Urban Transportation" (Ph.D. diss., University of Chicago, 1967), 1, 31, and 62 for the quotations.

22. Peter L. Watson, *The Value of Time: Behavioral Models of Modal Choice* (Lexington, MA: Lexington Books, 1974), about the author.

23. Peter Robert Stopher, "Factors Affecting Choice of Mode of Transport" (Ph.D. diss., University College [London], 1967).

24. Warren T. Adams, "Factors Influencing Mass-Transit and Automobile Travel in Urban Areas," *Public Roads* 30, no. 11 (December 1959): 256–260 (published also in *Highway Research Board Bulletin*, no. 230 [1959]: 101–111).

25. As Stopher put it in his comments on the draft of this book: "my Ph.D. was actually the first logit model, even though I did not know that name for it" (on the logit model see below).

26. In 1970–1971 Stopher was at McMaster University (Canada), and from 1971–1973 at Cornell University. He returned to Northwestern University in 1973, which he left definitively in 1979.

27. Peter R. Stopher, "Lecture Notes of Urban Models, 1970/71," typescript, copyright © P. R. Stopher 1970, 1973; Peter R. Stopher, "Transportation Analysis Methods," typescript, copyright © P. R. Stopher, 1971 (the preface dated September 1970).

28. Just one example: Daniel Brand, then at Harvard, corresponded with Stopher in the early 1970s. I was able to examine this correspondence when I interviewed Brand in Hanover, New Hampshire, on June 27, 2012.

29. Among the first master's theses and Ph.D. dissertations supervised by Stopher that dealt with the new modeling approach were: Paul F. Inglis, "A Multimodal Logit Model of Modal Split for a Short Journey" (master's thesis, McMaster University, 1971); Gregory C. Nicolaïdis, "Quantification of the Comfort Variable. An Application of Multidimensional Scaling Techniques: Use in a Binary Disaggregate Mode Choice Model" (Ph.D. diss., Cornell University, 1973); Bruce Spear, "The Development of a Generalized Convenience Variable for Models of Mode Choice" (Ph.D. diss., Cornell University, 1974).

30. See, for example, Peter Stopher (principal investigator), *A Method for Understanding and Predicting Destination Choices*, prepared for U.S. Department of Transportation (Washington, DC: n.p., 1979).

31. Peter R. Stopher, "A Probability Model of Travel Mode Choice for the Work Journey," *Highway Research Record*, no. 283 (1969): 57–65. For this article, Stopher was awarded the Fred Burggraf Award by the Highway Research Board for researchers aged thirty-five or under.

32. Stopher and Lisco, "Modelling Travel Demand."

33. Arnim H. Meyburg, "An Empirical Study of the Relationship Between the Intercity Passenger Transportation System and the Social State of Metropolitan Areas in the Northeast Corridor" (Ph.D. diss., Northwestern University, 1971), biographical sketch.

34. Stopher and Meyburg, *Urban Transportation Modeling and Planning*.

35. Charles Arthur Lave, "Modal Choice in Urban Transportation: A Behavioral Approach" (Ph.D. diss., Stanford University, 1968), acknowledgments; 76 for the quotations.

36. Robert Gordon McGillivray, "Binary Choice of Transport Mode in the San Francisco Bay Area" (Ph.D. diss., University of California at Berkeley, 1969).

37. See, for example, Dan G. Haney, *The Value of Travel Time for Passenger Cars: A Preliminary Study* (Menlo Park, CA: Stanford Research Institute, January 1963); Thomas C. Thomas and Gordon I. Thompson, "Value of Time Saved by Trip Purpose with Discussion and Closure" (with discussion), *Highway Research Record*, no. 369 (1971): 104–117.

38. Karilyn Crockett, *People before Highways: Boston Activists, Urban Planners, and a New Movement for City Making* (Amherst, MA: University of Massachusetts Press, 2018).

39. Cambridge Task Force on Transportation Policy, "Untitled document (Confidential)," typescript, December 11, 1968, Manheim's Personal Records, MC-0330, Box 5, Folder 32. On Manheim, see chapter 2.

40. David Warsh, *Economic Principals: Masters and Mavericks of Modern Economics* (New York: The Free Press, 1993), 145; John R. Meyer and Gerald Kraft, "The Evaluation of Statistical Costing Techniques as Applied in the Transportation Industry," *The American Economic Review* 51, no. 2 (May 1961): 313–334.

41. The Brookings Institution, *Transport Research Program: Progress Report, June 12, 1964–December 12, 1964* (Washington, DC: The Brookings Institution, December 1964), 7.

42. Systems Analysis and Research Corporation, *Demand for Intercity Passenger Travel in the Washington-Boston Corridor* (Boston and Washington, DC: n.p., n.d., ca. 1963). Kraft was the main author of part V (see Kraft and Wohl, "New Directions for Passenger Demand Analysis and Forecasting," 217).

43. Gerald Kraft, "Criteria for Transportation Investment," typescript, n.d., Marvin Manheim's Personal Records, MC-0330, Box 5, Folder 32.

44. The other cofounder of CRA was Alan Rush Willens (1936–2016), with a master's degree in business administration from the University of Michigan in 1958 (information provided by Daniel Brand during an Internet meeting, dated July 1, 2021).

45. On the history of the firm from 1965–1980, see the brochure *Charles River Associates Incorporated, 1965–1980* (n.p.: n.p, n.d.). In 2009, the transportation planning practice of CRA was acquired by Steer Davies Gleave.

46. Charles River Associates, *A Model of Urban Passenger Travel Demand*; Thomas A. Domencich, Gerald Kraft, and Jean-Paul Valette, "Estimation of Urban Passenger Travel Behavior: An Economic Demand Model," *Highway Research Record*, no. 238 (1968): 64–78.

47. The authors referred in particular to the 1952 edition of *The Theory of Price* authored by G. J. Stigler, first published in 1946.

48. For 240 zonal interchanges, one hour of IBM 360/65 time, and six man-months of effort, exclusive of data collection, were required for the Boston case (the model had thirteen independent variables). This was led by Daniel Brand, then a PML staffer. For comparative purposes, in the early 1970s, calibration of the gravity model and the intervening opportunity model, exclusive of initial data preparation, normally required one to three man-weeks. See Peat, Marwick, Mitchell & Co., *A Review of Operational Urban Transportation Models* (Washington, DC: n.p., 1973), 118, 91, 94.

49. U.S. Department of Transportation, *Highway Transportation Research and Development Studies 1972* (Washington, DC: U.S. Government Printing Office, n.d.), 439.

50. Unless otherwise mentioned, I draw from McFadden, "Disaggregate Behavioral Travel Demand's RUM Side."

51. Phoebe Humphrey Cottingham, "Decision Rules in a Public Bureaucracy: An Examination of Highway Planning" (Ph.D. diss., University of California at Berkeley, 1969).

52. Robert Duncan Luce, *Individual Choice Behavior. A Theoretical Analysis* (New York: Wiley, 1959). On the origins of the axiom, see R. Duncan Luce, "Luce's Choice Axiom," *Scholarpedia* 3, no. 12 (2008): 8077.

53. I draw upon Ben-Akiva, "Structure of Passenger Travel Demand Models," 103.

54. Daniel McFadden, "The Revealed Preferences of a Government Bureaucracy: Theory," *Bell Journal of Economics* 6, no. 2 (Autumn 1975): 401–416; Daniel McFadden, "The Revealed Preferences of a Government Bureaucracy: Empirical Evidence," *Bell Journal of Economics* 7, no. 1 (Spring 1976): 55–72.

55. Charles River Associates, *A Disaggregated Behavioral Model of Urban Travel Demand*, final report prepared under Contract No. FH-11-7566 for FHWA/DOT (n.p.: n.p., March 1972), especially 1-2, 6-9–6-14, 1-5.

56. Thomas A. Domencich and Daniel McFadden, *Urban Travel Demand: A Behavioral Analysis* (*A Charles River Associates Research Study*) (Amsterdam and Oxford: North-Holland Publishing Company; New York: American Elsevier Publishing Company, 1975). The book was reprinted by *The Blackstone Company* in 1996.

57. I follow here the presentation provided by Joel L. Horowitz, "Travel and Location Behavior: State of the Art and Research Opportunities," *Transportation Research Part A* 19, nos. 5–6 (1985): 441–453.

58. The idea that randomness (probabilities) lies in the eye of the beholder-modeler has been the standard justification for the existence of the utility function's stochastic component. However, for some people randomness is, on the contrary, an inherent part of the decision-making process itself, and probabilities are properties of the world out there. For the objective/subjective issue about probabilities, see Lorraine Daston, "How Probabilities Came to Be Objective and Subjective," *Historia Mathematica* 21, no. 3 (August 1994): 330–344.

59. National Cooperative Highway Research Program, *Summary of Progress through 1988, Special Edition*, 125–127. For the outcomes of these projects, see William B. Tye, Leonard Sherman, Michael Kinnucan, David Nelson, and Timothy Tardiff, *Application of Disaggregate Travel Demand Models* (Washington, DC: Transportation Research Board, 1982).

60. Ben-Akiva, "Structure of Passenger Travel Demand Models," biographical summary.

61. Information provided by Daniel Brand in an interview conducted at the author's house in Hanover, New Hampshire on June 27, 2012.

62. Frank Sanford Koppelman, "Travel Prediction with Models of Individual Choice Behavior" (Ph.D. diss., MIT, 1975). A brief biographical sketch of Koppelman can be found on 330.

63. Steven Richard Lerman, "A Disaggregate Behavioral Model of Urban Mobility Decisions" (Ph.D. diss., MIT, 1975).

64. Charles Frederick Manski, "The Analysis of Qualitative Choice" (Ph.D. diss., MIT, 1973).

65. See Manski's CV: http://faculty.wcas.northwestern.edu/~cfm754/charles_manski_vita.pdf, accessed December 28, 2021.

66. See for example Steven R. Lerman and Charles F. Manski, "Sample Design for Discrete Choice Analysis of Travel Behavior: The State of the Art," *Transportation*

*Research Part A* 13, no. 1 (February 1979): 29–44; M. Ben-Akiva, Ch. F. Manski, and L. Sherman, "A Behavioural Approach to Modelling Household Motor Vehicle Ownership and Applications to Aggregate Policy Analysis," *Environment and Planning* 13A, no. 4 (April 1981): 399–411.

67. See, for example, Charles F. Manski, "Recent Advances in and New Directions for Behavioral Travel Modeling," in *New Horizons in Travel-Behavior Research*, ed. Peter R. Stopher, Arnim H. Meyburg, and Werner Brög (Lexington, MA: Lexington Books, 1981), 73–86; Richard D. Westin and Charles F. Manski, "Theoretical and Conceptual Developments in Demand Modelling," in *Behavioural Travel Modelling*, ed. David A. Hensher and Peter R. Stopher (London: Croom Helm, 1979), 378–392.

68. See especially Massachusetts Institute of Technology, "Description of Transportation Course Offerings, 1978–79," prepared by Center for Transportation Studies, n.d., in particular course no. 1.202 titled "Analysis of Transportation Demand (A)" (Prof. Ben-Akiva); course no. 1.205 titled "Advanced Travel Demand Modeling (A)" (Prof. Ben-Akiva), and course no. 1.206 titled "Transportation and Urban Activity Modeling (A)" (Prof. S. R. Lerman). Manheim's Personal Records, MC-0330, Box 1, Folder 11.

69. Marvin L. Manheim, "*Letter to the Editor*—Transportation Programs at the Center for Transportation Studies, MIT," *Transportation Science* 13, no. 1 (February 1979): 80–81.

70. For example, Ben-Akiva gave a lecture at Northwestern University during the academic year 1984–1985 (Transportation Center/Northwestern University, *Annual Report, 1984–85 Academic Year* [n.p.: n.dp., n.d.], 18). Koppelman coauthored with Lerman a manual on disaggregate modeling (more follows) and used the now classic book written by Ben-Akiva and Lerman in 1985 (more follows) in his teaching. On Koppelman and his contributions to travel demand modeling, see *Transportation Research Part B* 42, no. 3 (March 2008): 185–190 (the entire issue is a tribute to the career of Koppelman).

71. Formerly the Institute of Transportation and Traffic Engineering. See chapter 2.

72. Daniel L. McFadden, "The Path to Discrete-Choice Models," *Access*, no. 20 (Spring 2002): 2–7 (on 4).

73. McFadden, "The Path to Discrete-Choice Models," 5.

74. Daniel McFadden, Antti P. Talvitie, and Associates, *The Urban Travel Demand Forecasting Project, Phase 1 Final Report Series,* vol. 5: *Demand Model Estimation and Validation* (Berkeley: Institute of Transportation Studies, 1977), iv.

75. Daniel L. McFadden, Fred A. Reid, Antti P. Talvitie, Michael A. Johnson, and Associates, *The Urban Travel Demand Forecasting Project, Final Report Series,* vol. 1:

*Overview and Summary: Urban Travel Demand Forecasting Project* (Berkeley: Institute of Transportation Studies, 1979), xi.

76. McFadden et al., *Urban Travel Demand Forecasting Project, Final Report Series*, appendix 1.

77. Kenneth E. Train, University of California, Berkeley, http://elsa.berkeley.edu/~train/, accessed November 25, 2020.

78. Kenneth E. Train, *Discrete Choice Methods with Simulation* (Cambridge: Cambridge University Press, [2003] 2009). The author who characterized the book as "masterful" is David Hensher (more on him later) of the University of Sydney.

79. Rick Donnelly, "Lifelong Education as a Necessary Foundation for Success in Travel Modeling," in *Innovations in Travel Demand Modeling, Volume 2: Papers*, ed. Transportation Research Board (Washington, DC: Transportation Research Board, 2008), 121–123 (on 122).

80. Carlos Daganzo, *Multinomial Probit: The Theory and Its Application to Demand Forecasting* (New York: Academic Press, 1979).

81. McFadden, "The Theory and Practice," 21.

82. See, inter alia: the survey article by Steven R. Lerman, "Mathematical Models of Travel Demand: A State-of-the Art Review," in *Travel Analysis Methods for the 1980s*, ed. Transportation Research Board (Washington, DC: Transportation Research Board, 1983), 114–127; the survey article drawn up by Horowitz ("Travel and Location Behavior") for one of the workshops held at the James L. Allen Center at Northwestern University in March 1985.

83. A technique known as the "maximum likelihood estimation" has therefore been used.

84. See, for example, the special issue of *Transportation Research Part A* 16, nos. 5–6 (September–November 1982); the various proceedings of the conferences organized first by TRB Travel Behavior and Values Committee, and, later, by the International Association for Travel Behaviour Research.

85. The "nested logit," developed first by Ben-Akiva in his Ph.D. dissertation, was a (successful) attempt to overcome some of the IIA-related drawbacks. For further developments in discrete-choice modeling in general, its logit variant in particular, see M. Ben-Akiva and M. Bierlaire, "Discrete Choice Methods and Their Applications to Short Term Travel Decisions," in *Handbook of Transportation Science*, ed. M. Hall (Dordrecht: Kluwer Academic Publishers, 1999), 5–33; McFadden, "Disaggregate Behavioral Travel Demand's RUM Side"; Train, *Discrete Choice Methods with Simulation*.

86. Joel Horowitz to Charles Hedges, dated March 21, 1977, Manheim's Personal Records, MC-0330, Box 6, Folder 17.

87. See for example Janusz Supernak, "Transportation Modeling: Lessons from the Past and Tasks for the Future," *Transportation* 12, no. 1 (August 1983): 79–90; Janusz Supernak and Walter R. Stevens, "Urban Transportation Modeling: The Discussion Continues," *Transportation* 14, no. 1 (March 1987): 73–82. It is worth noting that in the early 1980s, a series of studies showed that, under certain conditions, disaggregate and aggregate models (of the gravity family) could be seen as two mathematically equivalent views of the same structure. See for example Andrew Daly, "Estimating Choice Models Containing Attraction Variables," *Transportation Research Part B* 16, no. 1 (February 1982): 5–15; Alex Anas, "Discrete Choice Theory, Information Theory and the Multinomial Logit and Gravity Models," *Transportation Research B* 17, no. 1 (February 1983): 13–23.

88. For the 1970s and the early 1980s, see for example Stopher and Meyburg, *Urban Transportation Modeling and Planning*; Manheim, *Fundamentals of Transportation Systems Analysis*; Kanafani, *Transportation Demand Analysis*; Michael D. Meyer and Eric J. Miller, *Urban Transportation Planning: A Decision-Oriented Approach* (New York: McGraw-Hill, 1984); John W. Dickey (senior author) et al., *Metropolitan Transportation Planning*, 2nd ed. (New York: McGraw-Hill, 1983).

89. Frank Satlow to Marvin Manheim, dated February 5, 1979, Manheim's Personal Records, MC-0330, Box 1, Folder 9.

90. Moshe Ben-Akiva and Steven R. Lerman, *Discrete Choice Analysis: Theory and Application to Travel Demand* (Cambridge, MA: MIT Press, 1985).

91. For a survey of the theoretical developments in discrete choice modeling after 1990, see Boyce and Williams, *Forecasting Urban Travel*, 229–239, 486–493.

92. Lerman, "Mathematical Models of Travel Demand"; Joel. L. Horowitz, "Evaluation of Discrete-Choice Random-Utility Models as Practical Tools of Transportation Systems Analysis," in Transportation Research Board, *Travel Analysis Methods for the 1980s*, 127–136.

93. Transportation Research Board, *Travel Analysis Methods for the 1980s*.

94. Peat, Marwick, Mitchell & Co., *A Review of Operational Urban Transportation Models*, 109–112 (a short bibliography is given on 112).

95. See also Paul O. Roberts Jr., "Disaggregate Demand Forecasting: Theoretical Tantalizer or Practical Problem Solver?," in Brown et al., *Emerging Transportation Planning Methods*, 29–45; Tye et al., *Application of Disaggregate Travel Demand Models*.

96. For the years 1987 and 1988, see the relevant information contained in Institute of Transportation Engineers, *Undergraduate and Graduate Transportation Programs Directory* (Washington, DC: ITE, 1988).

97. Robert Henri Binder, "Major Issues in Travel Demand Forecasting," in Brand and Manheim, *Urban Travel Demand Forecasting*, 13–16 (quotation on 13).

98. Brand and Manheim, *Urban Travel Demand Forecasting*, 313–315.

99. Peter R. Stopher and Arnim H. Meyburg, eds., *Behavioral Demand Modeling and Valuation of Travel Time* (Washington, DC: Transportation Research Board, 1974).

100. Currently the Transportation Research Board ADB 10. On the committee's history, written by Stopher himself (October 2012), see http://depts.washington.edu/trbadb10/history.htm, accessed November 25, 2020. It is worth noting that the Committee on Passenger Travel Demand Forecasting (the reincarnation of the Origin & Destination Studies Committee of the Highway Research Board—see chapter 2), headed by Daniel Brand from 1973 to 1979, would also play an important role in promoting the disaggregate approach. On the various TRB committees, see Peterson, *The Transportation Research Board, 1920–2020*, chapter 11.

101. One of Hartgen's first contributions to the then emerging field of disaggregate modeling was David T. Hartgen and George H. Tanner, "Individual Attitudes and Family Activities: A Behavioral Model of Traveler Mode Choice," *High Speed Ground Transportation Journal* 4, no. 3 (1970): 439–467.

102. Stopher and Meyburg, *Behavioral Demand Modeling* (the composition of the committee for year 1974 is given on 234).

103. For the first session organized by the committee, see *Transportation Research Record*, no. 534 (1975). The issue, with a preface by Stopher, was titled "Travel Behavior and Values" and was comprised of six articles.

104. Cohen, McEvoy, and Hartgen, "Who Reads the Transportation Planning Literature?"

105. From the mid-1980s on, this organizing role was performed by International Association for Travel Behaviour Research, set up in April 1985 as "an international organization of scholars, researchers, practitioners, consultants, and public agency professionals dedicated to the advancement of travel behavior research" (http://www.iatbr.org/). On the conferences organized first by the TRB Travel Behavior and Values Committee and later by IATBR (seventeen conferences from 1973–2018), and the resulting publications, see International Conferences on Travel Behaviour Research Organized by IATBR, http://iatbr.weebly.com/iatbr-conferencesbooks.html, accessed November 25, 2020.

106. Peter R. Stopher and Arnim H. Meyburg, eds., *Behavioural Travel Demand Models* (Lexington, MA: Lexington Books, 1976); Hensher and Stopher, *Behavioural Travel Modelling*.

107. Ryuichi Kitamura's review of *Behavioural Research for Transport Policy* (Utrecht, The Netherlands: VNU Science Press, 1985), in *Transportation Science* 21, no. 3 (August 1987): 218–222 (on 219).

108. Brown et al., *Emerging Transportation Planning Methods*.

109. Bruce D. Spear, *Applications of New Travel Demand Forecasting Techniques to Transportation Planning: A Study of Individual Choice Models* (Washington DC: U.S. Government Printing Office, 1977).

110. Urban Mass Transportation Administration renamed Federal Transit Administration in 1991.

111. Joel L. Horowitz, Frank S. Koppelman, and Steven R. Lerman, *A Self-Instructing Course in Disaggregate Mode Choice Modeling* (Washington, DC: U.S. Department of Transportation, 1986), 3.

112. Frank S. Koppelman and Chandra Bhat (with technical support from Vaneet Sethi, Sriram Subramanian, Vincent Bernardin, and Jian Zhang), *A Self Instructing Course in Mode Choice Modeling: Multinomial and Nested Logit Models*, prepared for U.S. Department of Transportation/Federal Transit Administration (n.p.: n.p., January 2006; modified June 30, 2006), 5.

113. *The Urban Transportation Monitor* (November 10, 2006), 11.

114. For PMM's software, see Peat, Marwick, Mitchell & Co., *A Review of Operational Urban Transportation Models*, 110–112; CRA was able to use McFadden's "logit machinery." For Stopher's computer program, see Peter R. Stopher and John O. Lavender, "Disaggregate, Behavioral Travel Demand Models: Empirical Tests of Three Hypotheses," *Transportation Research Forum Proceedings—Thirteen Annual Meeting* 13, no. 1 (1972): 321–336 (on 325).

115. On Dial and UTPS, see also this book's introduction.

116. ITE Technical Council Committee 6F-34, "Refinement of Traffic Forecasts: Practitioners and Procedures (An Informational Report)," *ITE Journal* (February 1990): 43–47 (on 45).

117. *Journal of Research of the National Institute of Standards and Technology* 111, no. 2 (March–April 2006): iii–iv. On ULOGIT, see especially Urban Mass Transportation Administration/U.S. DOT, *UMTA Transportation Planning System Reference Manual* (Washington, DC: n.p., June 1975) (it is worth noting that ULOGIT is not featured in an older version of the UTPS Reference Manual dated April 1974). On the use of ULOGIT by practitioners, see for example Snehamay Khasnabis, Michael J. Cynecki, and Mark A. Flak, "Systematic Calibration of Multinomial Logit Models," *Journal of Transportation Engineering* 109, no. 2 (March 1983): 209–231. On UTPS in general in the early 1980s, see also Robert H. Watkins, William R. Wolfe, with the assistance of Jon Kowolaski and Henry Lieu, *A UTPS Lexicon* (n.p.: n.p., July 6, 1981; reprint May 1982).

118. QUAIL stands for the QUAlitative, Intermittent, and Limited dependent variable statistical program. On QUAIL, see McFadden et al., *The Urban Travel Demand Forecasting Project, Final Report Series*, chapter 3 and appendix C.

119. "Dial_hudhistory_8.15.07.doc"; see note 5 in this book's introduction.

120. Dial to Ben-Akiva, dated April 4, 1977, Manheim's Personal Records, MC-0330, Box 6, Folder 17.

121. One can mention: The ALOGIT package, the original version of which goes back to the mid-1980s. See ALOGIT's "Ownership and History," http://www.alogit.com/History_and_ownership.htm, accessed November 26, 2020; and chapter 6 of this book. See also National Bureau of Economic Research's TROLL package, which had LOGIT routines (see Tye et al., *Application of Disaggregate Travel Demand Models*, 40); the LIMDEP and NLOGIT computer suites, both marketed by Econometric Software, Inc. For these and a series of other available packages, see also Charles G. Renfro, ed., "A Compendium of Existing Econometric Software Packages," *Journal of Economic and Social Measurement* 29, nos. 1–3 (2004): 359–409; Joseph Molloy, Felix Becker, Basil Schmid, and Kay W. Axhausen, "Mixl: An Open-source R Package for Estimating Complex Choice Models on Large Datasets," *Journal of Choice Modelling* 39 (June 2021), https://www.sciencedirect.com/science/article/pii/S1755534521000178, accessed January 15, 2022.

122. *The Urban Transportation Monitor* (September 15, 2006): 12–16; *The Urban Transportation Monitor* (September 29, 2006): 11–15; *The Urban Transportation Monitor* 26, no. 7 (September 28, 2012): 17–31; *The Urban Transportation Monitor* 28, no. 8 (October 17, 2014): 16–30.

123. Interviews with Lance Neumann—a Ph.D. holder from MIT in 1976, the president and CEO of the firm from 1986 to 2011, and the Chairman of the firm's Board of Directors from 2011-2021; and Thomas Francis Rossi, leading travel modeler, at Cambridge Systematics premises on June 23, 2011 and June 28, 2012.

124. Manheim's Personal Records, MC-0330, Box 1, Folder 16.

125. On Ruiter, see chapter 2.

126. Leonard Sherman, "The Impacts of the Federal Aid Highway Program on State and Local Highway Expenditures" (Ph.D. diss., MIT, 1975).

127. Hanna P. H. Kollo and Charles L. Purvis, "Regional Travel Forecasting Model System for the San Francisco Bay Area," *Transportation Research Record*, no. 1220 (1989): 58–65 (especially 58–59).

128. Cambridge Systematics was part of a larger consultant team including COMSIS Corporation (see chapter 4) and Barton-Aschman Associates (chapter 2). On the San Francisco modeling system, see Earl R. Ruiter and Moshe Ben-Akiva, "Disaggregate Travel Demand Models for the San Francisco Bay Area: System Structure, Component Models, and Application Procedures," and Moshe E. Ben-Akiva, Len Sherman, and Brian Kullman, "Non-Home-Based Models," both in *Transportation Research Record*, no. 673 (1978): 121–128, and 128–133, respectively. The comments

by Frederick C. Dunbar (Charles River Associates) as well as by Gordon A. Shunk and Hanna P. H. Kollo (both from the Metropolitan Transportation Commission), also appeared in the aforementioned issue of the journal (on 133–136). Shunk and Kollo are the authors of the quotation, which can be found on 135. For a detailed presentation of this modeling endeavor, see Cambridge Systematics, *Travel Model Development Project, Phase 2: Final Report* (three volumes) (Cambridge, MA: n.p., 1980). On the undertaking of this modeling, see also Gregory Louis Newmark, "Implementing Innovation in Planning Practice: The Case of Travel Demand Forecasting" (Ph.D. diss., University of California at Berkeley, 2011).

129. See chapter 2.

130. Deen, "RDSA/AMV."

131. Another academic who played a part in spreading the disaggregate approach to practitioners was Peter Stopher, who worked in consulting from 1980 to 1990. In 1980 Stopher joined Schimpeler-Corradino Associates, a small transportation engineering and planning firm out of Kentucky, and while working with them he headed up a number of projects that included setting up disaggregate mode choice models in such places as Honolulu, San Juan, Miami, Florida in general, and Los Angeles.

132. Atlanta, Baltimore, Boston, Dallas-Fort Worth, Denver, Detroit, Philadelphia, Pittsburgh, St. Louis, and Washington, DC.

133. Institute of Transportation Engineers, *Travel Demand Forecasting Processes Used by Ten Large Metropolitan Planning Organizations* (Washington, DC: ITE, 1994).

134. William A. Martin and Nancy A. McGuckin, *Travel Estimation Techniques for Urban Planning* (Washington, DC: Transportation Research Board, 1998), chapter 6: "Mode-Choice Analysis."

135. Vanasse Hangen Brustlin (VHB), *Determination of the State of the Practice in Metropolitan Area Travel Forecasting: Findings of the Surveys of Metropolitan Planning Organizations* (n.p.: n.p., 2007), 55–56.

136. Institute of Transportation Engineers, *Travel Demand Forecasting Processes*; VHB, *Determination of the State of the Practice in Metropolitan Area Travel Forecasting.*

137. VHB, *Determination of the State of the Practice in Metropolitan Area Travel Forecasting*, 51. However, as of 2018, disaggregate destination choice models were used, instead of gravity family models, by almost all of the top twenty-five largest metropolitan areas in the country as well as by a growing number of small and mid-sized MPOs. See Vince Bernardin, Clint Daniels, and Jason Chen, *How-To: Model Destination Choice* (Washington, DC: Federal Highway Administration, 2018), 3.

138. It is worth noting that the growing interest in disaggregate mode choice models seemed also to impact on the techniques used in the trip generation step.

See for example Peter R. Stopher and Kathie G. McDonald, "Trip Generation by Cross-Classification: An Alternative Methodology," *Transportation Research Record*, no. 944 (1983): 84–91.

139. Peter R. Stopher, "Data Needs and Data Collection-State of the Practice," in Transportation Research Board, *Travel Analysis Methods for the 1980s*, 63–71 (on 67). One should note here that thanks to the adoption of more sophisticated sampling procedures, such as stratified sampling, even aggregate travel demand models could progressively rely on significantly smaller samples than those traditionally used in the past. According to a seasoned travel modeler such as David Hartgen, if the 1960s could be described as the age of the "dinosaurs" in transportation surveys, the 1970s should be thought of as the age of the "mammals" (David T. Hartgen, "Coming in the 1990s: The Agency-Friendly Travel Survey," *Transportation* 19, no. 2 (May 1992): 79–95). A lot of information on sample sizes used in travel surveys conducted from the late 1980s on is contained in the following works: Peter R. Stopher and Helen M. A. Metcalf, *Methods for Household Travel Surveys* (Washington, DC: Transportation Research Board, 1996); Cambridge Systematics, *Scan of Recent Travel Surveys* (Washington, DC: U.S. Department of Transportation, 1996), appendix B; and especially David T. Hartgen and Elizabeth San Jose (The Hartgen Group), *Costs and Trip Rates of Recent Household Travel Surveys* (n.p.: n.p., 2009). A first account of the shift toward smaller samples in urban traffic forecasting is provided by Chatzis, "La Modélisation des Déplacements Urbains aux États-Unis et en France," chapter 5.

140. Stopher and Metcalf, *Methods for Household Travel Surveys*, 13.

141. *Journal of Transport Economics and Policy* 22, no. 1 (January 1988) (the entire issue is devoted to "Stated Preference Methods in Transport Research"); David A. Hensher, "Stated Preference Analysis of Travel Choices: the State of Practice," *Transportation* 21, no. 2 (May 1994): 107–133; Jordan J. Louviere, David A. Hensher, and Joffre D. Swait (with a contribution of Wiktor Adamowicz), *Stated Choice Methods: Analysis and Applications* (Cambridge: Cambridge University Press, 2000); Mark Bradley, "Important Stated Preference Experimental Design Issues in Recent Transportation Applications," *Transport Reviews* 29, no. 5 (September 2009): 657–663; Joan L. Walker, Yanqiao Wang, Mikkel Thorhauge, and Moshe Ben-Akiva, "D-efficient or Deficient? A Robustness Analysis of Stated Choice Experimental Designs," *Theory and Decision* 84, no. 2 (March 2018): 215–238. For supplementary references (as well as a survey note), see also Boyce and Williams, *Forecasting Urban Travel*, 219–229.

142. The term "revealed preference" seems to have been introduced by Paul Samuelson (1915–2009) in his study of consumer behavior at the end of the 1940s. See for example D. Wade Hands, "Introspection, Revealed Preferences, and Neoclassical Economics: A Critical Response to Don Ross on the Robbins-Samuelson Argument Pattern," *Journal of the History of Economic Thought* 30, no. 4 (December 2008): 453–478.

143. See especially Ottilia Angela Morlok, "Attitudes and Mode Choice Behavior for Work Trips by Chicago Suburbanites" (master's thesis, Graduate School of Loyola University of Chicago, 1974), especially chapter 2, where the author provides a review of a series of relevant research studies carried out up until the early 1970s.

144. J. D. Davidson, "Forecasting Traffic on STOL," *Operational Research Quarterly* 24, no. 4 (December 1973): 561–569; and especially Jordan J. Louviere, Lynn L. Beavers, Kent L. Norman, and Frank C. Stetzer, "Theory, Methodology, and Findings in Mode Choice Behavior," Working Paper no. 11 (Iowa City: The Institute of Urban and Regional Research/The University of Iowa, 1973).

145. Jordan J. Louviere, "A Psychophysical-Experimental Approach to Modeling Spatial Behavior" (Ph.D. diss., University of Iowa, 1973). Chapter 11 of the dissertation is devoted to "Transportation Case Studies."

146. Louviere, Hensher, and Swait, *Stated Choice Methods*.

147. Steven R. Lerman and Jordan J. Louviere, "Using Functional Measurement to Identify the Form of Utility Functions in Travel Demand Models," *Transportation Research Record*, no. 673 (1978): 78–86.

148. George Kocur, Tom Adler, William Hyman, and Bruce Aunet, *Guide to Forecasting Travel Demand with Direct Utility Assessment*, prepared for Urban Mass Transportation Administration (n.p.: n.p., 1982), acknowledgments. In the early 1980s, the SP-based models were labeled "Direct Utility Assessment" (DUA) models.

149. Peter R. Stopher, "On the Application of Psychological Measurements Techniques to the Estimation of Travel Demand," *Environment and Behavior* 9, no. 1 (March 1977): 67–80; and especially Ricardo Dobson, "Data Collection and Analysis Techniques for Behavioral Transportation Planning," *Traffic Quarterly* 31, no. 1 (January 1977): 77–96. For the few initial applications of this approach in the area of transportation see Kocur et al., *Guide to Forecasting Travel Demand*; George Kocur, William Hyman, and Bruce Aunet, "Wisconsin Work Mode-Choice Models Based on Functional Measurement and Disaggregate Behavioral Data," *Transportation Research Record*, no. 895 (1982): 24–32.

150. Cambridge Systematics, *Travel Survey Manual*, 2 vol. (n.p.: n.p., 1996); Hensher, "Stated Preference Analysis of Travel Choices," 108.

151. Peter R. Stopher, "A Review of Separate and Joint Strategies for the Use of Data on Revealed and Stated Choices," *Transportation* 25, no. 2 (May 1998): 187–205 (especially 190–191).

152. Thomas A. Lynch (ed.), *High Speed Rail in the U.S. Super Trains for the Millennium* (London: CRC Press, 1998). For a recent high-speed rail project for which modeling work was based (also) on SP techniques, see Maren Outwater, Kevin Tierney, Mark

Bradley et al., "California Statewide Model for High-Speed Rail," *Journal of Choice Modelling* 3, no. 1 (2010): 58–73.

153. Daniel Brand, Thomas E. Parody, Poh Ser Hsu, and Kevin F. Tierney, "Forecasting High-Speed Rail Ridership," *Transportation Research Record*, no. 1341 (1992): 12–18.

154. See especially Frank S. Koppelman, Chandra R. Bhat, and Joseph L. Schofer, "Market Research Evaluation of Actions to Reduce Suburban Traffic Congestion: Commuter Travel Behavior and Response to Demand Reduction Actions," *Transportation Research Part A* 27, no. 5 (September 1993): 383–393, which contains numerous relevant references.

155. See for example Takayuki Morikawa, "Incorporating Stated Preferences Data in Travel Demand Analysis" (Ph.D. diss., MIT, 1989) (Ben-Akiva was the supervisor of this dissertation); M. Ben-Akiva, M. Bradley, T. Morikawa, J. Benjamin, T. Novak, H. Oppewal, and V. Rao, "Combining Revealed and Stated Preferences Data," *Marketing Letters* 5, no. 4 (October 1994): 335–350; M. A. Bradley and A. J. Daly, "Estimation of Logit Choice Models Using Mixed Stated Preference and Revealed Preference Information," in *Understanding Travel Behavior in an Era of Change*, ed. Peter Stopher and Martin Lee-Gossselin (Oxford: Pergamon Press, 1997), 209–232.

156. See the references cited in note 141 in this chapter.

157. Cambridge Systematics, *Scan of Recent Travel Surveys*, especially 4-2, C-6, C-7, C-12, C-18, C-19.

158. Anthony J. Richardson, Elizabeth S. Ampt, and Arnim H. Meyburg, *Survey Methods for Transport Planning* (Parkville, Victoria: Eucalyptus, 1995); Juan de Dios Ortuzar, *Stated Preference Modelling Techniques* (London: PTRC Education and Research Services Ltd, 2000); D. Pearmain, J. Swanson, E. Kroes, and M. Bradley, *Stated Preference Techniques: A Guide to Practice* (London: Steer Davies Gleave and Hague Consulting Group, [1990] 1991). Both Steer Davies Gleave and Hague Consulting Group are consultancies.

159. In the first half of the 1990s, J. J. Louviere, D. A. Hensher, and A. Shocker offered, for example, in Australia and the United States an annual course titled "Conjoint Measurement: A Short Course" (Hensher, "Stated Preference Analysis of Travel Choices," 108, 132); J. J. Louviere and D. A. Hensher, "Stated Preference Analysis: Applications in Land Use, Transportation Planning and Environmental Economics" was another short course delivered in the United States (Portland, OR), Australia, and Sweden in the 1990s. Louviere also taught stated preference choice modeling and design of choice experiments in the annual MIT summer short course in travel demand modeling from 1984–2006.

160. See chapter 2.

161. Peter Stopher, "The Travel Survey Toolkit: Where to from Here?" in *Transport Survey Methods: Keeping up with a Changing World*, ed. Patrick Bonnel, Martin Lee-Gossselin, Johanna Zmud, and Jean-Loup Mardre (Bingley: Emerald, 2009), 15–46.

162. Werner Brög and Elisabeth Ampt, "State of the Art in the Collection of Travel Behavior Data," in Transportation Research Board, *Travel Analysis Methods for the 1980s*, 48–62; Transportation Research Board, *Conference on Household Travel Surveys: New Concepts and Research Needs* (Washington, DC: National Academy Press, 1996); Transportation Research Board, *Information Needs to Support State and Local Transportation Decision Making into the 21th Century* (Washington, DC: National Academy Press, 1997).

163. Such as Cambridge Systematics, *Travel Survey Manual*. It is worth noting that the manual's appendix K mentions three travel survey textbooks and two transportation planning journals only, while it features eight general surveying textbooks and three marketing research periodicals.

164. Douglas R. Berdie and John F. Anderson, *Questionnaires: Design and Use* (Metuchen, NJ: John Scarecrow Press, 1974); Doug R. Berdie, John F. Anderson, and Marsha A. Niebuhr, *Questionnaires: Design and Use* (Metuchen, NJ: John Scarecrow Press, 1986).

165. In 2007 SBRI was acquired by Abt Associates, the well-known research and consulting firm set up in 1965. Abt SRBI has been fully integrated into Abt Associates as of April 1, 2017.

166. NuStats was bought by the German transportation consultancy and proprietary software firm PTV in 2006 (on PTV, see chapter 4).

167. Carlos Humberto Arce, "Historical, Institutional, and Contextual Determinants of Black Enrollment in Predominantly White Colleges and Universities, 1946 to 1974" (Ph.D. diss., University of Michigan, 1976).

168. Johanna P. Zmud, "Ethnic Identity, Language, and Mass Communication: An Empirical Investigation of Assimilating among U.S. Hispanics" (Ph.D. diss., University of Southern California, 1992).

169. See for example W. Brög, A. H. Meyburg, P. R. Stopher, and M. J. Wermuth, "Collection of Household Travel and Activity Data: Development of a Survey Instrument," in *New Survey Methods in Transport*, ed. E. S. Ampt, A. J. Richardson, and W. Brög (Utrecht: VNU Science Press, 1985), 151–172.

170. A series of other consultancies involved in travel surveys by the mid-1990s are mentioned in Cambridge Systematics, *Travel Survey Manual*.

171. According to a survey published in 1996, the use of consultants was the choice of over 81 percent of the MPOs and states surveyed (Stopher and Metcalf, *Methods for Household Travel Surveys*, 16, 25–26).

172. Faced with the problems of trip misreporting and trip underreporting, travel survey specialists have pinned high hopes on methods that can *automatically* generate data information on travel patterns and practices. No wonder that the UTDM social world has been enamored with GPS from its inception and invested much effort into its potential uses in the design and implementation of travel surveys. See, for example, Jean Wolf, William Bachman, Marcelo Simas Oliveira, Joshua Auld, Abolfazl (Kouros) Mohammadian, Peter Vovsha, *Applying GPS Data to Understand Travel Behavior,* vol. I: *Background, Methods, and Tests* (Washington, DC: Transportation Research Board, 2014); Jean Wolf, William Bachman, Marcelo Oliveira, Joshua Auld, Abolfazl (Kouros) Mohammadian, Peter Vovsha, and Johanna Zmud, *Applying GPS Data to Understand Travel Behavior,* vol. II: *Guidelines* (Washington, DC: Transportation Research Board, 2014). See also Peter Stopher, ed., *New Technologies and Transportation Research: Applications of GPS in Travel Surveys* (Cheltenham, UK: Edward Elgar Publishing Ltd., 2017). For a first historical account of the usage of GPS techniques in travel surveys in the United States, see Chatzis, "La Modélisation des Déplacements Urbains aux États-Unis et en France," chapter 5.

173. On Wolf and GeoStats, bought by the giant Westat (founded in 1963) in 2012, see Chatzis, "La Modélisation des Déplacements Urbains aux États-Unis et en France," chapter 5.

## Chapter 4

1. Larry Joseph LeBlanc, "Mathematical Programming Algorithms for Large Scale Network Equilibrium and Network Design Problems" (Ph.D. diss., Northwestern University, 1973). On LeBlanc's work, see also Larry J. LeBlanc, Edward K. Morlok, and William P. Pierskalla, "An Accurate and Efficient Approach to Equilibrium Traffic Assignment on Congested Networks," *Transportation Research Record,* no. 491 (1974): 12–23.

2. See introduction and chapter 3.

3. See chapter 2.

4. "Dial_hudhistory_8.15.07.doc"; see note 5 in this book's introduction; Hamburg's Personal Records, C0073, Box 6, Folders 2 and 3.

5. Mertz, "Memories of 499." For more on Mertz, see chapter 2.

6. Bureau of Public Roads, *Calibrating and Testing a Gravity Model for Any Size Urban Area* (Washington DC: U.S. Government Printing Office, November 1968, reprint), A-48.

7. On UROAD, see for example Urban Mass Transportation Administration/U.S. DOT, *UMTA Transportation Planning System Reference Manual,* 30; UMTA and FHWA, *Highway Network Analysis: HNET/UROAD Lecture Guide* (Washington, DC: n.p.,

February 1986). It is worth noting that, despite the incorporation of LeBlanc's work into UTPS, few applications of the new method were being made in the 1970s as most practitioners kept sticking to the older assignment algorithms that were devised for the most part in the early 1960s; see R. W. Eash, B. N. Janson, and D. E. Boyce, "Equilibrium Trip Assignment: Advantages and Implications for Practice," *Transportation Research Record*, no. 728 (1979): 1–8 (on 1).

8. See, for example, "Memorandum to Robert Dial from Marvin L. Manheim, subject: 'Proposed Direct Equilibrium Approach Using Exponential Functions,'" typescript, September 24, 1973 (Manheim's Personal Records, MC-0330, Box 8, Folder 16).

9. Note that a two-way street is modeled by two arcs (links) in the opposite direction. On the coding issue, see, for example: Creighton, *Urban Transportation Planning*, 114–120; Peter Stopher, "Spatial Data Issues: A Historical Perspective," in *Handbook of Transport Geography and Spatial Systems*, ed. David A. Hensher, Kenneth J. Button, Kingsley E. Haynes, and Peter R. Stopher (Amsterdam: Elsevier, 2004), 293–308.

10. It is worth noting that empirical studies conducted by Bureau of Public Roads staffers in the first half of the 1960s seemed to show that it is the "total disutility"—a composite measure of travel time, tension (stress), and other factors—and not the sole time, that drivers seek to minimize. See, for example, Richard M. Michaels (reported by), "Attitudes of Drivers Determine Choice between Alternate Highways," *Public Roads* 33, no. 11 (December 1965): 225–236.

11. By the early 1970s transit assignment algorithms were often extensions of algorithms first applied to highway networks, such as the Moore algorithm (see chapter 2). The extension accommodated the peculiarities of transit systems by considering, for example, transfer time from one type of mass transportation to another or to an automobile, the waiting time, and the walking time. For an early transit assignment procedure, see Robert B. Dial, "Transit Pathfinder Algorithm," *Highway Research Record*, no. 205 (1967): 67–85. For a recent comprehensive review of transit assignment models, see Ektoras Chandakas, "Modelling Congestion in Passenger Transit Networks" (Ph.D. diss., Université Paris-Est, 2014).

12. See chapter 2.

13. In order to use the "capacity restraint" principle in assignment techniques, the *link performance function*—i.e., a function relating the speed, and therefore the travel time over a link, to the traffic flow (volume) assigned to the link—must be determined for the various types of transportation facilities. For many years, the most commonly used function was the formula designed by the staff of the BPR, or a variant of it. On the link performance function and its history, see among others Frederick W. Cron, "Highway Design for Motor Vehicles—A Historical Review," *Public Roads* 39, no. 3 (December 1975): 96–108; and, especially, Roger P. Roess, and

Elena S. Prassas, *The Highway Capacity Manual: A Conceptual and Research History, Volume 1: Uninterrupted Flow* (New York: Springer, 2014).

14. For a presentation of the various "constraint capacity" assignment techniques, which can be grouped into two broad classes: the "iterative-based," and the "incremental-based" ones, by the early 1970s, see Martin, Memmott, and Bone, *Principles and Techniques*, 172–191; E. C. Matsoukis, "Road Traffic Assignment—A Review: Part I. Non-Equilibrium Methods," *Transportation Planning and Technology* 11, no. 1 (1986): 69–79; and, especially, Patriksson, *The Traffic Assignment Problem*.

15. Brian V. Martin, *Minimum Path Algorithms for Transportation Planning*, Research Report R63–52 (Cambridge, MA: Department of Civil Engineering/MIT, 1963); Brian V. Martin and Marvin L. Manheim, "A Research Program for Comparison of Traffic Assignment Techniques," *Highway Research Record*, no. 88 (1965): 69–84.

16. National Cooperative Highway Research Program, *Summary of Progress through 1988, Special Edition*, 115.

17. Matthew J. Huber, Harvey B. Boutwell, and David K. Witheford, *Comparative Analysis of Traffic Assignment Techniques with Actual Highway Use* (n.p.: Highway Research Board, 1968), 59.

18. The paper by Robert W. Antonisse, Andrew J. Daly, and Moshe Ben-Akiva, "Highway Assignment Method Based on Behavioral Models of Car Drivers' Route Choice," *Transportation Research Record*, no. 1220 (1989): 1–11, contains a good discussion of the various factors affecting drivers' route choice preferences.

19. Dial, "Probabilistic Assignment", 12.

20. Wallace Alvin McLaughlin, *Traffic Assignment by Systems Analysis*, Joint Highway Research Project, no. 7 (Lafayette, IN: Purdue University, 1965). Dial himself referred to a model devised by Howard Bevis, the former CATS staffer, and his then fellow modeler at Alan M. Voorhees & Associates (Dial, "Probabilistic Assignment," 21).

21. The Briton J. E. Burrell, while working with Freeman, Fox, Wilbur Smith and Associates (chapter 2), also proposed a multipath stochastic assignment method in the second half of the 1960s. Burrell's work is not mentioned in Dial's thesis, however.

22. Dial, "Probabilistic Assignment," 27.

23. Memorandum from Manheim to Dial, "Proposed Direct Equilibrium Approach Using Exponential Functions; MLM [obviously Marvin L. Manheim], "Behavioral Basis of Path Choice," typescript, October 1975; Earl Ruiter to Manheim, dated January 4, 1976 (both in Manheim's Personal Records, MC-0330, Box 8, Folder 16).

24. Manheim and Ruiter, "DODOTRANS I."

25. Urban Mass Transportation Administration/U.S. DOT, *UMTA Transportation Planning System Reference Manual*: "All-or-nothing," "capacity-restraint," and "probabilistic multipath" assignment techniques were all part of the UROAD module. UTP, the large computer suite released by the Bureau of Public Roads at the end of the 1960s (see chapter 2), also included a capacity-restraint assignment program, named CAPRES, written by Alan M. Voorhees & Associates. More follows on CDC and its urban traffic forecasting computer battery.

26. Ross B. Emmett, *Frank Knight and the Chicago School in American Economics* (New York: Routledge, 2009).

27. Frank H. Knight, "Some Fallacies in the Interpretation of Social Cost," *Quarterly Journal of Economics* 38, no. 4 (August 1924): 582–606 (quotation on 585); emphasis mine. Regarding Pigou's formulation, see Arthur Cecil Pigou, *The Economics of Welfare* (London: Macmillan and Co., 1920), 194.

28. T. R. Lakshmanan, "Martin J. Beckmann: A Retrospective," in *Structure and Change in the Space Economy: Festschrift in Honor of Martin J. Beckmann*, ed. T. R. Lakshmanan and Peter Nijkamp (Berlin: Springer-Verlag, 1993), 3–10; David Boyce and Anna Nagurney, "In Memoriam: Martin Beckmann (1924–2017)," *Transportation Science* 53, no. 6 (November–December 2019): 1798–1799.

29. Martin Beckmann, C. Bartlett McGuire, and Christopher B. Winsten, *Studies in the Economics of Transportation*, U.S. Air Force Project RAND RM-1488 (Santa Monica: The RAND Corporation, May 1955) (the passage in question is on 3.3; the authors do not refer to the 1924 article, but to its reprinted version in Kenneth E. Boulding and George S. Stigler, eds., *Readings in Price Theory* [Homewood, IL: Richard D. Irwin, 1952], 160–179); Martin Beckmann, C. Bartlett McGuire, and Christopher B. Winsten, *Studies in the Economics of Transportation* (New Haven: Yale University Press, 1956), 60. We owe the best analyses of the book and its reception to Professor David Boyce. See, for example, David Boyce, "Forecasting Travel on Congested Urban Transportation Networks: Review and Prospects for Network Equilibrium Models," *Networks and Spatial Economics* 7, no. 2 (June 2007): 99–128; Boyce and Williams, *Forecasting Urban Travel*, 301–310.

30. Beckmann, McGuire, and Winsten, *Studies in the Economics of Transportation*, chapter 3 (1956 edition).

31. On the mathematization of American economics after World War II, see inter alia Gerard Debreu, "The Mathematization of Economic Theory," *American Economic Review* 81, no. 1 (March 1991): 1–7.

32. Beckmann, McGuire, and Winsten, *Studies in the Economics of Transportation*, 73 (1956 edition).

33. Boyce, "Forecasting Travel on Congested Urban Transportation Networks," 105.

34. Vernon Webster, "John Glen Wardrop: Obituary," *Transportation* 16, no. 1 (March 1989): 1–2.

35. John Glen Wardrop, "Some Theoretical Aspects of Road Traffic Research" (with discussion), *Proceedings of the Institution of Civil Engineers (Part II)* 1, no. 3 (1952): 325–378 (on 345).

36. Concerning staffers of the Chicago Area Transportation Study, see, for example, *CATS Research News* 1, no. 2 (1957): 15–16 (ref. to Wardrop); 1, no. 11 (1957): 16–20 (ref. to Beckmann et al.); 1, no. 19 (1957): 12–16 (ref. to Wardrop); 2, no. 9 (1958): 8–12 (ref. to Beckmann, McGuire, and Wardrop). In a practice-oriented review article, K. Rask Overgaard of Denmark also referred to Wardrop: Overgaard, "Urban Transportation Planning: Traffic Estimation," *Traffic Quarterly* 21, no. 2 (April 1967): 197–218 (on 215). According to Overgaard the "Wayne Arterial Assignment Method," developed for the Detroit Area Traffic Study in the early 1960s, converges on a situation in which the requirements of the "equal travel times" principle, namely Wardrop's state of equilibrium, are met (on 215–216). On the "Wayne method," see especially Robert Smock, "An Iterative Assignment Approach to Capacity Restraint on Arterial Networks," *Highway Research Board Bulletin*, no. 347 (1962): 60–66.

37. Jorgensen, "Some Aspects of the Urban Traffic Assignment Problem," 3, 37.

38. Styliani-Stella Constantine Dafermos, "Traffic Assignment and Resource Allocation in Transportation Networks" (Ph.D. diss., Johns Hopkins University, 1968). On Dafermos (1940–1990), see, for example, Anna Nagurney, "Equilibrium Modeling, Analysis and Computation: The Contributions of Stella Dafermos," *Operations Research* 39, no. 1 (January–February 1991): 9–12.

39. Stella C. Dafermos and Frederick T. Sparrow, "The Traffic Assignment Problem for a General Network," *Journal of Research of the National Bureau of Standards—B. Mathematical Sciences* 73B, no. 2 (April–June 1969): 91–118.

40. Dafermos, "Traffic Assignment," iii.

41. The authors associated the second "Wardrop criterion"—today named system equilibrium (SE) with the existence of a "central authority" that would force or incite drivers to cooperate in order to obtain a collective optimum, such as minimizing the sum of their individual journey times. This was a design principle, oriented toward planners and engineers whose job is to manage a network system, and it recently resurfaced with the advent of ITS technology (chapter 7).

42. See especially the survey article by Earl R. Ruiter, "Implementation of Operational Network Equilibrium Procedures," *Transportation Research Record*, no. 491 (1974): 40–51.

43. T. Leventhal, G. Nemhauser, and L. Trotter Jr., "A Column Generation Algorithm for Optimal Traffic Assignment," *Transportation Science* 7, no. 2 (May 1973):

168–176. See also the older article by Dennis F. Wilkie and Robert G. Stefanek, "Precise Determination of Equilibrium in Travel Forecasting Problems Using Numerical Optimization Techniques," *Highway Research Record*, no. 369 (1971): 239–252.

44. LeBlanc, "Mathematical Programming Algorithms," 18 (for Dafermos), 24 (for the quotation).

45. Marguerite Frank and Philip Wolfe, "An Algorithm for Quadratic Programming," *Naval Research Logistics Quarterly* 3, nos. 1–2 (March–June 1956): 95–110.

46. On Wolfe, see Alan J. Hoffman, "Philip Starr Wolfe," in *Profiles in Operations Research: Pioneers and Innovators*, ed. Arjang A. Assad and Saul I. Gass (New York: Springer, 2011), 627–642.

47. Sang Nguyen, "Une Approche Unifiée des Méthodes d'Équilibre pour l'Affectation du Trafic" (Ph.D. diss., University of Montréal, 1973).

48. Nguyen, "Une Approche Unifiée," 21, 101.

49. Willard I. Zangwill, "The Convex Simplex Method," *Management Science* 14, no. 3 (November 1967): 221–238.

50. Sang Nguyen, "An Algorithm for the Traffic Assignment Problem," *Transportation Science* 8, no. 3 (August 1974): 203–216 (on 213–214).

51. See, for instance: Patriksson, *The Traffic Assignment Problem*; Yosef Sheffi, *Urban Transportation Networks: Equilibrium Analysis with Mathematical Programming Methods* (Englewood Cliffs, NJ: Prentice-Hall, Inc., 1985); E. C. Matsoukis and P. C. Michalopoulos, "Road Traffic Assignment—A Review. Part II: Equilibrium Methods," *Transportation Planning and Technology* 11, no. 2 (1986): 117–135; Michael Florian and Donald Hearn, "Network Equilibrium Models and Algorithms," in *Handbooks in Operations Research & Management Science, Volume 8: Network Routing*, ed. M. O. Ball, T. L. Magnanti, C. L. Monma, and G. L. Nemhauser (New York: Elsevier Science, 1995), 485–550; Juan de Dios Ortúzar and Luis G. Willumsen, *Modelling Transport* (West Sussex: John Wiley & Sons, 2011), chapter 10.

52. See for example Michael Florian and Sang Nguyen, "An Application and Validation of Equilibrium Trip Assignment Methods," *Transportation Science* 10, no. 4 (November 1976): 374–390 (city of Winnipeg, Canada); the relevant references in Sheffi, *Urban Transportation Networks*.

53. See for example Sheffi, *Urban Transportation Networks*; Carolyn Frank, "A Study of Alternative Approaches to Combined Trip Distribution-Assignment Modeling" (Ph.D. diss., University of Pennsylvania, 1978).

54. For these research studies, see the following survey works: Sheffi, *Urban Transportation Networks*; Patriksson, *The Traffic Assignment Problem*; David Boyce, "Urban Transportation Network-Equilibrium and Design Models: Recent Achievements and

Future Prospects," *Environment and Planning A* 16, no. 11 (November 1984): 1445–1474; David E. Boyce, Larry J. LeBlanc, and Kyung S. Chon, "Network Equilibrium Models of Urban Location and Travel Choices: A Retrospective Survey," *Journal of Regional Science* 28, no. 2 (May 1988): 159–183; Ortúzar and Willumsen, *Modelling Transport*; Boyce and Williams, *Forecasting Urban Travel*.

55. Yosef Sheffi, "Transportation Networks Equilibration with Discrete Choice Models" (Ph.D. diss., MIT, 1978), 40; emphasis in original. Needless to say, SUE is a generalization of the traditional user equilibrium definition since, if the perceived travel times are assumed to be entirely accurate, all motorists would perceive the same travel time, and the stochastic user equilibrium will be identical to the (deterministic) user equilibrium (Sheffi, *Urban Transportation Networks*, 20).

56. On Daganzo, see chapter 3.

57. Sheffi, "Transportation Networks Equilibration"; Carlos F. Daganzo and Yosef Sheffi, "On Stochastic Models of Traffic Assignment," *Transportation Science* 11, no. 3 (August 1977): 253–274.

58. Sheffi, "Transportation Networks Equilibration," 3.

59. Sheffi, *Urban Transportation Networks*, preface.

60. Those of the city of Winnipeg, Milwaukee, Chicago, and Pittsburgh in particular. See Florian and Nguyen, "An Application"; Eash, Janson, and Boyce, "Equilibrium Trip Assignment"; D. E. Boyce, B. N. Janson, and R. W. Eash, "The Effect on Equilibrium Trip Assignment of Different Link Congestion Functions," *Transportation Research Part A* 15, no. 3 (1981): 223–232; Bruce N. Janson, Selwyn P. T. Thint, and Chris T. Hendrickson, "Validation and Use of Equilibrium Network Assignment for Urban Highway Reconstruction Planning," *Transportation Research Part A* 20, no. 1 (1986): 61–73.

61. "Heuristic" in the sense that they cannot guarantee an optimal and/or rigorous analytic solution to the stated problem.

62. Let us recall here that the Federal Highway Administration decided around 1976–1977 to join with UMTA in promoting the latter's computer battery. As a result, the FHWA stopped maintaining and developing its own computer program system, which it had inherited from the Bureau of Public Roads, a system known in the 1970s as PLANPAC/BACKPAC. For the last version of PLANPAC/BACKPAC, see, for example, Federal Highway Administration, *Computer Programs for Urban Transportation Planning: PLANPAC/BACKPAC General Information Manual* (Washington, DC: U.S. Department of Transportation, 1977).

63. In 1985, commenting on the "Reagan effect," a scholar heavily involved in urban traffic forecasting spoke of a "reduction of resources for applied research on short-term problems and the virtual elimination of any basic research orientation by

the U.S. Department of Transportation"; see David E. Boyce, foreword, *Transportation Research Part A* 19, nos. 5–6 (1985): 349–350 (on 349). However, it should be noted that after the first years of Reagan's presidency the U.S. university system received significant financing again, since during "the two Reagan Administrations, real federal funding for academic research increased by 45 percent, while total research grew by 62 percent"; Roger L. Geiger, *Research and Relevant Knowledge: American Research Universities since World War II* (New York: Oxford University Press, 1993), 311.

64. See notes 62 and 65 in this chapter.

65. Richard B. Robertson, "Opening Session: Keynote Address," in Transportation Research Board, *Travel Analysis Methods for the 1980s*, 4–7 (on 6). For the status of UTPS and PLANPAC/BACKPAC computer programs in 1979 and 1980, see UMTA and FHWA, *Highlights of the Transportation Planning Computer Users' Symposium, Detroit May 23–24, 1979, San Francisco Aug. 22–23, 1979* (n.p.: n.p., n.d.); UMTA and FHWA, *Highlights of the Transportation Planning Computer Users' Symposium, Philadelphia May 20–21, 1980, Los Angeles June 25–26, 1980, St. Louis September 24–25, 1980* (n.p.: n.p., n.d.).

66. UMTA and FHWA, *The Urban Transportation Planning System (UTPS): An Introduction for Management* (Washington, DC: U.S. Government Printing Office, 1986), 14.

67. UMTA and FHWA, *Sources of Information on Urban Transportation Planning Methods* (n.p.: n.p., 1983), 13.

68. Center for Microcomputers in Transportation, *McTrans* 11, no. 1 (September 1996): 2. Though the FHWA contract ended in May 1988, since then McTrans Center has been a full-service, entirely user-supported center.

69. "McTrans Turns 30: A Graphical Timeline," May 4, 2016, https://www.transportation.institute.ufl.edu/2016/05/timeline-for-mctrans-center/, accessed January 26, 2021; A. D. Wilbur, "The McTrans Center," *Public Roads* 52, no. 2 (1988–89): 44–46.

70. Let me recall here that as early as 1955 several of IBM's major scientific and engineering customers set up a user group named Share. See Akera, *Calculating a Natural World*, chapter 7.

71. See, for example, Robert E. Stammer Jr. and Mark D. Abkowitz, eds., *Microcomputer Applications within the Urban Transportation Environment* (New York: Published by the American Society of Civil Engineers, 1987); Robert E. Stammer Jr. and Mark D. Abkowitz, eds., *Microcomputer Applications in Transportation II* (New York: Published by the American Society of Civil Engineers, 1987).

72. On the software industry, its history and historiography, see for example M. Campbell-Kelly, *From Airline Reservations to Sonic the Hedgehog: A History of the Software Industry* (Cambridge, MA: MIT Press, 2003); M. Campbell-Kelly, "The History of the History of Software," *IEEE Annals of the History of Computing* 29, no. 4

(2007): 40–51. On econometric software and its history, which shared a number of features with the history of traffic forecasting-related software, see the survey article by Charles G. Renfro, "Econometric Software: The First Fifty Years in Perspective," *Journal of Economic and Social Measurement* 29, nos. 1–3 (2004): 9–107.

73. ITE Technical Council Committee 6F-34, "Refinement of Traffic Forecasts: Practitioners and Procedures (An Informational Report)": 45; ITE Technical Council Committee 6F-41, "Traffic Forecasting on State and Provincial Highway Segments," *ITE Journal* 61 (1991): 25–30.

74. On the various urban traffic software systems available on microcomputers in the 1980s see Earl Ruiter and Mike Waller, "Microcomputers Software for Transportation Planning," *Transportation Research Record*, no. 932 (1983): 3–8; and, especially, the successive (annual) editions of *Microcomputers in Transportation: Software and Source Book*, released by the FHWA and UMTA in 1982.

75. Unless otherwise mentioned, information about COMSIS and other similar firms has been drawn from the various editions of the *Consultants and Consulting Organizations Directory* (several updated editions).

76. MBE designates a business that is at least 51 percent owned, operated, and controlled by one or more ethnic minority American citizens. On COMSIS see for example Lawrence Berkeley Laboratory, *1988 Directory of Small Disadvantaged Businesses and Woman-Owned Businesses* (California: University of California Berkeley, n.d.), 15.

77. Lee would later be associated with several Silicon Valley-based start-ups, including Solectron, Komag, and IMR (on Lee, see http://taiwaneseamericanhistory.org/blog/ourjourneys19/, accessed November 26, 2020). On Taiwanese American entrepreneurs, see Claudia Bird Schoonhoven and Elaine Romanelli, eds., *The Entrepreneurship Dynamic: Origins of Entrepreneurship and the Evolution of Industries* (Stanford: Stanford University Press, 2001).

78. On Sosslau's modeling and code-writing work when he was at BPR, see Bureau of Public Roads, *Traffic Assignment Manual for Application with a Large, High Speed Computer*, nos. 10, 13, 15; Bureau of Public Roads, *Calibrating and Testing a Gravity Model with a Small Computer*, XII-1.

79. Bureau of Public Roads, *Calibrating & Testing a Gravity Model for Any Size Urban Area*, A-62; *Newsletter of the Transportation Planning Computer Program Exchange Group* 1, no. 1 (1963): 2; and 1, no. 4 (1964): 14; Transportation Planning Program Exchange Group, *T-PEG Newsletter* (Spring 1967): 2; (Fall 1967): 9; and (Spring 1968): 2.

80. See the many reports authored by COMSIS, which can be found in the catalogs of WorldCat and HathiTrust.

81. Unless otherwise mentioned, all information about the proprietary microcomputer suites in the 1980s are drawn from the various editions of *Microcomputers in Transportation: Software and Source Book*.

82. UMTA and FHWA, *The Urban Transportation Planning System (UTPS)*, 11.

83. Like PRC Engineering, MVA Systematica was closely related to the consultancy founded by Alan M. Voorhees in the United States (chapter 2). In 1968, the latter opened a subsidiary in London, which a few years later was renamed Martin and Voorhees Associates (MVA). In 1980, MVA established MVA Systematica.

84. Introduction to this book.

85. Among the early users of TRANPLAN (for the traffic assignment step) was the Bay Area Transportation Study Commission; see Bay Area Transportation Study Commission, *Bay Area Transportation Report* (Berkeley: n.p., 1969).

86. *Newsletter of the Transportation Planning Computer Program Exchange Group* 2, no. 1 (1964): 5; 2, no. 2 (1964): 1–3; 2, no. 3 (1965): 19–20; *Minutes of the T-PEG Meeting* (December 5 and 6, 1966): 3; Transportation Planning Program Exchange Group, *T-PEG Newsletter* (Autumn 1967): 7; *Newsletter of the Transportation Planning Program Exchange Group* (October 1, 1971): 4.

87. Another person closely associated with TRANPLAN was Ralf Kulunk, a longtime CDC staffer.

88. *Traffic Engineering* 40 (August 1970): 12.

89. *The Professional Geographer* 40, no. 1 (February 1988): 114–115 (on 114).

90. The Urban Analysis Group, Inc. (distributor), "URBAN/SYS Version 9.0 User Manuals for TRANPLAN and NIS plus Related Programs Miscellaneous Utilities DBC Interfaces" (copyright 1988–1998 © The Urban Analysis Group, Inc.).

91. Ruiter and Waller, "Microcomputers Software," and, especially, the successive editions of *Microcomputers in Transportation: Software and Source Book* provide much useful information on these now-forgotten software packages.

92. For an in-depth historical account of INRO and its flagship software, now called Emme, see Konstantinos Chatzis, "Managing Traffic Complexity. Canadian Transport Planning Software Package Emme, 1970s–2010s," *The Journal of Transport History* 42, no. 3 (December 2021): 444–466. Unless otherwise stated, all information about INRO presented herein is drawn from this study.

93. Let us recall here that the outputs of the mainframe-based UTPS computer packages sponsored by the federal administration were in the form of matrices, trip tables, and line reports. They were not interpretable until transferred to a node link or network map. This was done by hand posting the output onto a network map, which was tedious and time-consuming, or by an offline plotter.

94. Heinz Spiess, "Contributions à la Théorie et aux Outils de Planification des Réseaux de Transport Urbain" (Ph.D. diss., University of Montreal, 1984); see "Sommaire."

95. Matthias Hans Rapp, "Planning Demand-Adaptive Urban Public Transportation Systems: The Man-Computer Interactive Graphic Approach" (Ph.D. diss., University of Washington [Seattle], 1972). On the various applications of computer graphics to transportation planning issues and other engineering problems in the 1970s and the early 1980s, see among others Jerry B. Schneider, "A Review of a Decade of Applications of Computer Graphics Software in the Transportation Field," *Computers, Environment and Urban Systems* 9, no. 1 (1984): 1–20; "Urban Transportation Planning Using Interactive Graphics," *Transportation Research Record*, no. 866 (1982) (the entire issue); "Application of Interactive Graphics to Transportation Systems Planning," *Highway Research Record*, no. 455 (1973) (the entire issue).

96. Kenneth J. Dueker, Rishinath L. Rao, Andrew Cotugno, Keith Lawton, and Richard Walker, *The Impact of EMME/2 on Urban Transportation Planning: A Portland Case Study* (Portland, OR: School of Urban and Public Affairs/Portland State University, 1985), 5.

97. In 2014, Emme prices started at $9,000 per license and varied by network size, CUBE (which included the following software packages: TRANPLAN, TRIPS, MINUTP, and TP+, the MS-Window-based successor to MINUTP; see note 117) prices ranged from $1,500 to $12,500, TransCAD (follows) from $4,000 to $12,000, and VISUM (follows) from $6,000 to $40,000; see *The Urban Transportation Monitor* 28, no. 8 (October 2014): 16.

98. As has been seen, the first name of the software package was EMME while its second version was labeled EMME/2; however, starting with the third version of the software, the name changed to Emme.

99. In addition to INRO's Support Center in Canada, there was a second one in Switzerland managed by Spiess until 2005.

100. The 1st International EMME/2 Users' Conference took place in Vancouver in July 1986.

101. Unless explicitly mentioned otherwise, information on Caliper is based on an author interview with Howard Slavin at Caliper's premises in June 2011, and an email from Howard Slavin to the author, dated January 20, 2021.

102. Howard L. Slavin, "The Transport of Goods and Urban Spatial Structure" (Ph.D. diss., University of Cambridge, 1979).

103. Now John A. Volpe National Transportation Systems Center, and, more commonly, Volpe Center. See Marian Ott, Howard Slavin, and Donald Ward, "Behavioral Impacts of Flexible Working Hours," *Transportation Research Record*, no. 767 (1980): 1–6 (on 1).

104. Dan Kärreman, "Knowledge-Intensive Firms," in *International Encyclopedia of Organization Studies*, vol. 1, ed. Stewart R. Clegg and James R. Bailey (Thousand Oaks, CA: Sage Publications, 2008), 755–758.

105. Eric Adam Ziering, "Framework for the Development of a Massachusetts State Highway Access Policy" (master's thesis, MIT, 1979).

106. Zvi Tarem, "Evaluation of a Sampling Optimization Method for Discrete Choice Models" (master's thesis, MIT, 1983).

107. Let us recall here that GIS, that is, a computerized database management system for the capture, storage, retrieval, analysis, and display of spatial data, traces its roots back to a handful of research initiatives in the United States, Canada, and Europe during the late 1950s. On the use of GIS in urban planning from a social sciences perspective, see for example Shannon Jackson, "The City from Thirty Thousand Feet: Embodiment, Creativity, and the Use of Geographic Information Systems as Urban Planning Tools," *Technology and Culture* 49, no. 2 (April 2008): 325–346.

108. As one can expect, the GIS component of the tool is for data required or produced by the urban traffic forecasting model. This provides analysts with the ability to quickly and graphically evaluate several alternatives. See for example Howard L. Slavin, "The Role of GIS in Land Use and Transport Planning," in Hensher et al., *Handbook of Transport Geography and Spatial Systems*, 329–356. Examples of early combinations between distinct GIS systems, such as Arc/INFO and MAPINFO, and traffic forecasting software packages, including TMODEL2, QRS II, and TRANPLAN, at the end of the 1980s and the early 1990s, are given in Michael D. Anderson and Reginald R. Souleyrette, "Geographic Information System-Based Transportation Forecast Model for Small Urbanized Areas," *Transportation Research Record*, no. 1551 (1996): 95–104. See also Keechoo Choi and Tschangho John Kim, "Integrating Transportation Planning Models with GIS: Issues and Prospects," *Journal of Planning Education and Research* 13, no. 3 (April 1994): 199–207; and especially Bruce Spear, "Linking Spatial and Transportation Data," in Hensher et al., *Handbook of Transport Geography and Spatial Systems*, 309–326.

109. *The Urban Transportation Monitor* (September 29, 2006): 10–15 (on 15).

110. Caliper website (https://www.caliper.com/, accessed January 14, 2019).

111. In 2014, QRS II prices ranged from free (fifteen zones) to $975 (6,000 zones). See *The Urban Transportation Monitor* 28, no. 8 (October 2014): 16.

112. "QRS II Page," http://ajhassoc.com/, accessed November 26, 2020.

113. Arthur B. Sosslau, Amin B. Hassam, Maurice M. Carter, and George V. Wickstrom, *Travel Estimation Procedures for Quick Response to Urban Policy Issues* (Washington, DC: Transportation Research Board, 1978); Arthur B. Sosslau, Amin B. Hassam, Maurice M. Carter, and George V. Wickstrom, *Quick-Response Urban Travel Estimation*

*Techniques and Transferable Parameters: User's Guide* (Washington, DC: Transportation Research Board, 1978).

114. See, for example, Said M. Easa, "Urban Trip Distribution in Practice. II: Quick Response and Special Topics," *Journal of Transportation Engineering* 119, no. 6 (November 1993): 816–834.

115. G. Scott Rutherford and Norma T. Pennock, "Travel Demand Forecasting with the Quick-Response Microcomputer System: Application and Evaluation of Use," *Transportation Research Record*, no. 1037 (1985): 44–51. On the development of QRS I, see the various updated editions of *Microcomputers in Transportation: Software and Source Book.*

116. Alan J. Horowitz, *Reference Manual Quick Response System II™ (Version 7)* (Milwaukee, WI: AJH Associates, July 2004), acknowledgments, 1.

117. In 1997 COMSIS faced dissolution. Parts of COMSIS went to the giant Westat, while Larry Seiders joined in the Urban Analysis Group and brought with him MINUTP (and another software package, called TP+). In 2001, the software division of MVA (TRIPS) and Urban Analysis Group (TRANPLAN, MINUTP, TP +) merged to create Citilabs, the flagship software product of which is now called CUBE (it has included TP+, TRANPLAN, TRIPS, and MINUTP). Citilabs has been directed since its creation by Michael Clarke, who holds a master's degree in transportation engineering and urban planning from the University of Michigan (1982). Clarke worked for COMSIS from 1982–1986, when MINUPT was first developed. Another Citilabs staffer who has played an important part in the company's computer package development (in VIPER, a graphical user interface, in particular) is Victor Siu, a master's degree holder in transportation engineering, and a former travel demand modeler at TJKM, a Californian transportation firm established in 1974.

118. One can mention here the example of Zhong Zhou, who joined Citilabs in 2008 after getting his Ph.D. from Utah State University. Citilabs and Caliper have incorporated, for example, in their software the "Bi-Conjugate Frank-Wolfe user equilibrium assignment," first proposed by scholars Per Olov Lindberg and Maria Daneva (now Mitradjieva) in 2003; see Per Olov Lindberg and Maria Mitradjieva, "The Stiff Is Moving—Conjugate Direction Frank-Wolfe Methods with Applications to Traffic Assignment," *Transportation Science* 47, no. 2 (May 2013): 280–293, especially 280. See Cambridge Systematics, with Gallop Corporation, *Fiscal Year 2010 Task Reports*, final report prepared for National Capital Region Transportation Planning Board [TPB] (n.p.: n.p., 2010), 1–5. See also Vanasse Hangen Brustlin (VHB), *State of the Art in Equilibrium Traffic Assignment* (Vienna, VA: n.p., 2007).

119. A civil engineer and a research and teaching assistant at Oregon State University, Robert Shull created Professional Solutions, Inc. (1982–1998) to develop TMODEL, a microcomputer software for travel demand forecasting. In 2004, TMODEL Corporation (1991–2003) merged with Innovative Transportation

Concepts (ITC) to form PTV America, with Shull as vice president (2004–2010). See https://www.linkedin.com/in/robertmshull/, accessed November 26, 2020; http://www.angelfire.com/punk/calymayor/_Files/Customer_Magazine_03_04.htm#01, accessed November 28, 2020.

120. Duk M. Chang, Vergil G. Stover, and George B. Dresser, *A Comparison of Microcomputer Packages for Network-Based Highway Planning* (College Station: Texas Transportation Institute, October 1988), especially 18–21.

121. The questionnaire, received from thirty-eight (out of forty-five) transportation agencies, was crafted by JHK & Associates on behalf of the Transportation Research Board. See Neil J. Pedersen and Donald R. Samdahl, *Highway Traffic Data for Urbanized Area Project Planning and Design* (Washington, DC: Transportation Research Board, 1982), 5, 9.

122. VHB, *Determination of the State of the Practice in Metropolitan Area Travel Forecasting*, 1, 66–67.

123. For a brief presentation of the software products, including the user equilibrium assignment techniques incorporated in them, developed and commercialized by AJH Associates, Caliper, Citilabs, INRO, and PTV, in 2006, 2012, and 2014, see *The Urban Transportation Monitor* (September 15, 2006): 12–16; *The Urban Transportation Monitor* (September 29, 2006): 11–15; *The Urban Transportation Monitor* 26, no. 7 (September 2012): 17–31; *The Urban Transportation Monitor* 28, no. 8 (October 2014): 16–30. Much information about these software packages can also be found in Boyce and Williams, *Forecasting Urban Travel*, chapter 10. For a comparative study of advanced UE assignment techniques provided by the "Big Four" at the end of the 2000s, see Cambridge Systematics, with Gallop Corporation, *Fiscal Year 2010 Task Reports*. Concerning post-traditional Frank-Wolfe (static) UE traffic assignment techniques as well as their incorporation into commercial software, see for example Boyce and Williams, *Forecasting Urban Travel*, chapter 7; Jun Xie and Chi Xie, "Origin-Based Algorithms for Traffic Assignment: Algorithmic Structure, Complexity Analysis, and Convergence Performance," *Transportation Research Record*, no. 2498 (2015): 46–55; VHB, *State of the Art in Equilibrium Traffic Assignment*.

124. Caliper Corporation, *Traffic Assignment and Feedback Research to Support Improved Travel Forecasting*, final report prepared for Federal Transit Administration (Newton, MA: Caliper Corporation, 2015), 2-2, and 2-13–2-14.

## Chapter 5

1. For an early interest in the connections between air pollution and transportation planning, see G. V. Wickstrom, "Air Pollution: Implications for Transportation Planning," *Highway Research Record*, no. 465 (1973): 46–54; and chapter 3 of this book.

2. On this trial, see especially Mark Garrett and Martin Wachs, *Transportation Planning on Trial: The Clean Air Act and Travel Forecasting* (London: Sage Publications, 1996). For a more recent similar trial, see Wisconsin Department of Transportation, "Accuracy of Traffic Projections at Center of Wisconsin Legal Challenge: Court Ruling Calls WisDOT Methodology into Question," *The Urban Transportation Monitor* 29, no. 6 (July 2015): 1 and 9. On the (recent) intersections of law and science (and engineering) in the United States, see for example the survey article by Sheila Jasanoff, "Making Order: Law and Science in Action," in Hackett et al., *The Handbook of Science and Technology Studies*, 761–786.

3. On the 1990 CAAA and 1991 ISTEA, and their rather dramatic impact on transportation planning, see for example Weiner, *Urban Transportation Planning in the United States*; "Transportation and Air Quality," special issue, *Transportation* 22, no. 3 (August 1995); John Felleman, "Clean Air Act and Amendments" and Juita-Elena (Wie) Yusuf and Kaitrin Mahar, "Intermodal Surface Transportation Efficiency Act," both in *Encyclopedia of Transportation. Social Science and Policy*, ed. Mark Garrett (Los Angeles: Sage, 2014), 371–373 and 829–832, respectively.

4. Mark Garrett, introduction, in Garrett, *Encyclopedia of Transportation: Social Science and Policy*, xxxi–xxxv (on xxxiv).

5. Including the distribution of transportation infrastructures and their negative externalities as well as the appropriateness of transportation policies for various groups. On these issues, see Kate J. Darby, "Environmental Justice" and Eric Petersen, "Social Equity and Discrimination in Transportation," both in Garrett, *Encyclopedia of Transportation. Social Science and Policy*, 515–517 and 1239–1241, respectively; Tierra Suzan Bills, "Enhancing Transportation Equity Analysis for Long-Range Planning and Decision Making" (Ph.D. diss., University of California at Berkeley, 2013); Evelyn Blumenberg, "Social Equity and Urban Transportation," in Giuliano and Hanson, *The Geography of Urban Transportation*, 332–358. See also the monograph produced by Mark Garrett, "The Struggle for Transit Justice: Race, Space, and Social Equity in Los Angeles" (Ph.D. diss., University of California at Los Angeles, 2006).

6. Bayarmaa Alexander, "Telecommuting and Telecommunications," in Garrett, *Encyclopedia of Transportation: Social Science and Policy*, 1319–1322.

7. Deakin, Harvey, Skabardonis, Inc. (prepared by), *Manual of Regional Transportation Modeling Practice for Air Quality Analysis*, sponsored by the National Association of Regional Councils, Washington, DC (n.p.: n.p., 1993).

8. Patricia Bass, Dennis G. Perkinson, Brigitta Keitgen, and George B. Dresser, *Travel Forecasting Guidelines* (College Station: Texas Transportation Institute, October 1994); JHK & Associates in association with Dowling Associates (prepared by), *State of California Department of Transportation Travel Forecasting Guidelines* (n.p.: n.p., 1992); Office of Air Quality Planning and Standards and Office of Mobile Sources,

*Issues and Approaches to Improving Transportation Modeling for Air Quality Analysis* (n.p.: U.S. Environmental Agency, 1993).

9. For an overview, see: Xueming Chen, "Travel Model Improvement Program," in Garrett, *Encyclopedia of Transportation: Social Science and Policy*, 1594–1596; Weiner, *Urban Transportation Planning in the United States*. See also the Travel Model Improvement Portal: https://tmip.org/.

10. COMSIS (prepared by), *Incorporating Feedback in Travel Forecasting: Methods, Pitfalls, and Common Concerns* (Washington, DC: Federal Highway Administration, 1996). In the first half of the 1990s, the feedback issue was about to become quite popular with urban traffic forecasters. See, for example, David Levinson and Ajay Kumar, "Integrating Feedback into Transportation Planning Model: Structure and Application," *Transportation Research Record*, no. 1413 (1993): 70–77; David E. Boyce, Yu-Fang Zhang, and Mary R. Lupa, "Introducing 'Feedback' into Four-Step Travel Forecasting Procedure Versus Equilibrium Solution of Combined Model," *Transportation Research Record*, no. 1443 (1994): 65–74; Robert A. Johnston and Raju Ceerla, "Travel Modeling with and without Feedback to Trip Distribution," *Journal of Transportation Engineering* 122, no. 1 (January/February 1996): 83–86; W. Thomas Walker and Haiou Peng, "Alternative Methods to Iterate a Regional Travel Simulation Model: Computational Practicality and Accuracy," *Transportation Research Record*, no. 1493 (1995): 21–28.

11. Martin Wohl, "Demand Cost, Price and Capacity Relationships Applied to Travel Forecasting," *Highway Research Record*, no. 38 (1963): 40–54; and especially N. A. Irwin, Norman Dodd, and H. G. von Cube, "Capacity Restraint in Assignment Programs," *Highway Research Board Bulletin*, no. 297 (1961): 109–127.

12. See, for example, the recollections of David Boyce in Boyce, Zhang and Lupa, "Introducing 'Feedback,'" 65. For the necessity of such feedback, see introduction, box 0.1.

13. COMSIS, *Incorporating Feedback*, "Appendix A (Regulatory Requirements for Feedback)."

14. Vanasse Hangen Brustlin (VHB), *Use of Feedback Loops Coupled with Nested Logit Mode Choice Models* (n.p.: MWCG/NCRTPB, 2007), 4.

15. See for example two recent studies: Phillip Reeder, Chandra Bhat, Karen Lorenzini, and Kevin Hall, *Positive Feedback: Exploring Current Approaches in Iterative Travel Demand Model Implementation* (College Station: Texas Transportation Institute, January 2012); Caliper Corporation, *Traffic Assignment and Feedback Research to Support Improved Travel Forecasting*.

16. On the Cold War in general, see the now classic John Lewis Gaddis, *The Cold War: A New History* (New York: The Penguin Press, 2005). On Cold War science and technology, see two recent survey works: Audra J. Wolfe, *Competing with the Soviets:*

*Science, Technology, and the State in Cold War America* (Baltimore: John Hopkins University Press, 2012); Michael Aaron Dennis, "Military, Science, and Technology and the," in *The Oxford Encyclopedia of the History of American Science, Medicine, and Technology*, vol. 2, ed. Hugh Richard Slotten (New York: Oxford University Press, 2014), 95–109. On the connections between urban planning and the military in the United States from 1950–1970, see especially Jennifer S. Light, *From Warfare to Welfare: Defense Intellectuals and Urban Problems in Cold War America* (Baltimore: Johns Hopkins University Press, 2003); the author does not mention urban travel demand modeling at all.

17. Laura Agnes McNamara, "Ways of Knowing about Weapons: The Cold War's End at the Los Alamos National Laboratory" (Ph.D. diss., University of New Mexico, 2001). For another example of such a shift, see Thomas Heinze and Olof Hallonsten, "The Reinvention of the SLAC National Accelerator Laboratory, 1992–2012," *History of Technology* 33, no. 3 (2017): 300–332.

18. On TRANSIMS and its connection with the 1990 CAAA and 1991 ISTEA see especially Jack Morrison and Verne Loose, *TRANSIMS Model Design Criteria as Derived from Federal Legislation* (Washington, DC: U.S. Department of Transportation, 1995). See also Office of Intermodalism (prepared by), *Intermodal Technical Assistance Activities for Transportation Planners* (n.p.: U.S. Department of Transportation, 1993), 66.

19. Daniel S. Metzger, "A New Approach to Public-Private Cooperation in Transportation Research," *Public Roads* 57, no. 1 (Summer 1993), https://highways.dot.gov/public-roads/summer-1993/new-approach-public-private-cooperation-transportation-research, accessed January 27, 2021.

20. The cover reads: "Rough Draft of TRANSIMS: Transportation Analysis and Simulation System," prepared by Los Alamos National Laboratory and New Mexico State Highway and Transportation Department, Albuquerque, NM, Alliance for Transportation Research, n.d. (ca. April/May 1992) (hereafter "TRANSIMS White Paper"). The copy I have consulted is part of the Personal Records of Thomas B. Deen. See Deen's Personal Records, C0106, Box 50, Folder 8.

21. Letter from Albright to Deen, dated May 29, 1992, Deen's Personal Records, C0106, Box 50, Folder 8.

22. "TRANSIMS White Paper," 4.

23. Letter from Albright to Deen, dated May 29, 1992.

24. Letter from Deen to Albright, dated June 10, 1992, Deen's Personal Records, C0106, Box 50, Folder 8.

25. Unlike urban travel demand modeling, which is for transport planning purposes and usually operates over both long periods of time (several years) and large areas (a city or even a larger metropolitan region), the "traffic models," which were also

developed from the 1950s onward, model (simulate) car flows on a transportation network on a much smaller spatial scale and for short periods of time. Much more computationally demanding, they were used, at the same time as when the FSM was being developed, for local regulations (traffic lights, etc.). TRANSIMS was intended to be a comprehensive tool, which could be utilized both as a "traffic model" and as a "travel demand model."

26. Federal Highway Administration, *Federal Highway Administration Research and Technology Program Highlights 1993* (n.p.: n.p. n.d.), 1. See also Office of Intermodalism (prepared by), *Intermodal Technical Assistance Activities.*

27. Email from John L. Bowman to the author, dated January 23, 2021. John Bowman was among the session attendees as a late member of the Ben-Akiva team, which had prepared a commissioned white paper for the FHWA (see chapter 6).

28. A mathematician by training, Ducca became familiar with urban modeling at the time of his dissertation: Frederick W. Ducca Jr., "Retail and Service Employment Location: Simulation, Analysis and Projection" (Ph.D. diss., University of Pennsylvania, 1978).

29. *TMIP Connection—The Travel Model Improvement Program Newsletter*, no. 1 (August 1994).

30. Gordon A. Shunk and Patricia L. Bass (prepared by), *Travel Model Improvement Program Conference Proceedings, August 14–17, 1994, Fort Worth, Texas* (Washington, DC: U.S. Department of Transportation, n.d.). The list of attendees is on 63–65.

31. See, for example, Martin Wachs, "Introduction, Keynote Address: Evolution and Objectives of the Travel Model Improvement Program," in Shunk and Bass, *Travel Model Improvement Program Conference Proceedings*, 3–9, especially 9. Wachs was the first chairman of the TMIP Review Panel. See also Catherine T. Lawson, "Microsimulation for Urban Transportation Planning: Miracle or Mirage," *Journal of Urban Technology* 13, no. 1 (2006): 55–80 (on 59–60).

32. The Transportation Equity Act for the Twenty-First Century (TEA-21) was signed into law on June 9, 1998 and was built and expanded upon by the 1991 ISTEA. TEA-21 expired on September 30, 2003, and after a two-year debate in Congress, President Bush signed SAFETEA-LU into law in 2005. SAFETEA-LU was replaced in turn by the Moving Ahead for Progress in the 21st Century Act in 2012 (on these acts, see Weiner, *Urban Transportation Planning in the United States*).

33. Yet, according to several experts, however important, even this funding was not adequate. By comparison, in the late 1970s and early 1980s, federal highway and transit agencies spent about $5 million a year on travel demand modeling, an amount that equated to about $15 million in 2007 dollars (Committee for Determination of the State of the Practice in Metropolitan Area Travel Forecasting,

*Metropolitan Travel Forecasting: Current Practice and Future Direction* (Washington DC: Transportation Research Board, 2007), 7.

34. Committee for Determination of the State of the Practice in Metropolitan Area Travel Forecasting, *Metropolitan Travel Forecasting*, 98–99, and 6.

35. Given the criticism leveled against the TRANSIMS project by a significant part of the UTDM social world, it is no wonder that at the end of 1996 the promoters of the LANL project still felt, not without a dash of humor, the need to answer the charges. See *TMIP Connection—The Travel Model Improvement Program Newsletter*, no. 6 (November 1996).

36. Christopher Barrett, Darrell Morgeson, Kathy Berkbigler, Mike Brown, Brian Bush, John Davis, Deborah Kubicek, Verne Loose, Mike McKay, Jack Morrison, Kai Nagel, Rob Oakes, Steen Rasmussen, Marcus Rickert, Jay Riordan, Doug Roberts, Paula Stretz, Steve Sydoriak, Gary Thayer, Murray Wolinsky, LaRon Smith, Richard Beckman, Keith Baggerly, Doug Anson, and Michael Williams (see the publication titled: *TRANSIMS: TRansportation Analysis and SIMulation System* [Los Alamos: LANL; n.d., ca. 1995], cover and acknowledgments, 10). Unsurprisingly, the LANL group would be renewed following departures and new recruitments. The cover of another TRANSIMS-related text published in 2002 featured, in fact, several new people: See C. L. Barrett, R. J. Beckman, K. P. Berkbigler, K. R. Bisset, B. W. Bush, K. Campbell, S. Eubank, K. M. Henson, J. M. Hurford, D. A. Kubicek, M. V. Marathe, P. R. Romero, J. P. Smith, L. L. Smith, P. L. Speckman, P. E. Stretz, G. L. Thayer, E. Van Eeckhout, and M. D. Williams, *TRansportation ANalysis SIMulation System (TRANSIMS), Version TRANSIMS—3.0, Volume 4: Calibrations, Scenarios, and Tutorials* (Los Alamos: LANL, 2002).

37. Much more information on the TRANSIMS team can be found in Chatzis, "La Modélisation des Déplacements Urbains aux États-Unis et en France," chapter 6.

38. Sipress, "Lab Studying Science behind Traffic Patterns."

39. The German master's degree.

40. On the (relatively recent) close bonds formed between the physical sciences and computer science, see especially Peter J. Denning, "Computer Science: The Discipline," in *Encyclopedia of Computer Science*, ed. Anthony Ralston, Edwin D. Reilly, and David Hemmendinger (Chichester: John Wiley and Sons, 2003), 405–419.

41. Sipress, "Lab Studying Science behind Traffic Patterns."

42. See for example Darrel Morgeson, "Travel Model Improvement: TRANSIMS Presentation. Introductory Remarks," in Shunk and Bass, "*Travel Model Improvement Program Conference Proceedings*, 39–42 (on 39).

43. Kai Nagel, Paula Stretz, Martin Pieck, Shannon Leckey, Rick Donnelly, and Christopher L. Barrett, *TRANSIMS Traffic Flow Characteristics* (preprint) (Los Alamos: LALN, 1998).

44. K. S. Kurani and R. Kitamura, *Recent Developments and the Prospects for Modeling Household Activity Schedules*, report prepared for the Los Alamos National Laboratory (Davis: Institute of Transportation Studies/University of California at Davis, 1996).

45. Ram M. Pendyala, "Federal TMIP Prepares for Next Phase of TRANSIMS," *Florida Transportation Modeling* 5 (September 1997): 3–4.

46. Of the nineteen TRANSIMS-related publications by the Los Alamos team that appeared from 1994–1997, fourteen were published in 1995 (see Texas Transportation Institute, *Early Deployment of TRANSIMS, Issue Paper* [TMIP] [n.p.: US. DOT/ U.S. EPA, June 1999; revised August 1999], 27).

47. W. Edwards Deming and Frederick F. Stephan, "On a Least Squares Adjustment of a Sampled Frequency Table When the Expected Marginal Totals Are Known," *The Annals of Mathematical Statistics* 11, no. 4 (December 1940): 427–444.

48. McFadden et al., *The Urban Travel Demand Forecasting Project, Final Report Series*, vol. 1, chapter 7 and appendix B.

49. Richard J. Beckman, Keith A. Baggerly, and Michael D. McKay, *Creating Synthetic Baseline Populations* (Los Alamos: LANL, n.d.); Richard J. Beckman, Keith A. Baggerly, and Michael D. McKay, "Creating Synthetic Baseline Populations," *Transportation Research Part A* 30, no. 6 (November 1996): 415–429.

50. See, for example, Texas Transportation Institute (prepared by), *Activity-Based Travel Forecasting Conference, June 2–5, 1996, Summary, Recommendations and Compendium of Papers* (Arlington: Texas Transportation Institute, 1997); TRANSIMS in general and its submodule devoted to "synthetic population" in particular were also presented to over 125 people at the second annual conference sponsored by TMIP held in Daytona Beach, Florida, on December 4–6, 1995 (*TMIP Connection—The Travel Model Improvement Program Newsletter*, no. 5 [May 1996]).

51. For a recent review of the state of the "population synthesis" issue, see Abdoul-Ahad Choupani and Amir Reza Mamdoohi, "Population Synthesis Using Iterative Proportional Fitting (IPF): A Review and Further Research," *Transportation Research Procedia* 17 (2016): 223–233 (the authors list no fewer than fifty-three references).

52. *TMIP Connection—The Travel Model Improvement Program Newsletter*, no. 3 (July 1995).

53. Kai Nagel, *Particle Hopping Models, Traffic Flow Theory, and Traffic Jam Dynamics* (Los Alamos: LANL, November 1995); Christopher L. Barrett, Stephen Eubank, Kai Nagel, Steen Rasmussen, Jason Riodan, and Murray Wolinsky, *Issues in the Representation of Traffic Using Multiresolution Cellular Automata* (Los Alamos: LANL, n.d., ca. 1995); Kai Nagel, Steen Rasmussen, and Christopher L. Barret, *Network Traffic as a Self-Organized Critical Phenomena* (Los Alamos: LANL, 1995); Kai Nagel, Christopher L. Barrett, and Marcus Rickert, *Parallel Traffic Micro-simulation by Cellular Automata*

*and Application for Large Transportation Modeling* (Los Alamos: LANL, 1996). A series of other reports are mentioned in Texas Transportation Institute, *Early Deployment of TRANSIMS, Issue Paper*, 27.

54. A brief history of CA theory is provided in the following texts: Palash Sarkar, "A Brief History of Cellular Automata," *AMC Computing Surveys* 32, no. 1 (March 2000): 80–107; Wolfram Science, "Some Historical Notes," https://www.wolframscience.com/reference/notes/876b, accessed November 27, 2020.

55. See, for example, the seminal article by Stephen Wolfram, "Statistical Mechanics of Cellular Automata," *Reviews of Modern Physics* 55, no. 3 (July–September 1983): 601–644, as well as his book: Stephen Wolfram, *A New Kind of Science* (n.p.: Wolfram Media, 2002), https://www.wolframscience.com/nks/, accessed January 24, 2022.

56. For references, see among others Nagel, *Particle Hopping Models*; and, especially, Sven Maerivoet and Bart De Moor, "Cellular Automata Models of Road Traffic," *Physics Reports* 419, no. 1 (November 2005): 1–64.

57. Kai Nagel and Michael Schreckenberg, "A Cellular Automaton Model for Freeway Traffic," *Journal de Physique I France* 2, no. 12 (December 1992): 2221–2229.

58. For example, the driver is supposed to slow down when the distance to the car in front of him is about to decrease, or to change lane in order to go past slower cars.

59. Nicholas Lewis, "Purchasing Power: Rivalry, Dissent, and Computing Strategy in Supercomputer Selection at Los Alamos," *IEEE Annals of the History of Computing* 37, no. 3 (July–September 2017): 25–40.

60. On SUN workstations, see for example Ceruzzi, *History of Modern Computing*, chapter 9. On the different tests conducted by the LANL team, see among others K. Nagel and A. Schleicher, "Microscopic Traffic Modeling on Parallel High Performance Computers," *Parallel Computing* 20, no. 1 (January 1994): 125–146. On the choice of SUN workstations, see *TMIP Connection—The Travel Model Improvement Program Newsletter*, no. 3 (July 1995); no. 6 (November 1996); no. 7 (November 1997).

61. The many reports crafted by the Los Alamos team were made widely available through the Travel Model Improvement Program. The "Traffic microsimulation" component was also presented at the 5th National Conference on Transportation Planning Methods Applications, held at Seattle in April 1995 as well as during the 2nd annual TMIP-sponsored conference held at Daytona Beach, Florida, on December 4–6, 1995. Several single-authored or coauthored papers were also published by Nagel and his colleagues in various academic journals. See, for example: Kai Nagel, "Particle Hopping Models and Traffic Flow Theory," *Physical Review E*, 53, no. 5 (May 1996): 4655–4670; Kai Nagel, "From Particle Hopping Models to Traffic Flow Theory," *Transportation Research Record*, no. 1644 (1998): 1–9; Peter Wagner, Kai Nagel, and Dietrich E. Wolf, "Realistic Multi-Lane Traffic Rules for Cellular

Automata," *Physica A: Statistical Mechanics and Its Applications* 234, nos. 3–4 (January 1997): 687–698.

62. Dallas-Fort Worth, Boston, Portland (OR), Oakland, Chicago, and Denver.

63. *TMIP Connection—The Travel Model Improvement Program Newsletter*, no. 2 (January 1995). It should be noted that before the first large Dallas-Fort Worth experiment, several other smaller traffic microsimulation studies had been conducted by the group on real sites in New Mexico and also in Germany.

64. Travel Model Improvement Program, *Transportation Analysis Simulation System (TRANSIMS): The Dallas Case Study*, prepared by Los Alamos National Laboratory and Texas Transportation Institute (n.p.: U.S. Department of Transportation, January 1998); Ken Cervenka, "Large-Scale Traffic Microsimulation from a MPO Perspective," in *Proceedings of the Sixth TRB Conference on the Application of Transportation Planning Methods, May 19–23, 1997, Dearborn (Michigan)*, ed. Rick Donnelly and Julie Dunbar (n.p.: TRB, December 1997), 314–320; Kai Nagel, Patrice M. Simon, Marcus Rickert, and Jörg Esser, *Iterated Transportation Simulation for Dallas and Portland* (Very Extended Abstract) (Los Alamos: LANL, 1998); *TMIP Connection—The Travel Model Improvement Program Newsletter*, no. 7 (November 1997).

65. *TMIP Connection—The Travel Model Improvement Program Newsletter*, no. 8 (July 1998).

66. *TMIP Connection—The Travel Model Improvement Program Newsletter*, no. 8 (July 1998); C. L. Barrett et al., *TRANSIMS User Notebook: Version 1.0* (Los Alamos: LANL, 1998).

67. *TMIP Connection—The Travel Model Improvement Program Newsletter*, no. 7 (November 1997).

68. *TMIP Connection—The Travel Model Improvement Program Newsletter*, no. 8 (July 1998); no. 10 (December 1999).

69. On METRO, its modeling resources, and achievements, see Rick Donnelly, Greg D. Erhardt, Rolf Moeckel, and William A. Davidson, eds., *Advanced Practices in Travel Forecasting. A Synthesis of Highway Practice* (Washington, DC: Transportation Research Board, 2010), 58–59. On the MPOs in the 2000s, see U.S. Government Accountability Office, *Metropolitan Planning Organizations: Options Exist to Enhance Transportation Planning Capacity and Federal Oversight* (n.p.: GAO, 2009).

70. Cambridge Systematics, *Data Collection in the Portland, Oregon Metropolitan Area Case Study* (Travel Model Improvement Program) (n.p.: n.p., 1996). See also chapter 6.

71. *TMIP Connection—The Travel Model Improvement Program Newsletter*, no. 8 (July 1998).

72. See for example Ken Cervenka and Mahmoud Ahmadi, "Getting Ready for TRANSIMS," *TMIP Connection—The Travel Model Improvement Program Newsletter*, no. 8 (July 1998) (without pagination).

73. K. P. Berkbigler, B.W. Bush, and J. F. Davis, *TRANSIMS Software Architecture for IOC-1* (Los Alamos: LANL, 1997), 3. On software architecture and engineering, and their history till the early 2000s, see for example Henry Muccini, Patricia Lago, Karthik Vaidyanathan, Francesco Osborne, and Eltjo Poort, "The History of Software Architecture—In the Eye of the Practitioner" (2018), https://arxiv.org/abs/1806.04055, accessed January 27, 2021; Leonor Barroca, Jon Hall, and Patrick Hall, "An Introduction and History of Software Architectures, Components, and Reuse," in *Software Architectures. Advances and Applications*, ed. Leonor Barroca, Jon Hall, and Patrick Hall (New York: Springer, 2000), 1–11; Niklaus Wirth, "A Brief History of Software Engineering," *IEEE Annals of the History of Computing* 30, no. 3 (July–September, 2008): 32–39.

74. *TMIP Connection—The Travel Model Improvement Program Newsletter*, no. 9 (March 1999).

75. *TMIP Connection—The Travel Model Improvement Program Newsletter*, no. 10 (December 1999); email from Howard Slavin to the author, dated February 7, 2021.

76. Kimberly M. Fisher, "TRANSIMS Is Coming," *Public Roads* 63, no. 5 (March/April 2000), https://www.fhwa.dot.gov/publications/publicroads/00marapr/transims.cfm, accessed November 27, 2020.

77. *TMIP Connection—The Travel Model Improvement Program Newsletter*, no. 11 (December 2000).

78. *TMIP Connection—The Travel Model Improvement Program Newsletter*, no. 13 (September 2001).

79. *TMIP Connection—The Travel Model Improvement Program Newsletter*, no. 10 (December 1999); no. 11 (December 2000); no. 12 (March 2001).

80. *TMIP Connection—The Travel Model Improvement Program Newsletter*, no. 9 (March 1999).

81. *TMIP Connection—The Travel Model Improvement Program Newsletter*, no. 12 (March 2001).

82. *The Charles E. Via, Jr. Department of Civil and Environmental Engineering, VirginiaTech*, no. 14 (2000): 32; no. 16 (2002): 33; no. 25 (2011): 31.

83. Antoine Hobeika, *TRANSIMS Fundamentals* (Blacksburg: Virginia Polytechnic Institute and State University, July 2005); A. G. Hobeika, M. Jeihani, and C. Jeenanunta, *Transportation Analysis and Simulation Systems (TRANSIMS): Short Course Instructions Manual* (Blacksburg: Virginia Tech, 2002). On this course, and the first

TRANSIMS-related research studies, including master's theses and Ph.D. dissertations, at Virginia Tech, see, for example, *TMIP Performance Report 2003* (Draft) (n.p.: n.p., n.d.), 11–12; *The Charles E. Via, Jr. Department of Civil and Environmental Engineering, VirginiaTech*, no. 18 (2004): 29; Mansoureh Jeihani, "A Review of Dynamic Traffic Assignment Computer Packages," *Journal of the Transportation Research Forum* 46, no. 2 (Summer 2007): 35–46 (the article contains also several relevant references); Kwang-Sub Lee, Antoine G. Hobeika, Hong Bo Zhang, and HeeJin Jung, "Traveler's Response to Value Pricing: Application of Departure Time Choices to TRANSIMS," *Journal of Transportation Engineering* 136, no. 9 (September 2010): 811–817.

84. Laurence R. Rilett, "Transportation Planning and TRANSIMS Microsimulation Model: Preparing for the Transition," *Transportation Research Record*, no. 1777 (2001): 84–92; see also box 5.1 in this chapter.

85. In the first years of the 2000s, thirty-one copies of versions 1.0 and 1.1 as well as one copy of version 3.1 were acquired by academic institutions. See Lawson, "Microsimulation for Urban Transportation Planning," 68.

86. For some of the research studies authored by academics by 2006, see Lawson, "Microsimulation for Urban Transportation Planning," 68 and bibliography.

87. *TMIP Performance Report 2003* (Draft), 5.

88. Ronald Eash, longtime head of CATS (chapter 2), and Thomas Rossi from Cambridge Systematics were joined by Eric Miller, professor at the University of Toronto, Rilett, Joseph L. Schofer from Northwestern University, and David L. Kurth, then with the Parsons Transportation Group. See *TMIP Connection—The Travel Model Improvement Program Newsletter*, no. 17 (August 2003); email from Thomas Rossi to the author, dated January 28, 2021.

89. In October 2002, IBM bought PwC Consulting, thus inheriting the commercialization process of TRANSIMS.

90. On AECOM, see chapter 2.

91. David B. Roden, "Impact of Regional Simulation on Emission Estimates," *Transportation Research Record*, no. 1941 (2005): 81–88 (on 81). An ex-staffer at JHK & Associates, Roden has been well versed in computerized transportation planning since the early 1980s. See, for example, David B. Roden, "Modeling MultiPath Transit Networks," *Transportation Research Record*, no. 1064 (1986): 35–41. It is worth noting that JHK & Associates would develop a comprehensive travel demand modeling computer suite in the 1980s, labeled System II. See M. Nazrul Islam, Prianka N. Seneviratne, and Koti R. Kalakota, "Traffic Management During Road Closure," *Transportation Research Record*, no. 1509 (1995): 46–54.

92. Donnelly et al., *Advanced Practices in Travel Forecasting*, 16.

93. *TMIP Connection—The Travel Model Improvement Program Newsletter*, no. 25 (Winter 2007); Donnelly et al., *Advanced Practices in Travel Forecasting*, 58–59.

94. AECOM, *Using Traditional Model Data as Input to TRANSIMS Microsimulation (Support for the Implementation of TRANSIMS in Portland, Oregon)*, prepared for USDOT Federal Highway Administration (n.p.: n.p., 2006); AECOM, *Revisiting the Portland GEN2 Modeling Process with TRANSIMS Version 4.0 Software Methods (Revised TRANSIMS Implementation Plan for Portland)*, prepared for USDOT Federal Highway Administration) (n.p.: n.p., 2009).

95. AECOM, *Revisiting the Portland GEN2*; Donnelly et al., *Advanced Practices in Travel Forecasting*, 16. It is worth noting that the first versions of TRANSIMS had run on UNIX (*TMIP Connection—The Travel Model Improvement Program Newsletter*, no. 8 [July 1998]).

96. By the end of fiscal year 2003, holders of commercial software licenses included USDOT, Portland METRO, AECOM/PB Consult, Virginia Polytechnic University, and the University of Idaho; see *TMIP Performance Report 2003* (Draft), 15.

97. *TMIP Connection—The Travel Model Improvement Program Newsletter*, no. 24 (Spring 2006); Dr. Ing Hubert Ley, *TRANSIMS Training Course at TRACC, Part 13: The Open Source Concept and TRANSIMS Resources on the Web* (n.p.: Argonne National Laboratory, last updated April 21, 2008).

98. Donnelly et al., *Advanced Practices in Travel Forecasting*, appendix C (list of the projects by fall 2009).

99. Renamed Noblis in 2007.

100. *TMIP Connection—The Travel Model Improvement Program Newsletter*, no. 25 (Winter 2007); no. 26 (Spring 2007); and especially Ley, *TRANSIMS Training Course at TRACC*. The website was supported and maintained by USDOT and could be found at http://www.transims-opensource.org/; it has ceased to be operational as of writing (2020).

101. Jack M. Holl (with the assistance of Richard G. Hewlett and Ruth R. Harris), *Argonne National Laboratory, 1946–96* (Urbana and Chicago: University of Illinois Press, 1997).

102. *TMIP Fiscal Year 2006 Annual Report* (n.p.: n.p., October 2007), 10.

103. Including the Illinois Institute of Technology (IIT), the Northern Illinois University, and the University of Illinois in Urbana-Champaign.

104. Chicago Department of Transportation and Chicago Metropolitan Agency for Planning, to name just two.

105. Hubert Ley, *Transportation Research and Analysis Computing Center (TRACC) Year 6 Quarter 4 Progress Report (Final) (Argonne National Laboratory, Energy Systems Division)*

(Oak Ridge, TN: U.S. Department of Energy, 2013), 39–40. TRANSIMS Studio could be freely downloaded at https://sourceforge.net/projects/transimsstudio/, accessed March 29, 2019.

106. Thus, for the Portland Study two types of parallel computing were used: Linux cluster resorting to Intel processors from VA Linux and a shared memory multiprocessor system from Sun Microsystems. See C. L. Barrett et al., *TRansportation ANalysis SIMulation System (TRANSIMS): Portland Study Reports*, vol. 1: *Introduction/Overview* (Los Alamos: Los Alamos National Laboratory, 2002), 13.

107. See especially Ley, *Transportation Research and Analysis Computing Center (TRACC) Year 6 Quarter 4 Progress Report (Final)*; Hubert Ley, *Transportation Research and Analysis Computing Center (TRACC) Year 5 Quarter 4 Progress Report (Argonne National Laboratory, Energy Systems Division)* (n.p.: n.p., 2011). A wealth of documents and other archival sources (videos, audios, etc.) chronicling the work done on TRANSIMS at TRACC can be found at https://www.tracc.anl.gov/joomla/?searchword=TRANSIMS&searchphrase=any&limit=100&ordering=newest&view=search&option=com_search, accessed October 7, 2022.

108. Argonne National Laboratory, *RTSTEP, Final Report* (n.p.: n.p., 2011); CMAP, *UWP for Transportation Unified Work Program, Northeastern Illinois Fiscal Year 2012* (n.p.: n.p., n.d.), 153–154.

109. On TRANSIMS training courses organized by TRACC staffers, see Ley, *Transportation Research and Analysis Computing Center (TRACC) Year 6 Quarter 4 Progress Report (Final)*, and, especially "Transportation Systems Modeling Training," https://www.tracc.anl.gov/joomla/index.php/transportation-systems-modeling-training, accessed October 7, 2022.

110. *TMIP Fiscal Years 2008–2009 Report* (n.p.: n.p., n.d.), 5; *TMIP Fiscal Years 2010–2011 Report* (available at https://www.fhwa.dot.gov/planning/tmip/publications/annual_reports/fys10-11/, accessed January 26, 2021).

111. *TMIP Annual Report May–September 2012* (n.p.: n.p., n.d.), 21; *TMIP Annual Report October 2012–September 2013* (n.p.: n.p., n.d.), 17.

112. Two categories of TRANSIMS initiatives should be mentioned here. The "Application studies" were targeted to technology development and demonstration. They were limited to one year, with funding levels below $100,000. "Deployment case studies" were geared toward analyses involving local planning issues, typically limited to eighteen months, with funding levels below $400,000. See *TMIP Fiscal Years 2008–2009 Report* (n.p.: n.p., n.d.), 8.

113. See Donnelly et al., *Advanced Practices in Travel Forecasting*, appendix C; the various *TMIP Annual Reports* in the 2000s, https://www.fhwa.dot.gov/planning/tmip/publications/annual_reports/, accessed November 26, 2020; and the materials available at https://www.tracc.anl.gov/joomla/?searchword=TRANSIMS&searchphrase

=any&limit=100&ordering=newest&view=search&option=com_search, accessed October 7, 2022.

114. From August 2009–December 2013: see https://apps.trb.org/cmsfeed/TRBNetProjectDisplay.asp?ProjectID=2829, accessed January 28, 2021.

115. RSG, AECOM, Mark Bradley Research and Consulting, John Bowman Research and Consulting, Mohammed Hadi, Ram Pendyala, Chandra Bhat, Travis Waller, and North Florida Transportation Planning Organization, *Dynamic, Integrated Model System: Jacksonville-Area Application*, 2 vols. (Washington, DC: Transportation Research Board, 2014).

116. See chapter 4.

117. *TMIP Fiscal Year 2007 Report* (n.p.: n.p., n.d.), 12.

118. See, for example, RSG et al., *Dynamic, Integrated Model System: Jacksonville-Area Application*; Konstadinos G. Goulias, Nathanael A. Isbell, Daimin Tang, Michael Balmer, Yali Chen, Chandra Bhat, and Ram Pendyala, *TRANSIMS and MATSIM Experiments in SimAGENT* (Draft), Phase 2 Report Submitted to Southern California Association of Governments (Santa Barbara, CA: GeoTrans Laboratory, 2012).

119. See, for example, the literature review by Kwang Sun Lee, Jin Ki Eom, and Daeseop Moon, "Applications of TRANSIMS in Transportation: A Literature Review," *Procedia Computer Science* 32 (2014): 769–773.

120. "Transportation Analysis and Simulation," https://sourceforge.net/projects/transims/, accessed November 26, 2020; https://www.linkedin.com/groups/43594/, accessed November 26, 2020 (around forty members).

121. Cambridge Systematics, *Status of Activity-Based Models and Dynamic Traffic Assignment at Peer MPOs*, prepared for MWCG/NCRTPB (n.p.: n.p., 2015), 24–25.

122. Cambridge Systematics, *Status of Activity-Based Models and Dynamic Traffic Assignment at Peer MPOs*.

123. See chapters 6 and 7.

124. Joshua A. Auld, "Agent-Based Dynamic Activity Planning and Travel Scheduling Model: Data Collection and Model Development" (Ph.D. diss., University of Illinois at Chicago, 2011).

125. On POLARIS, see Joshua Auld, Michael Hope, Hubert Ley, Vadim Sokolov, Bo Xu, and Kuilin Zhang, "POLARIS: Agent-Based Modeling Framework Development and Implementation for Integrated Travel Demand and Network and Operations Simulations," *Transportation Research Part C* 64 (March 2016): 101–116; Joshua Auld, Vadim Sokolov, and Thomas S. Stephens, "Analysis of the Effects of Connected-Automated Vehicle Technologies on Travel Demand," *Transportation Research Record*, no. 2625 (2017): 1–8; https://www.anl.gov/es/polaris-transportation-system-simula

tion-tool, accessed November 26, 2020; https://www.anl.gov/topic/polaris, accessed January 26, 2021; https://www.anl.gov/search-public#q=POLARIS&sort=relevancy &site=Argonne%20National%2BLaboratory, accessed November 26, 2020.

## Chapter 6

1. For example, the amount and type of travel required to fulfill their activities, the destinations of these activities, the mode of travel used to access activity locations, and the timing of this travel.

2. It is obvious that possibilities for travel are for example constrained by the location of facilities where activities may be undertaken and the times of day when such activities are possible and desirable.

3. Since activities (and thus travel patterns) of different individuals within a single household interact with each other, it is *household*, and not individual-level analysis that gives better potential for understanding *individual* travel decisions.

4. "Activity Based Models," https://tfresource.org/topics/Activity_based_models.html, accessed January, 24, 2022.

5. Ian G. Heggie, *Transport Engineering Economics* (London: McGraw Hill, 1972).

6. On the research work carried out at the TSU in the 1970s and early 1980s, see David Banister, "The TSU Approach—The First Ten Years of the Transport Studies Unit, Oxford," *Transport Reviews* 4, no. 3 (1984): 299–330.

7. P. M. Jones, M. C. Dix, M. I. Clarke, and I. G. Heggie, *Understanding Travel Behaviour* (Aldershot, UK: Gower Publishing Company, 1983), 245. The book first appeared as a research report in 1980.

8. Ian G. Heggie, "Putting Behaviour into Behavioural Models of Travel Choice," *Journal of the Operational Research Society* 29, no. 6 (June 1978): 541–550 (on 541).

9. Jones would get his Ph.D. in 1983: Peter Malcolm Jones, "A New Approach to Understanding Travel Behaviour and Its Implications for Transportation Planning" (Ph.D. diss., University of London, 1983).

10. Jones et al., *Understanding Travel Behaviour*, ix (quotation on 247).

11. The opening sentence of an article coauthored by Chapin in 1965 reads as follows: "Most land uses and transportation routes exist not for *their own sake*, but because they are means for the *accomplishment of some desired activity*" [emphasis mine], in F. Stuart Chapin Jr. and Henry C. Hightower, "Household Activity Patterns and Land Use," *Journal of the American Institute of Planners* 31, no. 3 (1965): 222–231 (on 222). In their book, which appeared in 1983, the TSU authors quote from two of Chapin's works: F. Stuart Chapin Jr., *Urban Land Use Planning*, 1st ed. (Urbana: University of Illinois Press, [1957] 1965); and F. Stuart Chapin Jr., *Human Activity*

*Patterns in the City: Things People Do in Time and in Space* (London: John Wiley & Sons, 1974) (see Jones et al., *Understanding Travel Behaviour*, 276). On Chapin see David R. Godschalk, "ACSP Distinguished Educator, 1986: F. Stuart Chapin Jr.," *Journal of Planning Education and Research* 36, no. 1 (March 2016): 119–120; *The Center for Urban & Regional Studies. The First 50 Years* (n.p.: n.p., 2008).

12. Heggie and the TSU group are quoted the following article: Torsten Hägerstrand, "What about People in Regional Science," *Papers of the Regional Science Association* 24, no. 1 (January 1970): 7–24 (Hägerstrand's article is cited by Jones et al., *Understanding Travel Behaviour*, 277).

13. In fairness to history, it should be mentioned that an even earlier explicit account of travel as derived demand can be found in Oi and Shuldiner, *An Analysis of Urban Travel Demands*. But this book is not mentioned by the TSU group.

14. Jones et al., *Understanding Travel Behaviour*; M. I. Clarke, M. C. Dix, P. M. Jones, and I. G. Heggie, "Some Recent Developments in Activity-Travel Analysis and Modeling," *Transportation Research Record*, no. 794 (1981): 1–8.

15. S. Carpenter and P. M. Jones, eds., *Recent Advances in Travel Demand Analysis* (Aldershot, UK: Gower Publishing Company, 1983).

16. Ryuichi Kitamura, "An Evaluation of Activity-Based Travel Analysis," *Transportation* 15, nos. 1–2 (March 1988): 9–34 (quotation on 9).

17. The proceedings of the conference have been frequently quoted since their publication in 1983.

18. See, for example, Clarke et al., "Some Recent Developments"; P. M. Jones, "Methodology for Assessing Transportation Policy Impacts," *Transportation Research Record*, no. 723 (1979): 52–58; P. M. Jones, "'HATS': A Technique for Investigating Household Decisions," *Environment and Planning A*, 11, no. 1 (January 1979): 59–70; Heggie, "Putting Behaviour into Behavioural Models"; Ian G. Heggie and Peter M. Jones, "Defining Domains for Models of Travel Demand," *Transportation* 7, no. 2 (June 1978): 119–125; Ian G. Heggie "Consumer Response to Public Transport Improvements and Car Restraint: Some Practical Findings," *Policy and Politics* 5, no. 4 (1977): 47–69.

19. Hensher and Stopher, *Behavioural Travel Modelling*.

20. The two conferences were held in May 1978 and January 1979, respectively. See K. Patricia Burnett, R. Q. Hanham, and Allen Cook, "Choice and Constraints-Oriented Modeling: Alternative Approaches to Travel Behavior," in *Directions to Improve Urban Travel Demand Forecasting: Conference Summary and White Papers*, ed. Louise E. Skinner (Washington, DC: FHWA and UMTA, 1978), 343–411.

21. Joseph Henry Bain, "Activity Choice Analysis Time Allocation and Disaggregate Travel Demand Modeling," (master's thesis, MIT, 1976), acknowledgments.

22. David Damm, "Theory and Empirical Results: A Comparison of Recent Activity-Based Research," in Carpenter and Jones, *Recent Advances in Travel Demand Analysis*, 137–162.

23. David Damm, "Toward a Model of Activity Scheduling Behavior" (Ph.D. diss., MIT, 1979); Jesse Jacobson, "Models of Non-Work Activity Duration" (Ph.D. diss., MIT, 1979); Ilan Salomon, "Life Style as a Factor in Explaining Travel Behavior" (Ph.D. diss., MIT, 1980). One could also mention here the research carried out by Thomas Adler for his own doctoral dissertation: Thomas Adler, "Modeling Non-Work Travel Patterns" (Ph.D. diss., MIT, 1976). An early review of the activity approach-related modeling work conducted at MIT in the 1970s can be found in J. M. Golob and Thomas F. Golob, "Classification of Approaches to Travel-Behavior Analysis," in Transportation Research Board, *Travel Analysis Methods for the 1980s*, 83–107.

24. Eric Ivan Pas, "Toward the Understanding of Urban Travel Behaviour through the Classification of Daily Urban Travel/Activity Patterns" (Ph.D. diss., Northwestern University, 1980). On Pas's Ph.D. work, and especially Koppelman's involvement in ABM, see Chandra R. Bhat, Laurie A. Garrow, and Patricia L. Mokhtarian, "Frank Koppelman's Contributions and Legacy to the Travel Demand Modeling Field," *Transportation Research Part B* 42, no. 3 (March 2008): 185–190 (the entire issue is a tribute to the career of Koppelman).

25. A year after his death, in 1998, the International Association for Travel Behaviour Research established the Eric Pas Dissertation Prize Award.

26. Hani Sobhi Mahmassani, "Methodological Aspects of a Decision Aid for Transportation Choices under Uncertainty" (Ph.D. diss., MIT, 1981). Mahmassani was awarded his master's degree by Purdue University in 1978.

27. Chandra R. Bhat, "Toward a Model of Activity Program Generation" (Ph.D. diss., Northwestern University, 1991).

28. Kara Maria Kockelman, "A Utility-Theory-Consistent System-of-Demand-Equations Approach to Household Travel Choice" (Ph.D. diss., University of California at Berkeley, 1998).

29. John Polak, "A Tribute to Ryuichi Kitamura: An IATRB Perspective," *Transportation* 36, no. 6 (November 2009): 649–650 (on 650).

30. Satoshi Fujii and Toshiyuki Yamamoto, "Tribute to Kitamura *Sensei*: A Great Teacher," *Transportation* 36, no. 6 (November 2009): 647–648 (on 647).

31. Ryuichi Kitamura, "Urban Travel Demand Forecasting by Stratified Choice Models" (Ph.D. diss., University of Michigan, 1978).

32. Ryuichi Kitamura, "A Stratification Analysis of Taste Variations in Work-Trip Mode Choice," *Transportation Research Part A* 15, no. 6 (November 1981): 473–485 (on 473).

33. Ryuichi Kitamura, Lidia P. Kostyniuk, and Michael J. Uyeno, "Basic Properties of Urban Time-Space Paths: Empirical Tests," *Transportation Research Record*, no. 794 (1981): 8–19.

34. Kitamura chaired the Transportation Research Board Committee on Travel Behavior and Values from 1989 to 1995, and the International Association for Travel Behaviour Research between 1992 and 1994.

35. Patricia L. Mokhtarian, "What about People in Behavioral Modeling? Ryuichi Kitamura (1949–2009)," *Journal of Choice Modelling* 2, no. 1 (2009): 1–7 (on 1). The paper provides an overall presentation of Kitamura's work. See also Eric J. Miller, "Articulating the Activity-Based Paradigm: Reflections on the Contributions of Ryuichi Kitamura," *Transportation* 36, no. 6 (November 2009): 651–655.

36. Ram M. Pendyala, Konstadinos G. Goulias, and Cynthia Chen, "Remembering a Teacher," *Transportation* 36, no. 6 (November 2009): 643–645 (on 643).

37. Konstadinos G. Goulias, "Long Term Forecasting with Dynamic Microsimulation" (Ph.D. diss., University of California at Davis, 1991).

38. Ram M. Pendyala, "Causal Modeling of Travel Behavior Using Simultaneous Equations Systems. A Critical Examination" (Ph.D. diss., University of California at Davis, 1992).

39. W. W. Recker and R. Kitamura, "Activity Based Travel Analysis," in *Transportation and Mobility in an Era of Transition*, ed. G. R. M. Jansen, P. Nijkamp, and G. J. Ruijgrok (Amsterdam: North Holland/Elsevier, 1984), 157–183.

40. See, for example, Wilfred W. Recker and Thomas F. Golob, "An Attitudinal Modal Choice Model," *Transportation Research* 10, no. 5 (October 1976): 299–310.

41. Wilfred W. Recker and Lidia P. Kostyniuk, "Factors Influencing Destination Choice for the Urban Grocery Shopping Trip," *Transportation* 7, no. 1 (March 1978): 19–33.

42. Lidia P. Kostyniuk, "Tribute to Ryuichi Kitamura: A Reflection," *Transportation* 36, no. 6 (November 2009): 641–642.

43. Michael Greyson McNally, "On the Formation of Household Travel/Activity Patterns: A Simulation Approach" (Ph.D. diss., University of California at Irvine, 1986). See also W. W. Recker, M. G. McNally, and G. S. Root, "A Model of Complex Travel Behavior: Part I—Theoretical Development," *Transportation Research Part A* 20, no. 4 (July 1986): 307–318; W. W. Recker, M. G. McNally, and G. S. Root, "A Model of Complex Travel Behavior: Part II—An Operational Model," *Transportation Research Part A* 20, no. 4 (July 1986): 319–330.

44. McNally, "On the Formation of Household Travel/Activity Patterns," 182; emphasis mine.

45. For the period 1973–1985, see especially the review of the book *Behavioural Research for Transport Policy. The 1985 International Conference on Travel Behaviour* (Utrecht, The Netherlands: VNU Science Press, 1986), authored by Kitamura himself. The review appeared in *Transportation Science* 21, no. 3 (August 1987): 218–222.

46. Golob and Golob, "Classification of Approaches."

47. The proceedings were published in a special issue of *Transportation* 15, nos. 1–2 (March 1988).

48. Ryuichi Kitamura, "Life-Style and Travel Demand," in *A Look Ahead: Year 2020*, ed. Transportation Research Board (Washington, DC: Transportation Research Board, 1988), 149–189.

49. Peter M. Allaman, Timothy J. Tardiff, and Frederick C. Dunbar, *New Approaches to Understanding Travel Behavior* (Washington, DC: Transportation Research Board, 1982).

50. Susan Hanson and Perry Hanson, "The Impact of Married Women's Employment on Household Travel Patterns: A Swedish Example," *Transportation* 10, no. 2 (June 1981): 165–183.

51. Kitamura, "Life-Style and Travel Demand," 149.

52. Kenneth S. Kurani and Martin E. H. Lee-Gosselin, "Synthesis of Past Activity Analysis Applications," in Texas Transportation Institute, *Activity-Based Travel Forecasting Conference, June 2–5, 1996*, 51–77.

53. Ryuichi Kitamura and Chandra R. Bhat, "Guest Editorial: Special Issue Dedicated to the Memory of Eric Pas," *Transportation* 26, no. 2 (May 1999): 113–115 (on 114); emphasis mine.

54. Chandra R. Bhat and Frank S. Koppelman, "A Retrospective and Prospective of Time-Use Research," *Transportation* 26, no. 2 (May 1999): 119–139 (on 121).

55. John Urry, "The Sociabilities of Travel," in *The Expanding Sphere of Travel Behaviour Research: Selected Papers from the 11th International Conference on Travel Behaviour Research*, ed. Ryuichi Kitamura, Toshio Yoshii, and Toshiyuki Yamamoto (Bingley: Emerald Group Publishing, 2009), 3–15.

56. See especially the various contributions in *Transportation* 15, nos. 1–2 (March 1988) as well as the two following articles: Eric I. Pas, "State of the Art and Research Opportunities in Travel Demand: Another Perspective," *Transportation Research Part A* 19, nos. 5–6 (1985): 460–464; and Eric I. Pas, "Is Travel Demand Analysis and Modelling in the Doldrums?," in *Developments in Dynamic and Activity-Based Approaches to Travel Analysis*, ed. Peter Jones (Aldershot, UK: Avebury, 1990), 3–27.

57. Peter M. Jones, "Viewpoint," *Transportation* 15, nos. 1–2 (March 1988): 49–51.

58. Lidia P. Kostyniuk, "Viewpoint," *Transportation* 15, nos. 1–2 (March 1988): 43–45.

59. David T. Hartgen, "Viewpoint," *Transportation* 15, nos. 1–2 (March 1988): 47–48.

60. Hani S. Mahmassani, "Some Comments on Activity-Based Approaches to the Analysis and Prediction of Travel Behavior," *Transportation* 15, nos. 1–2 (March 1988): 35–40 (on 36–37).

61. Kitamura, "An Evaluation of Activity-Based Travel Analysis."

62. Eric I. Pas, "Recent Advances in Activity-Based Travel Demand Modeling," in Texas Transportation Institute, *Activity-Based Travel Forecasting Conference, June 2–5, 1996*, 79–102 (quotations on 79–80 and 81); emphasis mine.

63. Pas, "Is Travel Demand Analysis and Modelling in the Doldrums?"

64. In addition to the references provided in chapter 5, see also Scott Le Vine and Martin Lee-Gosselin, "Transportation and Environmental Impacts and Policy"; David L. Greene, "Transportation and Energy"; Evelyn Blumenberg, "Social Equity and Urban Transportation," all of them in Giuliano and Hanson, *The Geography of Urban Transportation*, 273–301, 302–331, 332–358, respectively.

65. Peter R. Stopher, "Use of an Activity-Based Diary to Collect Household Travel Data," *Transportation* 19, no. 2 (May 1992): 159–176.

66. Including Wasatch Front (1993), Southeast Michigan (1994), Oregon and Southwest Washington (1994), Oahu (1995), and the New York Metropolitan Area (1996), to mention a few. See Peter R. Stopher and Stephen P. Greaves, "Household Travel Surveys: Where Are We Going?," *Transportation Research Part A* 41, no. 5 (June 2007): 367–381 (on 370); and especially Cambridge Systematics, *Scan of Recent Travel Surveys*.

67. Like the development of SP techniques, the approach of using an activity diary was first experimented with in Europe and Australia.

68. Remember that in the traditional Home Interview O&D traffic survey, trained interviewers visited sampled dwellings and asked a person, sometimes whoever happened to open the door, questions about the travels the various members of the household had undertaken on the *previous* day. With the introduction of *prospective* gathering procedures, the "previous" day was superseded by a day in the "future," on which travel was to be recorded *individually* by all members of the household able to fill in the questionnaire. The prospective approach was probably first used in the United States in southeast Michigan and Oahu in the early 1980s; both surveys were administered by the private firm Schimpeler-Corradino Associates. In the 1990s, the prospective method had become by far the preferred one within the American community of travel survey professionals (Stopher and Metcalf, *Methods for Household Travel Surveys*, 19 and 22).

69. Pas, "Recent Advances in Activity-Based Travel Demand Modeling," 95–96. For a thorough discussion regarding a series of issues related to data needs, data quality

requirements, and methods of data collection pertinent to the further development of ABM at the end of the 1990s, see especially Theo Arentze, Harry Timmermans, Frank Hofman, and Nelly Kalfs, "Data Needs, Data Collection, and Data Quality Requirements of Activity-Based Transport Demand Models," in *Transportation Research E-Circular, no. E-C008*, ed. Transportation Research Board (Washington, DC: Transportation Research Board, August 2000), II-J/1–II-J/35.

70. T. Keith Lawton, "Activity and Time Use Data for Activity-Based Forecasting," and Martin Lee-Gosselin and John Polak, "Summary of Workshop One: Activity and Time Use Data Needs, Resources and Survey Methods," both in Texas Transportation Institute, *Activity-Based Travel Forecasting Conference, June 2–5, 1996*, 103–118 and 173–181, respectively.

71. Thus, the first two components of TRANSIMS, Population Synthesizer and Activity Generator, directly correspond to the same components in ABM. On the relationships between the most recent versions of TRANSIMS and the general ABM framework, see especially Parsons Brinckerhoff, University of Texas, Northwestern University et al., *Assessing Highway Tolling and Pricing Options and Impacts*, vol. 2: *Travel Demand Forecasting Tools* (Washington, DC: Transportation Research Board, 2012), 92–94.

72. See especially Pas, "Recent Advances in Activity-Based Travel Demand Modeling," 94.

73. Eric J. Miller, "Microsimulation and Activity-Based Forecasting," in Texas Transportation Institute, *Activity-Based Travel Forecasting Conference, June 2–5, 1996*, 151–172 (on 151).

74. Bob Sicko and Hani Mahmassani, "Summary of Workshop Three: Microsimulation in Activity Analysis," in Texas Transportation Institute, *Activity-Based Travel Forecasting Conference, June 2–5, 1996*, 193–197.

75. Bruce D. Spear, *New Approaches to Travel Forecasting Models: A Synthesis of Four Research Proposals* (Washington, DC: U.S. Department of Transportation and U.S. Environmental Protection Agency, 1994), without pagination. The report was also published as Bruce D. Spear, "New Approaches to Transportation Forecasting Models: A Synthesis of Four Research Proposals," *Transportation* 23, no. 3 (August 1996): 215–240 (quotation on 229–230). The same issue of *Transportation* contains the summaries of the results of the four proposals that were awarded a contract.

76. Ryuichi Kitamura, Eric I. Pas, Clarisse V. Lula, T. Keith Lawton, and Paul E. Benson, "The Sequenced Activity Mobility Simulator (SAMS): An Integrated Approach to Modeling Transportation, Land Use and Air Quality," *Transportation* 23, no. 3 (August 1996): 267–291.

77. Konstadinos G. Goulias and Ryuichi Kitamura, "Travel Demand Forecasting with Dynamic Microsimulation," *Transportation Research Record*, no. 1357 (1992):

8–17. It is worth noting that the authors refer to, among others, Guy H. Orcutt, Amihai Glazer, Robert Harris, and Richard Wertheimer II, "Microanalytic Modeling and the Analysis of Public Transfer Policies," in *Microeconomic Simulation Models for Public Policy Analysis*, vol. 1, ed. R. Haveman and K. Hollenbeck (New York: Academic Press, 1980), 81–106. Microsimulation was first introduced by Orcutt in an article published in 1957: Guy H. Orcutt, "A New Type of Socio-Economic System," *Review of Economics and Statistics* 39, no. 2 (May 1957): 116–123, and has been subsequently applied to a series of modeling tasks, including travel behavior, demographic change, spatial diffusion, health, and land use. On Orcutt's pioneering work, see the recent work by Chung-Tang Cheng, "Guy H. Orcutt's Engineering Microsimulation to Reengineer Society," *History of Political Economy* 52, no. S1 (2020): 191–217.

78. Kitamura et al. cited the seminal article by Herbert A. Simon, "A Behavioral Model of Rational Choice," *The Quarterly Journal of Economics* 69, no. 1 (February 1955): 99–118.

79. Peter R. Stopher, David T. Hartgen, and Yuanjun Li, "SMART: Simulation Model for Activities, Resources and Travel," *Transportation* 23, no. 3 (August 1996): 293–312 (quotations on 310, 298, and 293).

80. Moshe Ben-Akiva, John L. Bowman, and Dinesh Gopinath, "Travel Demand Model System for the Information Era," *Transportation* 23, no. 3 (August 1996): 241–266.

81. Dinesh Ambat Gopinath, "Modeling Heterogeneity in Discrete Choice Processes: Application to Travel Demand" (Ph.D. diss., MIT, 1995).

82. John L. Bowman, "Activity Based Travel Demand Model System with Daily Activity Schedules" (master's thesis, MIT, 1995).

83. Ben-Akiva et al., "Travel Demand Model System for the Information Era," 241.

84. Tours are travel events that start at one location and return to that same location.

85. See, for example, John L. Bowman and Moshe E. Ben-Akiva, "Activity-Based Disaggregate Travel Demand Model System with Activity Schedules," *Transportation Research Part A* 35, no. 1 (January 2001): 1–28 (especially 3).

86. The Metropolitan Washington Council of Governments (MWCOG), DOT, and EPA.

87. RDC, *Activity-Based Modeling System for Travel Demand Forecasting* (n.p.: n.p., 1995).

88. R. M. Pendyala, R. Kitamura, A. Kikuchi, T. Yamamoto, and S. Fujii, "Florida Activity Mobility Simulator, Overview and Preliminary Validation Results," *Transportation Research Record*, no. 1921 (2005): 123–130.

89. Bowman and Ben-Akiva expounded on the differences between their own econometric approach and the approach put forward by Kitamura et al. in John L. Bowman and Moshe Ben-Akiva, "Activity-Based Travel Forecasting," in Texas Transportation Institute, *Activity-Based Travel Forecasting Conference, June 2–5, 1996*, 3–36.

90. Donnelly et al., *Advanced Practices in Travel Forecasting*, 58.

91. On the aforementioned household travel survey in Portland, see Lawton, "Activity and Time Use Data for Activity-based Forecasting"; Catherine Theresa Lawson, "Household Travel/Activity Decisions" (Ph.D. diss., Portland State University, 1998), chapter 5 and passim.

92. Bowman, "Activity Based Travel Demand Model System."

93. Also contributing through a concurrent congestion pricing project was a group of three other consultants (see Mark Bradley Research and Consulting, Portland Metro, John Bowman [MIT], and Cambridge Systematics, *A System of Activity-Based Models for Portland, Oregon*, prepared for TMIP [n.p.: U.S. DOT and U.S. EPA, 1998]).

94. Shiftan, a Ph.D. holder from MIT in the early 1990s—Yoram Shiftan, "Transportation Workforce Planning in the Transit Industry: Incorporating Absence, Overtime, and Reliability Relationships" (Ph.D. diss., MIT, 1992)—had been already involved in the tour-based model developed for Boise, Idaho, and along with Rossi—a MIT civil engineer (1981) and a master's degree holder from the same institution in 1987 (Thomas Francis Rossi, "Traffic Allocation Methods for Use in Impact Fee Assessment" [master's thesis, MIT, 1987])—in the modeling framework incorporated in the New Hampshire statewide model.

95. John L. Bowman, "The Day Activity Schedule Approach to Travel Demand Analysis" (Ph.D. diss., MIT, 1998), 6.

96. Bowman, "The Day Activity Schedule Approach," 5.

97. Hague Consulting Group would merge with RAND Europe in 2001.

98. John L. Bowman, Mark Bradley, Yoram Shiftan, T. Keith Lawton, and Moshe Ben-Akiva, "Demonstration of an Activity-Based Model System for Portland," in *World Transport Research Proceedings of 8th World Conference on Transport Research*, vol. 3, ed. Hilde Meersman, Eddy van de Voorde, and Willy Winkelmans (Amsterdam: Pergamon, 1999), 171–184.

99. Bradley et al., "A System of Activity-Based Models," 5.

100. Donnelly et al., *Advanced Practices in Travel Forecasting*, 58.

101. Chandra Bhat, "Welcome," in *Innovations in Travel Demand Modeling, Volume 1: Session Summaries*, ed. Transportation Research Board (Washington, DC: Transportation Research Board, 2008), 1–2 (on 1).

102. Chandra R. Bhat, Jessica Y. Guo, Sivaramakrishnan Srinivasan, and Aruna Sivakumar, "A Comprehensive Econometric Microsimulator for Daily Activity-Travel Patterns," *Transportation Research Record*, no. 1894 (2004): 57–66.

103. The metropolitan regions examined included Portland (Oregon); San Francisco, Sacramento, and the Bay Area of California; New York; Columbus; Atlanta; and Denver. See Mark Bradley and John L. Bowman, "Design Features of Activity-Based Microsimulation Models for U.S. Metropolitan Planning Organizations: A Summary," in Transportation Research Board, *Innovations in Travel Demand Modeling, Volume 2: Papers*, 11–20; John L. Bowman, "Historical Development of Activity-Based Model Theory and Practice," *Traffic Engineering and Control* 50, no. 2 (February 2009): 59–62.

104. Donnelly et al., *Advanced Practices in Travel Forecasting*, 7, 59–61; for a summary of projects conducted by SFCTA with the help of its activity-based modeling suite by 2010, see Association of Metropolitan Planning Organizations, *Advanced Travel Modeling Study*, final report prepared by Vanasse Hangen Brustlin (VHB), Resource Systems Group (RSG), Shapiro Transportation Consulting, LLC, Urban Analytics (n.p.: n.p., 2011), 17–19. On the SFCTA model, see also Maren L. Outwater and Billy Charlton, "The San Francisco Model in Practice: Validation, Testing, and Application," in Transportation Research Board, *Innovations in Travel Demand Modeling, Volume 2: Papers*, 24–29; Lisa Zorn, Elizabeth Sall, and Daniel Wu, "Incorporating Crowding into the San Francisco Activity-Based Travel Model," *Transportation* 39, no. 4 (July 2012): 755–771.

105. On the New York and Columbus activity-based models and ABM experience, see Peter Vovsha and Kuo-Ann Chiao, "Development of New York Metropolitan Transportation Council Tour-Based Model"; Kuo-Ann Chiao, Ali Mohseni, and Sangeeta Bhowmick, "Lessons Learned from the Implementation of New York Activity-Based Travel Model"; Rebekah S. Anderson, "Development of Mid-Ohio Regional Planning Commission Tour-Based Model"; Rebekah S. Anderson, Zhuojun Jiang, and Chandra Parasa, "Hardware Requirements and Running Time for the Mid-Ohio Regional Planning Commission Travel Forecasting Model," all in Transportation Research Board, *Innovations in Travel Demand Modeling, Volume 2: Papers*, 21–23, 173–176, 30–32, and 181–184, respectively.

106. The committee was composed of thirteen specialists ranging from university and other knowledge-related organizations to consultants and government officials. It supplemented its own expertise by seeking technical guidance from three large consultancies heavily involved in travel demand modeling: Parsons Brinckerhoff, Cambridge Systematics, and AECOM. See Committee for Determination of the State of the Practice in Metropolitan Area Travel Forecasting, *Metropolitan Travel Forecasting*, 114–119.

107. Cambridge Systematics, *Status of Activity-Based Models and Dynamic Traffic Assignment at Peer MPOs*.

108. NYMTC (New York), MTC (San Francisco), SANDAG (San Diego), SACOG (Sacramento), MORPC (Columbus), and DRCOG (Denver).

109. CMAP (Chicago), Metropolitan Council (Twin Cities of Minneapolis and Saint Paul), BRTB/BMC (Baltimore), ARC (Atlanta), Portland Metro, DVRPC (Philadelphia), PSRC (Seattle), SCAG (Los Angeles), H-GAC (Houston-Galveston), and MAG (Phoenix).

110. Association of Metropolitan Planning Organizations, *Advanced Travel Modeling Study*, 11.

111. Outwater and Charlton, "The San Francisco Model in Practice," 29; Association of Metropolitan Planning Organizations, *Advanced Travel Modeling Study*, 14, 30.

112. Unless otherwise stated, the bulk of information about RSG has been drawn from the company's site: https://rsginc.com/history/, accessed November 28, 2020.

113. John P. Gliebe, "Models of Household Joint Decision Making in Activities and Travel" (Ph.D. diss., Northwestern University, 2004).

114. John Gliebe joined Cambridge Systematics in November 2019 and Joe Castiglione returned to San Francisco County Transportation Authority in February 2015 (email from Thomas Rossi to the author, dated January 28, 2021).

115. See, for example, Greg M. Spitz, Frances L. Niles, and Thomas J. Adler (RSG), *Web-Based Survey Techniques: A Synthesis of Transit Practice* (Washington, DC: Transportation Research Board, 2006).

116. This proof-of-concept study lasted more than three years, from October 2009–February 2012, while its total funding amounted to $1,266,223. The marketing research firm Abt SRBI was appointed project leader, and, in addition to Cambridge Systematics and Resource Systems Group, the university travel modeling and data collection specialist Peter Stopher was also part of the project. See Creg Giaimo, Rebekah Anderson, Laurie Wargelin, and Peter Stopher, "Will It Work? Pilot Results from First Large-Scale Global Positioning System-Based Household Travel Survey in the United States," *Transportation Research Record*, no. 2176 (2010): 26–34 (for the pilot study); Peter Stopher, Laurie Wargelin, Jason Minser, Kevin Tierney, Mindy Rhindress, Sharon O'Connor, *GPS-Based Household Interview Survey for the Cincinnati, Ohio Region* (n.p.: Abt SRBI/FHWA/R&D Ohio DOT, 2012).

117. Concerning the university segment of the UTDM social world, in addition to MIT, the University of California at Berkeley, another academic institution already heavily involved in urban traffic forecasting, has recently developed the first open-source smartphone platform for collecting travel data, called OpenPATH: Open Platform for Agile Trip Heuristics (formerly known as e-mission): https://www.nrel.gov/transportation/openpath.html#/home, accessed January 24, 2022; K. Shankari, Mohamed Amine Bouzaghrane, Samuel M. Maurer et al., "e-mission: An

Open-Source, Smartphone Platform for Collecting Human Travel Data," *Transportation Research Record* 2672, no. 42 (December 2018): 1–12.

118. rMove™ (https://rmove.rsginc.com/what-is-rmove/, accessed October 7, 2022) was first developed by the consultancy for a pilot travel study conducted in Anderson, Indiana, and sponsored by the Federal Highway Administration. Following an approximately five-month planning, testing, and application development period, rMove was published in the Android and iOS app stores in late April 2015. While the smartphone's sensors passively collect a wealth of travel temporal and spatial data, rMove "asks" the smartphone-holders in close to real-time at each trip destination a short set of survey questions, including trip purpose, travel party makeup, travel mode, specific household vehicle (if auto), and travel costs. See Resource Systems Group, *In-The-Moment Travel Study* (Revised Report) (White River Junction, VT: RSG, 2015). By mid-2018, rMove had recorded over two hundred thousand days of data, been used in over twenty projects and had collected over two million trips.

119. On Cambridge Systematics, see, for example, Cambridge Systematics, with Massachusetts Institute of Technology, *Cell Phone Location Data for Travel Behavior Analysis* (Washington, DC: Transportation Research Board, 2018).

120. Including the origin, the destination and the route of travel, accurate trip start and end times, the exact location of the various activities performed by the tripmaker, and the length of the trips.

121. See, for example, Soora Rasouli and Harry Timmermans, eds., *Mobile Technologies for Activity-Travel Data Collection and Analysis* (Hershey, PA: Information Science Reference, 2014); Bat-hen Nahmias-Biran, Yafei Han, Shlomo Bekhor et al., "Enriching Activity-Based Models Using Smartphone-Based Travel Surveys," *Transportation Research Record* 2672, no. 42 (December 2018): 280–291.

122. For an overview of the work done by Parsons Brinckerhoff in ABM by 2007, see William Davidson, Robert Donnelly, Peter Vovsha, Joel Freedman, Steve Ruegg, Jim Hicks, Joe Castiglione, and Rosella Picado, "Synthesis of First Practices and Operational Research Approaches in Activity-Based Travel Demand Modeling," *Transportation Research Part A* 41, no. 5 (June 2007): 464–488.

123. See, for example, Gordon W. Schultz, "Development of a Travel-Demand Model Set for the New Orleans Region," *Transportation Research Record*, no. 944 (1983): 128–134.

124. Rosella Picado, "Non-Work Activity Scheduling Effects in the Timing of Work Trips" (Ph.D. diss., University of California at Berkeley, 1999).

125. William Davidson, Peter Vovsha, Joel Freedman, Richard Donnelly, "CT-RAMP Family of Activity-Based Models" (Australian Transport Research Forum 2010, September 29–October 1, 2010, Canberra, Australia), https://www.researchgate.net/publication/267862957_CT-RAMP_family_of_Activity-Based_Models, accessed January

24, 2022; Peter Vovsha, Joel Freedman, Vladimir Livshits, and Wu Sun, "Design Features of Activity-Based Models in Practice: Coordinated Travel-Regional Activity Modeling Platform," *Transportation Research Record*, no. 2254 (2011): 19–27.

126. Parsons Brinckerhoff and The Corradino Group, *Southeast Florida Regional Planning Model—SERPM 7.0, Coordinated Travel—Regional Activity Based Modeling Platform (CT-RAMP): Model User Guide* (n.p.: n.p., 2016).

127. Unless otherwise stated, information about DaySim is drawn from RSG's GitHub wiki: https://github.com/RSGInc/DaySim/wiki/John-Bowman-Website-Archive, accessed January 27, 2021.

128. Ben Stabler, Mark Bradley, Dan Morgan, Howard Slavin, and Khademul Haque, *Volume 2: Model Impacts of Connected and Autonomous/Automated Vehicles (CAVs) and Ride-Hailing with an Activity-Based Model (ABM) and Dynamic Traffic Assignment (DTA)—An Experiment* (TMIP) (Washington, DC: Federal Highway Administration, 2018), 8.

129. "Daysim Activity Based Model," https://github.com/RSGInc/DaySim/wiki, accessed November 27, 2020.

130. University of Maryland and Cambridge Systematics, *An Activity-Based Maryland Statewide Transportation Model—MSTM Version 2* (Draft Final Report) (n.p.: n.p., 2018), 29; http://mitams.org/Main_Page, accessed November 27, 2020.

131. Association of Metropolitan Planning Organizations, *Advanced Travel Modeling Study*, A/53–A/57 (Portland Case); A/37–A/40 (Los Angeles); Chandra R. Bhat, Konstadinos G. Goulias, Ram M. Pendyala, Rajesh Paleti, Raghuprasad Sidharthan, Laura Schmitt, Hsi-Hwa Hu, "A Household-Level Activity Pattern Generation Model with an Application for Southern California," *Transportation* 40, no. 5 (September 2013): 1063–1086.

132. Mobility Futures Collaborative, "Research Areas," https://mfc.mit.edu/simmobility, accessed January 24, 2022; Isabel Viegas de Lima, Mazen Danaf, Arun Akkinepally, Carlos Lima De Azevedo, and Moshe Ben-Akiva, "Modeling Framework and Implementation of Activity—and Agent-Based Simulation: An Application to the Greater Boston Area," *Transportation Research Record* 2672, no. 49 (2018): 146–157.

133. "Activity Sim," https://activitysim.github.io/, accessed November 27, 2020; https://rsginc.com/project/activitysim-consolidated-travel-model-software-platform-development-and-enhancement/, accessed November 27, 2020.

134. Citilabs, *Discover Activity-Based Modeling Using Cube Voyageur* (n.p.: n.p., 2011), 1.

135. Citilabs, "Simplified Tour-Based Model," TMIP Webinar, July 26, 2017, https://www.fhwa.dot.gov/planning/tmip/community/webinars/summaries/index.cfm, accessed November 27, 2020.

136. For an overview, see *The Urban Transportation Monitor* 26, no. 7 (September 2012): 16–31 (on 21); *The Urban Transportation Monitor* 28, no. 8 (October 2014): 15–30 (on 20); each firm's website also contains a lot of relevant information.

137. See, for example, the five-day Short Course on Activity-Based Modeling of Transport Network Demand and Performance, organized by the Center for Transportation Research at the University of Texas at Austin in May 2016 and directed by three ABM academic specialists (Bhat, Pendyala, and Goulias).

138. Joe Castiglione, Mark Bradley, and John Gliebe, *Activity-Based Travel Demand Models: A Primer* (Washington, DC: Transportation Research Board, 2015), 66–67. On the transferability issue, see especially John Gliebe, Mark Bradley, Nazneen Ferdous, Maren Outwater, Haiyun Lin, and Jason Chen, *Transferability of Activity-Based Model Parameters* (Washington, DC: Transportation Research Board, 2014). In this project, the DaySim model developed first in Sacramento was transferred to both the Jacksonville and Tampa regions. For FHWA-sponsored studies of transferability, see, for example, John L. Bowman and Mark Bradley, "Testing Spatial Transferability of Activity-Based Travel Forecasting Models," *Transportation Research Record*, no. 2669 (2017): 62–71 (the authors quote a number of other studies conducted in the early 2010s). On the transferability issue, see also Sujan Sikder, "Spatial Transferability of Activity-Based Travel Forecasting Models" (Ph.D. diss., University of South Florida, 2013).

139. Donnelly et al., *Advanced Practices in Travel Forecasting*, 8–14, 32–35, and passim.

140. Castiglione, Bradley, and Gliebe, *Activity-Based Travel Demand Models*.

141. ABM specialist John Bowman was, for example, the presenter of a two-hour webinar that took place on June 18, 2009 and dealt with "Activity Model Development Experiences." Unless otherwise stated, information about ABM-related webinars is drawn from the TMIP site: https://www.fhwa.dot.gov/planning/tmip/community/webinars/summaries/index.cfm, accessed November 27, 2020.

142. For example, from February to September 2012, a twelve-session webinar on ABM was held by TMIP and conducted by travel forecasters from Parsons Brinckerhoff and Resource Systems Group.

143. For examples, see Donnelly et al., *Advanced Practices in Travel Forecasting*, 49–50.

144. TMIP, *Peer Review Program at a Glance* (n.p.: n.p., 2013). Thus, on April 14 and 15, 2011, the New York activity-based model system was reviewed by eight people outside the host agency, including Vovsha from PB, Rossi from Cambridge Systematics, and Bhat from the University of Texas at Austin; see TMIP, *New York Metropolitan Transportation Council (NYMTC) Travel Model Peer Review Report* (n.p.: n.p., 2011). Vovsha along with the academic Pendyala and four other panelists also reviewed the Baltimore ABM system on March 3, 2016, at a meeting that was a follow-up to a peer review conducted on December 6, 2013; see TMIP, *Baltimore Metropolitan*

*Council (BMC) Peer Review* (n.p.: n.p., 2016); TMIP, *Baltimore Metropolitan Council (BMC) Activity-Based Travel Model Peer Review Report* (n.p.: n.p., 2014).

145. On this mix of competition and cooperation, which now characterizes several industrial sectors, see, for example, Myriam Cloodt, John Hagedoorn, and Nadine Roijakkers, "Trends and Patterns in Interfirm R&D Networks in the Global Computer Industry: An Analysis of Major Developments, 1970–1999," *Business History Review* 80, no. 4 (Winter 2006): 725–746; Cyrus C. M. Mody, *The Long Arm of Moore's Law: Microelectronics and American Science* (Cambridge, MA: MIT Press, 2016).

146. See, for example: Vovsha et al., "Design Features of Activity-Based Models in Practice"; Bradley and Bowman, "Design Features of Activity-Based Microsimulation Models for U.S. Metropolitan Planning Organizations"; John Gliebe and Joel Freedman, Activity-Based Modeling; Session 4: Frameworks and Techniques, TMIP Webinar Series, April 5, 2012.

147. Regarding work and school location, auto ownership and transit pass holding, mode, or destination and time of day.

148. The Monte Carlo technique is a method of simulating a choice or action. The probabilities across all possible alternatives are used in a Monte Carlo approach to predict specific choices for each individual in the sample. The Monte Carlo approach predicts a single outcome per person, drawing randomly from the model probabilities. On the (relatively recent) marriage between disaggregate discrete choice models and simulation techniques, see especially Train, *Discrete Choice Methods with Simulation*. For a short history of this marriage, see Boyce and Williams, *Forecasting Urban Travel*, 207–213.

149. On the origins and the early history of the Monte Carlo simulation method, see among others Peter Galison, "Computer Simulation and the Trading Zone," in *From Science to Computational Science*, ed. Gabrielle Gramelsberger (Zürich: Diaphanes, 2011), 118–157. For early (European) applications of Monte Carlo simulation in transportation modeling, see Volker Kreibich, "Modelling Car Availability, Modal Split and Trip Distribution by Monte-Carlo Simulation: A Short Way to Integrated Models," *Transportation* 8, no. 2 (June 1979): 153–166; Peter Bonsall, "Microsimulation of Organized Car Sharing: Description of the Models and Their Calibration," *Transportation Research Record*, no. 767 (1980): 12–21; Kay W. Axhausen, and Raimund Herz, "Simulating Activity Chains: German Approach," *Journal of Transportation Engineering* 115, no. 3 (May 1989): 316–325.

## Chapter 7

1. See chapters 1 and 2.

2. Jeffrey A. Lindley, "Urban Freeway Congestion: Quantification of the Problem and Effectiveness of Potential Solutions," *ITE Journal* (January 1987): 27–32 (on 27).

3. The labor force participation rate of women stood at 34 percent in 1950, increasing to 60 percent by 2000. See Mitra Toossi, "A Century of Change: The U.S. Labor Force, 1950–2050," *Monthly Labor Review* (May 2002): 15–28 (on 15). See also Ruth Milkman, *On Gender, Labor, and Inequality* (Urbana: University of Illinois Press, 2016). On the travel patterns of American women, see, for example, the pioneering work conducted by Sandra Rosenbloom, "The Need for Study of Women's Travel Issues," *Transportation* 7, no. 4 (December 1978): 347–350; Transportation Research Board et al., eds., *Women's Issues in Transportation. Summary of the 4th International Conference, Volume 1: Conference Overview and Plenary Papers* (Washington, DC: Transportation Research Board, 2010); *Volume 2: Technical Papers* (Washington, DC: Transportation Research Board, 2011).

4. Robert Cervero, "Congestion Relief: The Land Use Alternative," *Journal of Planning Education and Research* 10, no. 2 (January 1991): 119–129 (the article contains a long list of relevant references); United States General Accounting Office, *Traffic Congestion: Trends, Measures, and Effects*, (n.p.: n.p., 1989). On the increasing suburbanization of the American city and urban jobs from the 1950s on, see among others Jon C. Teaford, *The Metropolitan Revolution: The Rise of Post-Urban America* (New York: Columbia University Press, 2006), chapters 3, 4, and 5. For a detailed analysis of the major shifts in the travel patterns of the American households after World War II, see in the first place the various updated reports prepared by Alan E. Pisarski. For two such reports, see Alan E. Pisarski (prepared by), *Commuting in America: A National Report on Commuting Patterns and Trends* (Westport: Eno Foundation for Transportation, 1987); Alan E. Pisarski and Steven E. Polzin, *Commuting in America 2013: The National Report on Commuting Patterns and Trends* (Washington, DC: American Association of State Highway and Transportation Officials, 2013).

5. Koppelman, Bhat, and Schofer, "Market Research Evaluation of Actions to Reduce Suburban Traffic Congestion," 383.

6. The yearly delay per auto commuter surged from eighteen hours in 1982 to thirty-seven hours in 2000; it was forty-two hours in 2006, forty in 2010, and forty-two again in 2014. See Texas A&M Transportation Institute and INRIX, *2015 Urban Mobility Scorecard* (n.p.: n.p., 2015), 1. The term "yearly delay per auto commuter" refers to the extra time spent during the year traveling at congested speeds rather than free-flow speeds by private vehicle driver and passengers who typically travel in the peak periods.

7. United States General Accounting Office, *Smart Highways: An Assessment of their Potential to Improve Travel* (n.p.: n.p., 1991).

8. The first name of ITS America was IVHS America. IVHS used to stand for intelligent vehicle-highway systems and for the Intelligent Vehicle-Highway Society. The term ITS was officially sanctioned by the U.S. Department of Transportation as a replacement for IVHS in 1994.

9. See chapter 5.

10. For a historical account of ITS technology in the United States, see, for example, Lyle Saxton, "Mobility 2000 and the Roots of IVHS," *IVHS Review* (Spring 1993): 11–26; Ashley Auer, Shelley Feese, and Stephen Lockwood, *History of Intelligent Transportation Systems* (Washington, DC: U.S. Department of Transportation, 2016).

11. Auer, Feese, and Lockwood, *History of Intelligent Transportation Systems*; see also Joseph M. Sussman, *Perspectives on Intelligent Transportation Systems (ITS)* (New York: Springer, 2005); Stephen Gordon and Jeff Trombly, *Deployment of Intelligent Transportation Systems: A Summary of the 2016 National Survey Results* (n.p.: n.p., 2018).

12. Including variable lane allocation, network access control, prescriptive rerouting, and variable tolls.

13. Remember that each cell of the origin and destination matrix represents the number of trips between a specific pair of an origin and destination. When the O&D matrix is *time-dependent*, an entry represents the volume of traffic departing from an origin i *in a specific time interval h* and destined for j.

14. Remember that after World War II, and for several decades, traffic and transportation engineers had treated network planning and operational analyses as two distinct functions (see chapter 2). However, the advent of ITS technology and the enactment of TEA-21, signed into law in 1998 and requiring transportation planners to promote efficient system management and operations as part of the transportation planning process, brought about the need (and the possibility) for tools that unify planning and operations analyses in a single, dynamic format. TRANSIMS (chapter 5) was designed to be such a tool.

15. On the historical links between the advent of ITS technology and the growing interest in DTA, see David E. Boyce, "Contributions of Transportation Network Modeling to the Development of a Real-Time Route Guidance System," in *Transportation for the Future*, ed. David F. Batten and Roland Thord (Berlin: Springer, 1989), 161–177; Srinivas Peeta and Athanasios K. Ziliaskopoulos, "Foundations of Dynamic Traffic Assignment: The Past, the Present and the Future," *Networks and Spatial Economics* 1, nos. 3–4 (September 2001): 233–265.

16. See, for example, Jaimison Sloboden, John Lewis, Vassili Alexiadis, Yi-Chang Chiu, and Eric Nava, *Traffic Analysis Toolbox Volume XIV: Guidebook on the Utilization of Dynamic Traffic Assignment in Modeling* (n.p.: FHWA, 2012), 1–3.

17. Samuel Yagar, "Analysis of the Peak Period Travel in a Freeway-Arterial Corridor" (Ph.D. diss., University of California at Berkeley, 1970).

18. Sam Yagar, "Emulation of Dynamic Equilibrium in Traffic Networks," in *Traffic Equilibrium Methods*, ed. Michael A. Florian (Berlin: Springer-Verlag, 1976), 240–264.

19. Deepak K. Merchant, "A Study of Dynamic Traffic Assignment and Control" (Ph.D. diss., Cornell University, 1974).

20. "George L. Nemhauser," http://www.informs.org/About-INFORMS/History-and-Traditions/Miser-Harris-Presidential-Portrait-Gallery/George-L.-Nemhauser, accessed November 28, 2020.

21. Chapter 4.

22. Deepak K. Merchant and George L. Nemhauser, "A Model and an Algorithm for the Dynamic Traffic Assignment Problems," *Transportation Science* 12, no. 3 (August 1978): 183–199 (on 198).

23. Deepak K. Merchant and George L. Nemhauser, "A Model and an Algorithm for the Dynamic Traffic Assignment Problem," in Florian, *Traffic Equilibrium Methods*, 265–273.

24. Merchant and Nemhauser, "A Model and an Algorithm for the Dynamic Traffic Assignment Problems"; Deepak K. Merchant and George L. Nemhauser, "Optimality Conditions for a Dynamic Traffic Assignment Model," *Transportation Science* 12, no. 3 (August 1978): 200–207.

25. Pierre Robillard, "Multipath Traffic Assignment with Dynamic Input Flows," *Transportation Research* 8, no. 6 (December 1974): 567–573.

26. On the state of the art in static assignment and DTA in the mid-1980s, see especially Terry L. Friesz, "Transportation Network Equilibrium, Design and Aggregation: Key Developments and Research Opportunities," *Transportation Research Part A* 19, nos. 5–6 (September–November 1985): 413–427, with an extensive bibliography.

27. Stella Dafermos, "Traffic Equilibrium and Variational Inequalities," *Transportation Science* 14, no. 1 (February 1980): 42–54. See, for example, Terry L. Friesz, David Bernstein, Tony E. Smith, Roger L. Tobin, and B. W. Wie, "A Variational Inequality Formulation of the Dynamic Network User Equilibrium Problem," *Operations Research* 41, no. 1 (January–February 1993): 179–191; Bin Ran and David Boyce, *Modeling Dynamic Transportation Networks: An Intelligent Transportation System Oriented Approach* (Berlin: Springer, [1994] 1996).

28. See, for example, M. Carey, "Optimal Time Varying Flows on Congested Networks," *Operations Research* 35, no. 1 (1987): 58–69 (system equilibrium); E. Codina and J. Barceló, "Dynamic Traffic Assignment: Considerations on Some Deterministic Modelling Approaches," *Annals of Operations Research* 60, no. 1 (December 1995): 1–58 (especially for the bibliography).

29. Here is a recent formulation of this extension: "Under equilibrium conditions in networks where congestion *varies over time* traffic arranges itself so that *at each instant* the costs incurred by drivers on those routes that are used are equal and no greater than those on any unused route" (Ortúzar and Willumsen, *Modelling*

*Transport*, 415). Another time-varying version of the first Wardrop principle reads as follows: "At equilibrium and *for a given start time*, no driver can unilaterally change paths and improve his or her travel time" in Cambridge Systematics, *Status of Activity-Based Models and Dynamic Traffic Assignment at Peer MPOs*, 18; emphasis in original.

30. Bruce N. Janson, "Dynamic Traffic Assignment for Urban Road Networks," *Transportation Research Part B* 25, nos. 2–3 (April–June 1991): 143–161. The article was received on May 29, 1989. See also Bruce N. Janson, "Network Design Effects of Dynamic Traffic Assignment," *Journal of Transportation Engineering* 121, no. 1 (January–February 1995): 1–13.

31. For an early investigation of the issue, see the now seminal article by William S. Vickrey, "Congestion Theory and Transport Investment," *The American Economic Review* 59, no. 2 (May 1969): 251–260.

32. Chris Hendrickson and George Kocur, "Schedule Delay and Departure Time in a Deterministic Model," *Transportation Science* 15, no. 1 (February 1981): 62–77; Chris Hendrickson and Edward Plank, "The Flexibility of Departure Times for Work Trips," *Transportation Research Part A* 18, no. 1 (January 1984): 25–36. See also, for example, Mark David Abkowitz, "The Impact of Service Reliability on Work Travel Behavior" (Ph.D. diss., MIT, 1980); Michèle Cyna, "Congestion and Schedule Delay" (master's thesis, MIT, 1981).

33. See, for example, Kraft and Wohl, "New Directions for Passenger Demand Analysis and Forecasting."

34. André de Palma, Moshe Ben-Akiva, Claude Lefèvre, and Nicolaos Litinas, "Stochastic Equilibrium Model of Peak Period Congestion," *Transportation Science* 17, no. 4 (November 1983): 430–453; Moshe Ben-Akiva, Michèle Cyna, and André de Palma, "Dynamic Model of Peak Period Congestion," *Transportation Research Part B* 18, nos. 4–5 (August-October 1984): 339–355; Moshe Ben-Akiva, André de Palma, and Pavlos Kanaroglou, "Dynamic Model of Peak Period Traffic Congestion with Elastic Arrival Rates," *Transportation Science* 20, no. 3 (August 1986): 164–181.

35. See chapter 6.

36. When Herman died in 1997, Mahmassani wrote a tribute to him, published in "Tributes to Robert Herman," *Transportation Science* 31, no. 2 (May 1997): 102–103.

37. On the general subject matter of modern (post–World War II) traffic science, see chapter 2. For a personal account of the history of modern traffic science, including the author's personal involvement in its development, see Robert Herman, "Technology, Human Interaction, and Complexity: Reflections on Vehicular Traffic Science," *Operations Research* 40, no. 2 (March–April 1992): 199–212. Herman's collaboration with Mahmassani is mentioned on 206–207.

38. Hani Mahmassani and Robert Herman, "Dynamic User Equilibrium Departure Time and Route Choice on Idealized Traffic Arterials," *Transportation Science* 18, no. 4 (November 1984): 362–384 (on 373); emphasis mine.

39. Mahmassani and Herman, "Dynamic User Equilibrium Departure Time," 383.

40. Gang-Len Ghang, "Departure Time Decision Dynamics in the Urban Transportation Network" (Ph.D. diss., University of Texas at Austin, 1985).

41. Hani S. Mahmassani and Gang-Len Chang, "Experiments with Departure Time Choice Dynamics of Urban Commuters," *Transportation Research B* 20, no. 4 (August 1986): 297–320 (on 298).

42. In the bibliography of his doctoral dissertation, Mahmassani lists no fewer than five works authored by Simon. See Mahmassani, "Methodological Aspects of a Decision Aid for Transportation Choices," 264.

43. Hani S. Mahmassani, Gang-Len Chang, and Robert Herman, "Individual Decisions and Collective Effects in a Simulated Traffic System," *Transportation Science* 20, no. 4 (November 1986): 258–271. See also Hani S. Mahmassani and Gang-Len Chang, "Dynamic Aspects of Departure-Time Choice Behavior in a Commuting System: Theoretical Framework and Experimental Analysis," *Transportation Research Record*, no. 1037 (1985): 88–101.

44. Hani S. Mahmassani, "Dynamic Models of Commuter Behavior: Experimental Investigation and Application to the Analysis of Planned Traffic Disruptions," *Transportation Research Part A* 24, no. 6 (1990): 465–484 (on 466). On the computer-based experiments conducted by Mahmassani and his collaborators in the 1980s, see also the survey chapter: Hani Mahmassani and Robert Herman, "Interactive Experiments for the Study of Tripmaker Behaviour Dynamics in Congested Commuting Systems," in Jones, *Developments in Dynamic and Activity-Based Approaches to Travel Analysis*, 272–298. For another experiment with "virtual" commuters, see Hani S. Mahmassani and R. Jayakrishnan, "Dynamic Analysis of Lane Closure Strategies," *ASCE Journal of Transportation Engineering* 114, no. 4 (July 1988): 476–496.

45. Hani S. Mahmassani and R. Jayakrishnan, "System Performance and User Response under Real-Time Information in a Congested Traffic Corridor," *Transportation Research Part A* 25, no. 5 (1991): 293–307.

46. On the behavioral foundations of the various user equilibrium traffic assignment methods, including BRUE, as well as their practical consequences, see, in the first place, Lei Zhang, "Behavioral Foundations of Route Choice and Traffic Assignment: Comparison of Principles of User Equilibrium Traffic Assignment under Different Behavioral Assumptions," *Transportation Research Record*, no. 2254 (2011): 1–10.

47. Hani S. Mahmassani and Gang-Len Chang, "On Boundedly Rational User Equilibrium in Transportation Systems," *Transportation Science* 21, no. 2 (May 1987): 89–99 (quotations on 89 and 91); emphasis mine.

48. See the introduction of this chapter.

49. Richard Arnott, André de Palma, and Robin Lindsey, "Does Providing Information to Drivers Reduce Traffic Congestion?" *Transportation Research Part A* 25, no. 5 (1991): 309–318; Moshe Ben-Akiva, André de Palma, and Isam Kaysi, "Dynamic Network Models and Driver Information Systems," *Transportation Research Part A* 25, no. 5 (1991): 251–266.

50. *Transportation Research Part B* 24, no. 6 (December 1990); *Transportation Research Part A* 25, no. 5 (1991).

51. "University Transportation Centers," https://www.transportation.gov/content/university-transportation-centers, accessed January 25, 2022.

52. MIT, *Reports to the President, 1989–1990* (n.p.: n.p., n.d.), 357; MIT, *Reports to the President for the Year Ended June 30, 1991* (n.p.: n.p., n.d.), 283–285.

53. IVHS America, *A Strategic Plan for Intelligent Vehicle-Highway Systems in the United States* (Washington, DC: IVHS, 1992).

54. MIT, *Reports to the President for the Year Ended June 30, 1991*, 283.

55. This monumental project captured the attention of the historian of technology Thomas Parke Hughes (1923–2014); see Thomas P. Hughes, *Rescuing Prometheus: Four Monumental Projects That Changed the Modern World* (New York: Pantheon Books, 1998), chapter 5. See also Michael R. Fein, "Tunnel Vision: 'Invisible' Highways and Boston's 'Big Dig' in the Age of Privatization," *Journal of Planning History* 11, no. 1 (February 2012): 47–69. Neither the book nor the article makes any reference to the involvement of Ben-Akiva and his team in the project.

56. Bechtel and Parsons Brinckerhoff, two large design and construction companies, along with Massachusetts Executive Office of Transportation and Construction, were the main actors involved in the Central Artery/Tunnel project. Bechtel and PB entered into a contract with MIT in the latter part of the academic year 1992–1993. See MIT, *Reports to the President for the Year Ended June 30, 1993* (n.p.: n.p., n.d.), 232, 291.

57. Moshe Ben-Akiva, David Bernstein, Antony Hotz, Haris Koutsopoulos, and Joseph Sussman, "The Case for Smart Highways," *MIT Technology Review* (July 1992): 38–47.

58. My reckoning based on the electronic catalog of the MIT Libraries.

59. Including travel times, queue lengths and fuel consumption. For a detailed presentation of the project, see especially Moshe Ben-Akiva, Haris Koutsopoulos, and Anil Mukundan, "A Dynamic Traffic Model System for ATMS/ATIS Operations," *IVHS Journal* 2, no. 1 (1994): 1–19.

60. Kalidas Ashok, "Dynamic Trip Table Estimation for Real Time Traffic Management Systems" (master's thesis, MIT, 1992); Kalidas Ashok, "Estimation and Prediction of Time-Dependent Origin-Destination Flows" (Ph.D. diss., MIT, 1996).

61. Work conducted by, among others, Isam Kaysi, another of Ben-Akiva's Ph.D. supervisees. See Isam Asnan Kaysi, "Framework and Models for the Provision of Real-Time Driver Information" (Ph.D. diss., MIT, 1992).

62. Amalia Polydoropoulou, "Modeling the Influence of Traffic Information on Drivers' Route Choice Behavior" (master's thesis, MIT, 1993); Amalia Polydoropoulou, "Modeling User Response to Advanced Traveler Information Systems (ATIS)" (Ph.D. diss., MIT, 1997).

63. See chapter 3.

64. Qi Yang, "A Microscopic Traffic Simulation Model for IVHS Applications" (master's thesis, MIT, 1993); Qi Yang, "A Simulation Laboratory for Evaluation of Dynamic Traffic Management System" (Ph.D. diss., MIT, 1997).

65. Microscopic, mesoscopic, macroscopic: Microscopic models simulate individual cars, mesoscopic models operate groups of vehicles and simulate the movement of these vehicle packages, while macroscopic models deal with traffic using macroscopic parameters, such as speed and density. More detailed than the two other kinds of traffic flow models, microscopic models are also much more demanding in computational terms.

66. MIT, *Reports to the President for the Year Ended June 30, 1997* (n.p.: n.p., n.d.), 252.

67. Called MesoTS, it is much quicker than MITSIM itself (Yang, "A Simulation Laboratory," 117.)

68. For the different areas for simulation in the early 1990s, see especially Averill M. Law and W. David Kelton, *Simulation Modeling & Analysis*, 2nd ed. (New York: McGraw-Hill, 1991), a book Yang himself referred to in his MIT doctoral dissertation.

69. For a series of examples and the corresponding references, see Yang, "A Simulation Laboratory," 23–24.

70. Yang, "A Simulation Laboratory," 20.

71. "60 Years of Great Science," special issue, *Oak Ridge National Laboratory Review* 36, no. 1 (2003).

72. Shaw-Pin Miaou, Rekha S. Pillai, Mike S. Summers, Ajay K. Rathi, and Henry C. Lieu "Laboratory Evaluation of Dynamic Traffic Assignment Systems: Requirements, Framework, and System Design," prepared by the Oak Ridge National Laboratory (Oak Ridge, Tennessee: n.p., 1996), https://rosap.ntl.bts.gov/view/dot/13553, accessed November 28, 2020; Henry Lieu, *A Roadmap for the Research, Development*

*and Deployment of Traffic Estimation and Prediction Systems for Real-Time and Off-Line Applications* (Revised Version) (n.p.: FHWA, 2003).

73. Ajay Kumar Rathi, "Development of a Microscopic Simulation Model for Freeway Lane Closures" (Ph.D. diss., Ohio State University, 1983).

74. MIT, *Reports to the President for the Year Ended June 30, 1996* (n.p.: n.p., n.d.), 241; MIT, *Development of a Deployable Real-time Dynamic Traffic Assignment System* (Cambridge, MA: Center for Transportation Studies, 1996).

75. Lieu, *A Roadmap for the Research*, 4.

76. U.S. Department of Transportation, *Department of Transportation's Intelligence Vehicle Highway Systems Projects* (n.p.: U.S. Government Printing Office, 1993).

77. R. Jayakrishnan, Michael Cohen, John Kim, Hani S. Mahmassani, and Ta-Yin Hu, *A Simulation-Based Framework for the Analysis of Traffic Networks Operating with Real-Time Information* (Berkeley: Institute of Transportation Studies, 1993).

78. R. Jayakrishnan, "In-Vehicle Information Systems for Network Traffic Control: A Simulation Framework to Study Alternative Guidance Strategies" (Ph.D. diss., University of Texas at Austin, 1992). On Jayakrishnan's contribution to DYNASMART through his thesis, see Jayakrishnan et al., *A Simulation-Based Framework*, 5.

79. Srinivas Peeta, "System Optimal Dynamic Traffic Assignment in Congested Networks with Advanced Information Systems" (Ph.D. diss., University of Texas at Austin, 1994); Athanasios Ziliaskopoulos, "Optimum Path Algorithms on Multidimensional Networks: Analysis, Design, Implementation and Computational Experience" (Ph.D. diss., University of Texas at Austin, 1994); Ta-Yin Hu, "Dynamic Analysis of Network Flows under Advanced Information and Control Systems" (Ph.D. diss., University of Texas at Austin, 1995).

80. Hani S. Mahmassani, Ta-Yin Hu, Srinivas Peeta, and Athanasios Ziliaskopoulos, *Development and Testing of Dynamic Traffic Assignment and Simulation Procedures for ATIS/ATMS Applications*, rev. ed. (Austin, TX: Center for Transportation Research, [June 1993] June 1994); see also R. Jayakrishnan, Hani S. Mahmassani, and Ta-Yin Hu, "An Evaluation Tool for Advanced Traffic Information and Management Systems in Urban Networks," *Transportation Research Part C* 2, no. 3 (September 1994): 129–147.

81. Mahmassani et al., *Development and Testing*, 8, 4; emphasis mine.

82. Kazi Iftekhar Ahmed, "Modeling Drivers' Acceleration and Lane Changing Behavior" (Ph.D. diss., MIT, 1999); Jon Alan Bottom, "Consistent Anticipatory Route Guidance" (Ph.D. diss., MIT, 2000); Constantinos Antoniou, "On-line Calibration for Dynamic Traffic Assignment" (Ph.D. diss., MIT, 2004); Ramachandran Balakrishna, "Off-line Calibration of Dynamic Traffic Assignment Models" (Ph.D.

diss., MIT 2006). On the MIT team and the characteristics of DynaMIT in the early 2000s, see among others Moshe Ben-Akiva, Michel Bierlaire, Didier Burton, Haris N. Koutsopoulos, and Rabi Mishalani, "Network State Estimation and Prediction for Real-Time Traffic Management," *Networks and Spatial Economics* 1, nos. 3–4 (September 2001): 293–318.

83. Yaser (El-Sayed) Hawas, "A Decentralized Architecture and Local Search Procedures for Real-Time Route Guidance in Congested Vehicular Traffic Networks" (Ph.D. diss., University of Texas at Austin, 1996); Nhan H. Huynh, "Development of a Graphical User Interface Framework for DYNASMART-X Real-Time Dynamic Traffic-Assignment System" (master's thesis, University of Texas at Austin, 1999); Akmal Saad Abdelfatah, "Time-Dependent Signal Control and System Optimal Traffic Assignment in Congested Vehicular Traffic Networks" (Ph.D. diss., University of Texas at Austin, 1999); Khaled F. Abddelghany, "Stochastic Dynamic Traffic Assignment for Intermodal Transportation Networks with Consistent Information Supply Strategies" (Ph.D. diss., University of Texas at Austin, 2001); Ahmed F. Abddelghany, "Dynamic Micro-Assignment of Travel Demand with Activity/Trip Chains" (Ph.D. diss., University of Texas at Austin, 2001); Yi-Chang Chiu, "Generalized Real-Time Route Guidance Strategies in Urban Networks" (Ph.D. diss., University of Texas at Austin, 2002); Xuesong Zhou, "Dynamic Origin-Destination Demand Estimation and Prediction for Off-line and On-line Dynamic Traffic Assignment Operation" (Ph.D. diss., University of Maryland, 2004). One can find the names of other collaborators (by 2001) in Hani S. Mahmassani, "Dynamic Network Traffic Assignment and Simulation Methodology for Advanced System Management Applications," *Networks and Spatial Economics* 1, nos. 3–4 (September 2001): 267–292 (on 290–291).

84. In technical terms, in the planning version the origin and destination traffic data are input *externally* and are fixed for the analysis period. In the online version, the O&D matrices used for network evaluation are *internally* estimated in the model, based on the surveillance traffic flows data, historical O&D data, and drivers' responses to the control and guidance measures and strategies (Lieu, *A Roadmap for the Research*, 8).

85. Lieu, *A Roadmap for the Research*, 17. See also the various reports prepared by Mahmassani's team in the author's CV (I consulted the 2018 version). For the MIT team, see among others the references in Ramachandran Balakrishna, "Calibration of the Demand Simulator in a Dynamic Traffic Assignment System" (master's thesis, MIT, 2002).

86. Lieu, *A Roadmap for the Research*, 17–25.

87. *McTrans Newsletter* 31 (Summer 2004): 1–2. For other DYNASMART training workshops held in 2006, see https://tmip.org/content/fhwa-dynasmart-p-training-opportunities-august, accessed November 29, 2020.

88. "Developing, Distributing, and Supporting Cost-Effective and Widely-Accessible Traffic Analysis Software," https://mctrans.ce.ufl.edu/featured/dynasmart/, accessed November 17, 2019.

89. M.T.A. (Mark) Roelofsen, "Dynamic Modelling of Traffic Management Scenarios Using Dynasmart" (master's thesis, University of Twente, 2012).

90. "Video Interview Transcript: Texas DOT & Texas Transportation Institute—DynaSMART," https://ops.fhwa.dot.gov/trafficanalysistools/videos/medina_transcript.htm, accessed November 29, 2020.

91. Chiu, "Generalized Real-Time Route Guidance Strategies in Urban Networks."

92. http://web.archive.org/web/20160327013820/http://ctr.utexas.edu/2015/05/01/alumni-spotlight-yi-chang-chiu/, accessed October 7, 2022.

93. Information on DynusT and its history is drawn from several Internet sources. See www.dynust.com, accessed November 29, 2020; www.metropia.com, accessed November 29, 2020. In-depth presentations of the DynusT modeling suite can be found in the following articles: Yi-Chang Chiu, Liang Zhou, and Houbing Song, "Development and Calibration of the Anisotropic Mesoscopic Simulation Model for Uninterrupted Flow Facilities," *Transportation Research Part B* 44, no. 1 (January 2010): 152–174; Cambridge Systematics (in association with), Sacramento Area Council of Governments, University of Arizona, University of Illinois (Chicago), Sonoma Technology, Fehr and Peers, *Dynamic, Integrated Model System: Sacramento-Area Application, Volume 2: Network Report* (Washington, DC: Transportation Research Board, 2014).

94. Xuesong Zhou and Jeffrey Taylor, "DTALite: A Queue-Based Mesoscopic Traffic Simulator for Fast Model Evaluation and Calibration," *Cogent Engineering* 1, no. 1 (October 2014) (961345, DOI: 10.1080/23311916.2014.961345); https://github.com/asu-trans-ai-lab/DTALite, accessed January 25, 2022; https://github.com/xzhou99?tab=repositories; https://code.google.com/archive/p/nexta/, accessed January 25, 2022.

95. Zhou, "Dynamic Origin-Destination Demand Estimation."

96. "Simmobility," https://www.its.mit.edu/link-platforms, accessed January 24, 2022 (the site contains a list of DynaMIT-related publications appeared by the mid-2010s).

97. Moshe Ben-Akiva, Haris N. Koutsopoulos, Constantinos Antoniou, and Ramachandran Balakrishna, "Traffic Simulation with DynaMIT," in *Fundamentals of Traffic Simulation*, ed. Jaume Barceló (New York: Springer, 2010), 363–398 (on 394).

98. Email from Ramachandran Balakrishna to Howard Slavin, dated January 24, 2021; I thank Howard Slavin for sharing with me the aforementioned email.

99. Xujun Eberlein, "Real-Time Control Strategies in Transit Operations: Models and Analysis" (Ph.D. diss., MIT, 1995).

100. Song Gao, "Optimal Adaptive Routing and Traffic Assignment in Stochastic Time-Dependent Networks" (Ph.D. diss., MIT, 2005).

101. On MITSIMLab, see, for example, Moshe Ben-Akiva, Haris N. Koutsopoulos, Tomer Toledo, Qi Yang, Charisma F. Choudhury, Constantinos Antoniou, and Ramachandran Balakrishna, "Traffic Simulation with MITSIMLab," in Barceló, *Fundamentals of Traffic Simulation*, 233–268; https://www.its.mit.edu/pd-fms-copy, accessed January 25, 2022. It is worth noting that MIT no longer provides support for the open-source MITSIMLab, the first version of which was released in 2004; however, questions may be posted to the larger computing community at the MITSIMLab user group website: http://sourceforge.net/projects/mitsim/, accessed November 29, 2019.

102. Srinivasan Sundaram, "Development of a Dynamic Traffic Assignment System for Short-Term Planning Applications" (master's thesis, MIT, 2002).

103. Email from Howard Slavin to the author, dated February 7, 2021. On TransModeler, in addition to the firm's website, see the recent article by Qi Yang, Ramachandran Balakrishna, Daniel Morgan, and Howard Slavin, "Large-Scale, High-Fidelity Dynamic Traffic Assignment: Framework and Real-World Case Studies," *Transportation Research Procedia* 25 (2017): 1290–1299. See also the list of publications authored by Caliper staffers at https://www.caliper.com/press/transportationlibrary.htm, accessed October 7, 2022. On the links between MITSIMLab and TransModeler, see Ben-Akiva et al., "Traffic Simulation with MITSIMLab," 235.

104. Carlos F. Daganzo, "The Cell Transmission Model: A Dynamic Representation of Highway Traffic Consistent with the Hydrodynamic Theory," *Transportation Research Part B* 28, no. 4 (1994): 269–287.

105. A. K. Ziliaskopoulos, S. T. Waller, Y. Li, and M. Byram, "Large-Scale Dynamic Traffic Assignments: Implementation Issues and Computational Analysis," *ASCE Journal of Transportation Engineering* 130, no. 5 (September–October 2004): 585–593.

106. Athanasios K. Ziliaskopoulos and S. Travis Waller, "An Internet-Based Geographic Information System That Integrates Data, Models and Users for Transportation Applications," *Transportation Research Part C* 8 (2000): 427–444 (on 428).

107. Jennifer C. Duthie, N. Nezamuddin, Natalia Ruiz Juri, et al., *Investigating Regional Dynamic Traffic Assignment Modeling for Improved Bottleneck Analysis: Final Report* (Austin: CTR/The University of Texas at Austin, 2012), 6.

108. On Dynameq, see, for example, Michael Mahut and Michael Florian, "Traffic Simulation with Dynameq," in Barceló, *Fundamentals of Traffic Simulation*, 323–361; https://www.inrosoftware.com/en/products/dynameq/, accessed January 25, 2022.

109. Michael Florian, "Models and Software for Urban and Regional Transportation Planning: The Contributions of the Center for Research on Transportation," *INFOR* 46, no. 1 (February 2008): 29–50, especially 43; V. Astarita, K. Er-Rafia, M. Florian, M. Mahut, and S. Velan, "Comparison of Three Methods for Dynamic Network Loading," *Transportation Research Record*, no. 1771 (2001): 179–190.

110. Michael Mahut, "A Discrete Flow Model for Dynamic Network Loading" (Ph.D. diss., University of Montreal, 2000). Thanks to his master's thesis, Mahut was already knowledgeable about static assignment modeling: Michael Mahut, "A Parametric Analysis of Arterial Travel Time" (master's thesis, University of Toronto, 1996). Michael Mahut shared with me his knowledge of DTA-related issues in Montréal on June 14, 2011.

111. Michael Mahut, Michael Florian, Nicolas Tremblay, Mark Campbell, David Patman, and Zorana Krnic McDaniel, "Calibration and Application of a Simulation-Based Dynamic Traffic Assignment Model," *Transportation Research Record*, no. 1876 (2004): 101–111.

112. Florian, "Models and Software for Urban and Regional Transportation Planning," 45.

113. *INTEGRATION Release 2.30 for Windows: User's Guide, Volume I: Fundamental Model Features*, copyright © M. Van Aerde & Assoc., Ltd., 1984–2010 (n.p.: n.p., February 2010): vi (the acknowledgments, on vi–viii, were written by Michel Van Aerde in July 1997); https://sites.google.com/a/vt.edu/hrakha/software, accessed January 25, 2022. On INTEGRATION, see also Michel Van Aerde and Sam Yagar, "Dynamic Integrated Freeway/Traffic Signal Networks: Problems and Proposed Solutions," *Transportation Research Part A* 22, no. 6 (November 1988): 435–443; Michel Van Aerde and Sam Yagar, "Dynamic Integrated Freeway/Traffic Signal Networks: A Routing-Based Modelling Approach," *Transportation Research Part A* 22, no. 6 (November 1988): 445–453; Hesham Rakha and Aly Tawfik, "Traffic Networks: Dynamic Traffic Routing, Assignment, and Assessment," in *Encyclopedia of Complexity and Systems Science*, ed. Robert A. Meyers (New York: Springer, 2009), 9429–9470.

114. On these firms' DTA products, see in addition to their websites *The Urban Transportation Monitor* 28, no. 8 (October 2014): 15–30 (on 20–21). On Citilabs' CUBE Avenue, see also Lehman Center for Transportation Research, URS Corporation and Citilabs, *Application of Dynamic Traffic Assignment to Advanced Managed Lane Modeling* (Miami: Lehman Center for Transportation Research, 2013).

115. Yi-Chang Chiu, Jon Bottom, Michael Mahut, Alex Paz, Ramachandran Balakrishna, Travis Waller, and Jim Hicks, *Dynamic Traffic Assignment: A Primer* (Washington, DC: Transportation Research Board, 2011).

116. Cambridge Systematics Inc., Vanasse Hangen Brustlin Inc., Gallop Corporation, Chandra R. Bhat, Shapiro Transportation Consulting LLC, Martin/Alexiou/

Bryson PLLC, *Travel Demand Forecasting: Parameters and Techniques* (Washington, DC: Transportation Research Board, 2012), chapter 6 (the 2012 report is an update to Martin and McGuckin, *Travel Estimation Techniques for Urban Planning.*

117. Krishna C. Patnam, David B. Roden, Dr. Weihao Yin, Li-Yang Feng, Sayeed Mallick, and Scott B. Smith, *A Practical Guide on DTA Model Applications for Regional Planning* (Washington, DC: FHWA, 2016).

118. TMIP/FMIP, "Webinars," https://tmip.org/webinars, accessed January 25, 2022.

119. Travel Forecasting Resource, "Dynamic Traffic Assignment," https://tfresource.org/topics/Dynamic_Traffic_Assignment.html, accessed January 25, 2022.

120. For a review of state of the art and applications of DTA by 2012, see Mohammed Hadi, Halit Ozen, Shaghayegh Shabanian, Yan Xiao, Wei Zhao, Frederick W. Ducca, and Jim Fennessy, *Use of Dynamic Traffic Assignment in FSUTMS in Support of Transportation Planning in Florida* (n.p.: n.p., 2012).

121. James Hicks, "Dynamic Traffic Assignment Model Breakdown," in Transportation Research Board, *Innovations in Travel Demand Modeling, Volume 2: Papers*, 101–108.

122. Parsons Brinckerhoff and San Francisco County Transportation Authority, *San Francisco Dynamic Traffic Assignment Project 'DTA Anyway': Final Methodology Report* (n.p.: SFCTA, 2012); https://github.com/sfcta/dta, accessed November 29, 2020.

123. DynusT (five cases), Dynameq (five), Citilab's Cube Avenue (four), DTALite (one), TRANSIMS (one) and Caliper's TransModeler (one). My reckoning is based on information drawn from Sloboden et al., *Traffic Analysis Toolbox Volume XIV*; see especially annexes A and B for a brief description of the various DTA projects.

124. Cambridge Systematics, *Status of Activity-Based Models and Dynamic Traffic Assignment at Peer MPOs*. In late 2011, despite some earlier attempts to apply DTA techniques to a large network, all the MPOs were still using static assignment models only. See Caliper Corporation, *Traffic Assignment and Feedback Research to Support Improved Travel Forecasting*, 1-3.

125. Chicago, Twin Cities, Baltimore, Atlanta, Detroit, San Diego, and Columbus. Of the seven DTA procedures under development, most models have been in development for one to two years and were expected to be completed between one and five years (Cambridge Systematics, *Status of Activity-Based Models and Dynamic Traffic Assignment at Peer MPOs*, 25).

126. At that time, Portland had also used Dynameq while Phoenix had resorted to TransModeler as well.

127. Scott Smith, Amy Fong, Ci Yang, and Brian Gardner, *TravelWorks Integrated Models: Final Report* (Washington, DC: FHWA, 2018).

128. In technical terms, integration means that a feedback loop is built between the "demand" (ABM) and the "supply" (DTA) model systems.

129. For an early attempt to integrate an activity-based model (CEMDAP) with DTA procedures (VISTA), see, for example, Dung-Ying Lin, Naveen Eluru, S. Travis Waller, and Chandra R. Bhat, "Integration of Activity-Based Modeling and Dynamic Traffic Assignment," *Transportation Research Record*, no. 2076 (2008): 52–61. For another case—the integration of openAMOS as the activity-based model, and DTALite as the DTA model—see http://pubsindex.trb.org/view/2017/C/1439702, accessed November 29, 2020.

130. On FAST-TrIPs, a recent dynamic transit model developed by a team from three public agencies plus several consultants, see the dedicated website: https://www.sfcta.org/shrp2-fast-trips, accessed January 25, 2022.

131. On the Sacramento project, see https://apps.trb.org/cmsfeed/TRBNetProjectDisplay.asp?ProjectID=2828, accessed November 29, 2020; Cambridge Systematics (in association with), Sacramento Area Council of Governments, University of Arizona, University of Illinois (Chicago), Sonoma Technology, Fehr and Peers, *Dynamic, Integrated Model System: Sacramento-Area Application, Volume 1: Summary Report* (Washington, DC: Transportation Research Board, 2014); Cambridge Systematics et al., *Dynamic, Integrated Model System: Sacramento-Area Application, Volume 2: Network Report*.

132. https://apps.trb.org/cmsfeed/trbnetprojectdisplay.asp?projectid=2829, accessed November 29, 2020; RSG et al., *Dynamic, Integrated Model System: Jacksonville-Area Application*.

133. The four awarded projects were: Two pilot projects, with Atlanta Regional Commission (ARC) and Ohio DOT, aiming to integrate an activity-based model of the CT-RAMP family (developed by Parsons Brinckerhoff) with DynusT in a highway setting; another pilot project, with Maryland State Highway Administration and Baltimore Metropolitan Council (BMC), to integrate the University of Maryland agent-based model (AgBM) with DTALite, as well as the BMC's activity-based model (InSITE), developed by Cambridge systematics, with a dynamic traffic assignment procedure based on DTALite software; a pilot project, with the Metropolitan Transportation Commission, San Francisco County Transportation Authority, and Puget Sound Regional Council (PSRC) to implement the FAST-TrIPs model, by joining it with their respective ABM procedures. These pilot projects led to several publications. See, for example, for the Baltimore case: Lei Zhang et al., "An Integrated, Validated, and Applied Activity-Based Dynamic Traffic Assignment Model for the Baltimore-Washington Region," *Transportation Research Record* 2672, no. 51 (2018): 45–55; for the Atlanta and Ohio pilot study, see http://pubsindex.trb.org/view/2018/C/1496801, accessed November 29, 2020.

134. Smith et al., *TravelWorks Integrated Models*, especially 14–17, plus appendices.

135. Including peer review meetings with selected experts who were not part of the projects, five Travel Model Improvement Program webinars, and presentations at conferences nationwide.

136. Mark Bradley, Ben Stabler, Khademul Haque, Howard Slavin, and Dan Morgan, *Volume 1: Integrated ABM-DTA Methods to Model Impacts of Disruptive Technology on the Regional Surface Transportation System—A Feasibility Study* (Washington, DC: Federal Highway Administration, 2017); Stabler et al., *Volume 2: Model Impacts of Connected and Autonomous/Automated Vehicles (CAVs).*

137. Travel Forecasting Resource, "Integrated Travel Demand and Network Models," https://tfresource.org/topics/Integrated_Travel_Demand_and_Network_Models.html, accessed January 25, 2022.

138. Douglass B. Lee Jr., "Requiem for Large-Scale Models," *Journal of the American Institute of Planners* 39, no. 3 (May 1973): 163–178 (on 163).

### Conclusion

1. See n. 34 in this book's introduction.

2. An in-depth analysis of these general changes characterizing the "intellectual" characteristics of UTDM can be found in the two following articles written by leading figures in the field: Moshe Ben-Akiva, Jon Bottom, Song Gao, Haris N. Koutsopoulos, and Yang Wen, "Towards Disaggregate Dynamic Travel Forecasting Models," *Tsinghua Science and Technology* 12, no. 2 (April 2007):115–130; Soora Rasouli and Harry Timmermans, "Activity-Based Models of Travel Demand: Promises, Progress and Prospects," *International Journal of Urban Sciences* 18, no. 1 (2014): 31–60.

3. Remember that problems for which solutions cannot be expressed *analytically*—when the size of the problem is too large or when no exact solution can be derived—can nonetheless be analyzed and resolved with the help of *simulation* techniques, such as the Monte Carlo method (chapter 6). Simulation techniques do not provide an exact solution to the problem, but they can offer "almost exact solutions."

4. See especially chapter 5.

5. See chapter 7.

6. On the differentiation of the roles of career scientist (researcher) and practitioner, see especially Joseph Ben-David, "Roles and Innovations in Medicine," *The American Journal of Sociology* 65, no. 6 (May 1960): 557–568.

7. Civil engineering, economics, statistics, computer science, operations research, geography, and urban planning, to name the more common disciplinary backgrounds today (see Joseph Y. J. Chow, R. Jayakrishnan, and Hani S. Mahmassani, "Is Transport Modelling Education *Too* Multi-disciplinary? A Manifesto on the Search

for Its Evolving Identity," in *Travel Behaviour Research: Current Foundations, Future Prospects*, ed. Matthew J. Roorda and Eric J. Miller (Morrisville, NC: Lulu Press, 2014), 231–250 (especially 234).

8. The withdrawal I refer to essentially concerns scientific and technical expertise. As has been seen, through the federal administration in particular, the "public" segment of the UTDM social world has remained an essential funder of this modeling field.

9. See the analyses and references in DiMento and Ellis, *Changing Lanes*.

10. For an early in-depth analysis of this interrelation, see especially Stephen H. Putman, *The Integrated Forecasting of Transportation and Land Use* (Washington, DC: U.S. Department of Transportation, n.d. [ca. 1976]).

11. On the land use modeling work performed at CATS, see, for example, Hamburg and Creighton, "Predicting Chicago's Land Use Pattern." On the so-called EMPIRIC land use model, which was first used in Boston in the second half of the 1960s, and was produced by TRC, see Daniel Brand, Brian Barber, and Michael Jacobs, "Technique for Relating Transportation Improvements and Urban Development Patterns," *Highway Research Record*, no. 207 (1967): 53–67.

12. In fairness to history, it should be noted that some members of the UTDM social world at least never lost sight of the global views that are necessary when urban travel demand modeling as a whole confronts general problems, such as congestion as well as environmental and societal issues. The history of urban traffic forecasting is also punctuated by attempts to integrate the increasingly specialized knowledge and practice produced within it into more *holistic* bodies. A detailed history of these attempts still remains to be written. One should mention here the federal government-sponsored pioneering work carried out by Stephen Putman and his collaborators in the 1970s: Stephen H. Putman et al., *The Interrelationships of Transportation Development and Land Development*, 2 vol. (n.p.: University of Pennsylvania, June 1973; revised September 1976). From the mid-1990s on, the same main forces driving new and dramatic developments in urban traffic forecasting (chapter 5) would rekindle interest in *integrated* land-use and transportation models (ILUT) as well. On the (re)encounter between transportation and land use modeling in the 2000s and the 2010s, see, for example, Ransford A. Acheampong, "Land Use–Transport Interaction Modeling: A Review of the Literature and Future Research Directions," *The Journal of Transport and Land Use* 8, no. 3 (2015): 11–38; Rolf Moeckel, *Integrated Transportation and Land Use Models* (Washington, DC: Transportation Research Board, 2018).

13. See especially Becker, *Art Worlds*. The notion of momentum, proposed by the historian Thomas P. Hughes (1923–2014), can also help explain why changes in the trajectory of a complex system are a rather tricky affair; see Thomas P. Hughes, "The

Evolution of Large Technological Systems," in Biagioli, *The Science Studies Reader*, 202–223.

14. Stopher, "History and Theory of Urban Transport Planning," 50.

15. Committee for Determination of the State of the Practice in Metropolitan Area Travel Forecasting, *Metropolitan Travel Forecasting*; Yoram Shiftan and Moshe Ben-Akiva, "A Practical Policy-Sensitive, Activity-Based, Travel-Demand Model," *The Annals of Regional Science* 47, no. 3 (December 2011): 517–541.

16. See, for example, Nazneen Ferdous, Chandra Bhat, Lakshmi Vana, David Schmitt, John L. Bowman, Mark Bradley, and Ram Pendyala, *Comparison of Four-Step Versus Tour-Based Models in Predicting Travel Behavior Before and After Transportation System Changes—Results Interpretation and Recommendations* (Austin: Center for Transportation Research/University of Texas at Austin, 2011). It is worth noting here that, to my surprise, the number of publications dealing with the "ex-post evaluation of forecast accuracy" issue—namely how well specific models actually used in existing transportation systems performed—as their main subject matter proved, all in all, very small compared to the total production within UTDM. See especially the recent publication by Greg Erhardt et al., *Traffic Forecasting Accuracy Assessment Research* (Washington, DC: Transportation Research Board, 2020), which refers to the relevant literature. One of the main results of the studies dealing with the "accuracy issue" is that highway traffic forecasts have been generally more accurate than transit traffic predictions.

17. I borrow this expression from Paul N. Edwards, *A Vast Machine: Computer Models, Climate Data, and the Politics of Global Warming* (Cambridge, MA: MIT Press, 2010), chapter 5.

18. A 2009 paper containing the results produced by an Austin, TX, application shows that a single run of the activity-based model was about forty times longer than that of its four-step counterpart. According to this study, the activity-based model required the estimation of 621 parameters across 43 submodels, while its aggregate competitor required just 132 parameters corresponding to 13 submodels (Cambridge Systematics, *Status of Activity-Based Models and Dynamic Traffic Assignment at Peer MPOs*, 34).

19. Regarding ABM, see chapter 6. On the (neglected) issue of maintenance in the history of technology, see the recent survey article by Andrew L. Russell and Lee Vinsel, "After Innovation, Turn to Maintenance," *Technology and Culture* 59, no. 1 (January 2018): 1–25.

20. Like financial practitioners, urban traffic forecasters are not "model dopes," uncritically accepting the outputs of their models. See Donald MacKenzie and Taylor Spears, "'A Device for Being Able to Book P & L': The Organizational Embedding of the Gaussian Copula," *Social Studies of Science* 44, no. 3 (June 2014): 418–440.

21. Back to the mid-1980s, Professor Briton Harris (1914–2005) explicitly referred to the "stabilization" efforts deployed by the BPR in the 1960s and underscored their mixed results in the long run. See Britton Harris, "Guest Editorial. How Should We Think about Theory?" *Environment and Planning A* 16, no. 2 (February 1984): 143–145. As a matter of fact, though well drafted "best practices" may provide an effective framework and guidelines for (good) local practices, the repeated use of the same "standards" is likely to make the people and the organizations employing them rigid and far less receptive to innovation. For a theoretical approach to the "standardization issue," see Mary Douglas, *How Institutions Think* (Syracuse: Syracuse University Press, 1986).

22. On the scale "issue" in historical studies, see, for example, Sebouth David Aslanian, Joyce E. Chaplin, Ann McGrath, and Kristin Mann, "AHR Conversation. How Size Matters: The Question of Scale in History," *American Historical Review* 118, no. 5 (December 2013): 1430–1472.

23. The (very long) careers of John Hamburg, Gordon Schultz, and David Roden are good cases in point (chapters 2, 5, 6, and 7).

24. On the continuity/discontinuity issue regarding the growth of scientific knowledge, see the perceptive comments addressed by Ben-David on Kuhn's paradigm-focused approach to scientific change (Joseph Ben-David, "Review Article, Scientific Growth: A Sociological View," *Minerva* 2, no. 4 [June 1964]: 455–476).

25. Becker, *Art Worlds*; Stephen Toulmin, *Human Understanding, Volume 1: General Introduction and Part I* (Princeton, NJ: Princeton University Press, 1972).

26. Think for example of Moshe Ben-Akiva, Frank Koppelman, and Peter Stopher, who had also been heavily involved in the creation and development of the utility-based disaggregate approach in the 1970s and the 1980s (chapters 3 and 6).

27. In order to identify these factors, I have adopted a comparative stance. After comparing the trajectories of urban travel demand modeling in the United States and France from the 1950s up to the present, I isolated the factors that have been strongly present in the American case and effectively absent in French urban traffic forecasting. On the history of French UTDM, see Konstantinos Chatzis, "Forecasting Urban Traffic in France, 1950s to 2000s: The Nation-State, Private Engineering Firms and the Globalization of an Area of Expertise" (Document de travail du LATTS—Working Paper, no. 14-02) (September 2014), http://hal-enpc.archives-ouvertes.fr/hal-01071139. On comparative history, see among others Heinz-Gerhard Haupt and Jürgen Kocka, eds., *Comparative and Transnational History: Central European Approaches and New Perspectives* (New York: Berghahn Books, 2009).

28. A largely similar pattern seems to be at work in other countries that can also boast sustained good performances in urban travel demand modeling over a significant span of time, such as Germany and, albeit to a lesser extent for the more recent

periods, Great Britain. See Chatzis, "Forecasting Urban Traffic in France, 1950s to 2000s," conclusion.

29. On the post-1945 university system in the United States see first Roger L. Geiger, *Research and Relevant Knowledge: American Research Universities since World War II*; Roger L. Geiger, *Knowledge and Money: Research Universities and the Paradox of the Marketplace* (Stanford: Stanford University Press, 2004). More recent references can be found in Emily J. Levine, "Baltimore Teaches, Gottingen Learns: Cooperation, Competition, and the Research University," *American Historical Review* 121, no. 3 (June 2016): 780–823.

30. As has been seen, although the Bureau of Public Roads did develop in-house expertise in urban traffic forecasting in the 1950s and the 1960s, it never considered itself a major research and development center in this domain. It did, however, disburse significant amounts of money to many universities, transportation agencies, and consulting firms in order for them to undertake such research and development work.

31. Joseph Ben-David, "Scientific Productivity and Academic Organization in Nineteenth-Century Medicine," *American Sociological Review* 25, no. 6 (December 1960): 828–843. See also Joel Mokyr, *A Culture of Growth: The Origins of the Modern Economy* (Princeton, NJ: Princeton University Press, 2017), especially the epilogue. A good case in point related to the subject matter of this book is the progressive transformation of Arizona University into a heavyweight in UTDM (chapters 6 and 7).

32. On the nonintegration of engineering departments within the university system in France, see Konstantinos Chatzis, "Theory and Practice in the Education of French Engineers from the Middle of the 18th Century to the Present," *Archives Internationales d'Histoire des Sciences* 60, no. 164 (2010): 43–78.

33. I explicitly refer here to the "role hybridization" process, identified and dissected by Ben-David, "Roles and Innovations in Medicine," especially 557. More recently, other scholars have spoken of "cultural borrowing" (Donald MacKenzie and Taylor Spears, "'The Formula that Killed Wall Street': The Gaussian Copula and Modelling Practices in Investment Banking," *Social Studies of Science* 44, no. 3 [June 2014]: 393–417 [on 396]).

34. See, for example, chapter 3.

35. Using Richard Whitley's classificatory scheme, we can state that researchers in urban travel demand modeling have mostly performed "explanatory instrumental research" (Richard Whitley, *The Intellectual and Social Organization of the Sciences*, 2nd ed. (Oxford: Oxford University Press, 2000).

36. Kleinman and Vallas, "Science, Capitalism, and the Rise of the 'Knowledge Worker.'"

37. The same remark seems to apply to financial modeling (MacKenzie, *An Engine, Not a Camera*).

38. Such as the Institute of Transportation Engineers.

39. Such as the *Transportation Research Record*.

40. Such as the many reports sponsored by the Transportation Research Board.

41. On the notion of "boundary organization," see Parker, "On Being All Things to All People"; Laurens K. Hessels, "Coordination in the Science System: Theoretical Framework and a Case Study of an Intermediary Organization," *Minerva* 51, no. 3 (September 2013): 317–339. Speaking of "boundary organizations," the Highway Research Board and its successor, the Transportation Research Board, with their various committees—one of the main places where both academics and practitioners could meet, according to one of the individuals involved, Daniel Brand (online meeting, July 1, 2021)—publication supports, meetings, and funding programs should be mentioned as a paramount example of such an organization in the field of urban traffic forecasting. The Travel Model Improvement Program has also worked as a "boundary entity since its establishment in the early 1990s, while the same applies also to the various travel modeling groups composed of academics and practitioners that gravitate toward open-source platforms.

42. As a useful introduction to an increasingly large literature one can select the following recent studies: MacKenzie and Spears, "'The Formula that Killed Wall Street'"; Armin W. Schulz, "The Heuristic Defense of Scientific Models: An Incentive-Based Assessment," *Perspectives on Science* 23, no. 4 (Winter 2015): 424–442; William Thomas and Lambert Williams, "The Epistemologies of Non-Forecasting Simulations, Part I: Industrial Dynamics and Management Pedagogy at MIT," *Science in Context* 22, no. 2 (June 2009): 245–270; William Thomas and Lambert Williams, "The Epistemologies of Non-Forecasting Simulations, Part II: Climate, Chaos, Computing Style, and the Contextual Plasticity of Error," *Science in Context* 22, no. 2 (June 2009): 271–310; Michel Armatte and Amy Dahan-Dalmedico, "Modèles et Modélisations, 1950–2000: Nouvelles Pratiques, Nouveaux Enjeux," *Revue d'histoire des sciences* 57, no. 2 (2004): 243–303. All these articles contain extensive bibliographies. See also the references in nn. 43 and 44 in this chapter.

43. Two such issues are whether modeling as a specific form of scientific knowledge departs from theory and experiment, and whether models are a sort of "intermediary" between theory and reality. See the now classic Eric Winsberg, *Science in the Age of Computer Simulation* (Chicago: University of Chicago Press, 2010); Stephan Hartmann, Carl Hoefer, and Luc Boven, eds., *Nancy Cartwright's Philosophy of Science* (New York: Routledge, 2008); several special issues of *Synthese* under the general heading "Models and Simulations" (190, no. 2 [January 2013]; 180, no. 1 [May 2011]; 169, no. 3 [August 2009]); "Simulation, Visualization, and Scientific standing," special issue, *Perspectives on Science* 22, no. 3 (Fall 2014).

44. On practice-oriented modeling used in decision-making, besides the relevant references in nn. 42 and 45 (in this chapter), see also Ekaterina Svetlova and Vanessa Dirksen, "Models at Work—Models in Decision Making," *Science in Context* 27, no. 4 (December 2014): 561–577; Ragna Zeiss and Stans van Egmond, "Dissolving Decision Making? Models and Their Roles in Decision-Making Processes and Policy at Large," *Science in Context* 27, no. 4 (December 2014): 631–657; "The Politics of Anticipatory Expertise: Plurality and Contestation of Futures Knowledge in Governance," special issue, *Science & Technology Studies* 32, no. 4 (2019).

45. One can nevertheless cite a number of studies that address weather and climate forecasting, financing modeling, and hydraulics. See, for example, Edwards, *A Vast Machine*; MacKenzie, *An Engine, Not a Camera*; MacKenzie and Spears, "'The Formula that Killed Wall Street'"; Kristine C. Harper, *Weather by the Numbers. The Genesis of Modern Meteorology* (Cambridge, MA: MIT Press, 2008); Martin Mahony and Mike Hulme, "Modelling and the Nation: Institutionalising Climate Prediction in the UK, 1988–92," *Minerva* 54, no. 4 (December 2016): 445–470; Catharina Landström and Sarah J. Whatmore, "Virtually Expert: Modes of Environment Computer Simulation Modeling," *Science in Context* 27, no. 4 (December 2014): 579–603 (flood risk management); Leon Wansleben, "Consistent Forecasting vs. Anchoring of Market Stories: Two Cultures of Modeling and Model Use in a Bank," *Science in Context* 27, no. 4 (December 2014): 605–630.

46. Using the terminology coined by Gibbons and his collaborators, one can argue that from the 1970s on, urban travel demand modeling has displayed several characteristics of the "Mode 2" science pattern. Be that as it may, like other historians, I lean toward the thesis that "Mode 1 and Mode 2 Knowledge production are coexisting, coevolving and interconnected" in history; see Andrea Bonaccorsi, "Search Regimes and the Industrial Dynamics of Science," *Minerva* 46, no. 3 (September 2008): 285–315 (on 287). See also Nordman, Radder, and Schiemann, *Science Transformed?*

47. The following works contain examples (and lists of references) of recent *disaggregate* techniques applied to different domains: David O'Sullivan and Mordechai Haklay, "Agent-Based Models and Individualism: Is the World Agent-Based?," *Environment and Planning A* 32, no. 8 (August 2000): 1409–1425; Michael Batty, *Cities and Complexity. Understanding Cities with Cellular Automata, Agent-Based Models, and Fractals* (Cambridge, MA: MIT Press, 2007); Yuna Huh Wong, "Ignoring the Innocent: Non-combatants in Urban Operations and in Military Models and Simulations" (Ph.D. diss., Pardee RAND Graduate School, 2006); Brian L. Heath, "The History, Philosophy, and Practice of Agent-Based Modeling and the Development of the Conceptual Model for Simulation Diagram" (Ph.D. diss., Wright State University, 2010); "Agent-Based Simulation: Principles and Applications to Social Phenomena," special issue, *Revue Française de Sociologie* 55, no. 4 (2014) (English edition); B. Edmonds and R. Meyer, eds., *Simulating Social Complexity: A Handbook* (New York:

Springer, 2013); Alison Heppenstall, Nick Malleson, and Andrew Crooks, "'Space, the Final Frontier': How Good are Agent-Based Models at Simulating Individuals and Space in Cities?," *Systems* 4, no. 1 (2016), https://doi.org/10.3390/systems4010009, accessed January 25, 2022.

48. *What-if* inferences determine by deduction what will happen (how the system will react) under the various "policies" we can institute.

49. For example: what will individual travel patterns become *if* mass-transit fares fall by 20 percent? Aggregate models have difficulty addressing these kinds of issues. For a nontechnical discussion of the general characteristics of disaggregate modeling and its advantage over aggregate techniques, see Petri Ylikoski, "Agent-Based Simulation and Sociological Understanding," *Perspectives on Science* 22, no. 3 (Fall 2014): 318–335; Gianluca Manzo, "The Potential and Limitations of Agent-Based Simulation: An Introduction," *Revue Française de Sociologie* 55, no. 4 (2014): 433–462 (English edition).

50. For a recent review of the strengths and limitations of Agent-Based Modeling and simulation in urban travel demand modeling, see Hong Zheng, Young-Jun Son, Yi-Chang Chiu, Larry Head, Yiheng Feng, Hui Xi, Sojung Kim, and Mark Hickman, *A Primer for Agent-Based Simulation and Modeling in Transportation Applications* (n.p.: Federal Highway Administration, 2013); the authors refer to several travel demand models associated with the ABM as well as to several dynamic traffic assignment models mentioned in this book. Recently, two characters of the story told here, Peter Vovsha and Kai Nagel, have contributed to an edited volume on agent-based modeling See Peter Vovsha, "Microsimulation Travel Models in Practice in the US and Prospects for Agent-Based Approach"; Michal Maciejewski, Joschka Bischoff, Sebastian Hörl, and Kai Nagel, "Towards a Testbed for Dynamic Vehicle Routing Algorithms," both in *Highlights of Practical Applications of Cyber-Physical Multi-Agent Systems*, ed. Javier Bajo et al. (Cham: Springer International Publishing AG, 2017), 52–68, and 69–79, respectively. It is also worth noting that a search within the TRB Publications Index database for the term "agent-based" shows 285 records from 2001 to 2020, many of which are concerned with urban traffic forecasting issues.

51. On the distinction between a model that *tests* a theory and a model rather than *applies* a theory, as well as for the "value" issue, see especially Winsberg, *Science in the Age of Computer Simulation*, especially 93–94, where the author comments on climate models.

52. Remember that without feedback loops between its steps, the four-step model is inconsistent.

53. See, for example, the case of climate models analyzed in Edwards, *A Vast Machine*.

54. On face-to-face interviews, see chapter 2. By the mid-1990s, computer-aided telephone Interviewing (CATI) systems, commercially introduced by Chilton Research Services in the early 1970, had grown into the most popular way of retrieving completed household travel data in the United States. (Stopher and Metcalf, *Methods for Household Travel Surveys*, 28.) On the early history of CATI, see the special issue of *Sociological Methods & Research* 12, no. 2 (November 1983).

55. For an overview of these techniques, see, for example, Li Shen and Peter R. Stopher, "Review of GPS Travel Survey and GPS Data-Processing Methods," *Transport Reviews* 34, no. 3 (2014): 316–334; Lara Montini, Nadine Rieser-Schüssler, Andreas Horni, and Kay W. Axhausen, "Trip Purpose Identification from GPS Tracks," *Transportation Research Record*, no. 2405 (2014): 16–23.

56. Edwards, *A Vast Machine*, 349, and passim; Mikaela Sundberg, "The Dynamic of Coordinated Comparisons: How Simulationists in Astrophysics, Oceanography and Meteorology Create Standards for Results," *Social Studies of Science* 41, no. 1 (February 2011): 107–125.

57. See chapter 2 (growth factor methods, gravity model, intervening opportunities model), chapter 4 (traffic assignment step), and conclusion (activity-based modeling versus four-step modeling framework).

58 For a list of such cities, with their coded networks, visit the site https://github.com/bstabler/TransportationNetworks, accessed January 27, 2022.

59. See, first, the excellent review article by Haigh, "History of Information Technology."

60. Including physics, chemistry, biology, geology, seismology, astronomy, oceanography, and meteorology.

61. Denning, "Computer Science: The Discipline."

62. On the growing interest in how scientists and engineers engage with visualization, see among others Regula Valérie Burri and Joseph Dumit, "Social Studies of Scientific Imaging and Visualization," in Hackett et al., *The Handbook of Science and Technology Studies*, 297–317; M. Friendly, "A Brief History of Data Visualization," in *Handbook of Data Visualization*, ed. C. Chen, W Härdle and A. Unwin (Berlin: Springer, 2008), 15–56; Henk W. de Regt, "Visualization as a Tool for Understanding," *Perspectives on Science* 22, no. 3 (Fall 2014): 377–396; Annamaria Carusi, Aud Sissel Hoel, Timothy Webmoor, and Steve Woolgar, eds., *Visualization in the Age of Computerization* (New York: Routledge, 2014). On visualization within transportation, see among others Charles L. Hixon III (consultant), *Visualization for Project Development. A Synthesis of Highway Practice* (Washington, DC: Transportation Research Board, 2006); Michael A. Manore et al., "Visualization in Transportation: Empowering Innovation," *TR News*, no. 252 (September–October 2007): 3–33; Nathan Higgins et al., *Data Visualization Methods for Transportation Agencies*

(Washington, D.C: Transportation Research Board, 2016); Eric J. Miller and Paul A. Salvini, "Workshop Report: Visualizing Travel Behaviour," in Roorda and Miller, *Travel Behaviour Research: Current Foundations, Future Prospects*, 81–102.

63. As Ann Johnson and Johannes Lenhard have cogently argued, these "images are tremendously powerful; they carry information more efficiently than do tables of numerical outputs," and "yield compelling results—sometimes in misleading ways"; see Ann Johnson and Johannes Lenhard, "Toward a New Culture of Prediction: Computational Modeling in the Era of Desktop Computing," in Nordman, Radder, and Schiemann, *Science Transformed?*, 189–199 (quotation on 196).

64. For two recent surveys on this topic, see Wolfe, *Competing with the Soviets*; Dennis, "Military, Science and Technology and the." See also Paul Erickson, Judy L. Klein, Lorraine Daston, Rebecca Lemov, Thomas Sturm, and Michael D. Gordin, *How Reason Almost Lost Its Mind. The Strange Career of Cold War Rationality* (Chicago: University of Chicago Press, 2013); Naomi Oreskes and John Krige, eds., *Science and Technology in the Global Cold War* (Cambridge, MA: MIT Press, 2014). On the all-in-all rather limited connections between urban planning and the military in the United States from 1950–1970, see especially Light, *From Warfare to Welfare*.

65. See chapter 4.

66. Significantly, neither the Bureau of Public Roads nor the various federal agencies within the Department of Transportation nor the Transportation Research Board are mentioned in the (otherwise excellent) review article by Steven W. Usselman, "Research and Development (R&D)," in Slotten, *The Oxford Encyclopedia of the History of American Science, Medicine, and Technology*, vol. 2, 369–387. Recently, Cyrus Mody has studied the increasing civilianization of the computer and electronics industry in the United States; Mody, *The Long Arm of Moore's Law*.

67. See, among others: Geiger, *Knowledge and Money*; Elizabeth Popp Berman, *Creating the Market University: How Academic Science Became an Economic Engine* (Princeton, NJ: Princeton University Press, 2012). For a recent in-depth case study, see Brian Dick and Mark Jones, "The Commercialization of Molecular Biology: Walter Gilbert and the Biogen Startup," *History and Technology* 33, no. 1 (2017): 126–151.

68. Henry Etzkowitz, "Innovation in Innovation: The Triple Helix of University-Industry-Government Relations," *Social Science Information* 42, no. 3 (September 2003): 293–337. On the university-industry relationships, see also the recent work by Mody, *The Long Arm of Moore's Law* (the book focuses on the microelectronics industry, but the author also discusses the relevant literature). For the commercialization of science in the nineteenth and twentieth centuries through the personal involvement of a series of academic teachers and researchers in consulting, patenting, and business entrepreneurship, see especially Joris Mercelis, Gabriel Galvez-Behar, and Anna Guagnini, eds., "Commercializing Science: Nineteenth- and

Twentieth- Century Academic Scientists as Consultants, Patentees, and Entrepreneurs," special issue, *History and Technology* 33, no. 1 (2017).

69. See, among others, Matthias Kipping and Timothy Clark, eds., *The Oxford Handbook of Management Consulting* (New York: Oxford University Press, 2012); and the recent article by Claire Dunning, "Outsourcing Government: Boston and the Rise of Public-Private Partnerships," *Enterprise and Society* 19, no. 4 (December 2018): 803–815.

70. James C. Scott, *Seeing Like a State: How Certain Schemes to Improve the Human Condition Have Failed* (New Haven: Yale University Press, 1998).

71. Edward Higgs, *The Information State in England: The Central Collection of Information on Citizens, 1500–2000* (Basingstoke: Palgrave Macmillan, 2004).

72. Gary Gerstle, "A State Both Strong and Weak," *American Historical Review* 115, no. 3 (June 2010): 779–785 (quotation on 780).

73. Alfred D. Chandler Jr. and James W. Cortada, eds., *A Nation Transformed by Information: How Information Has Shaped the United States from Colonial Times to the Present* (New York: Oxford University Press, 2000); James W. Cortada, *All the Facts: A History of Information in the United States since 1870* (New York: Oxford University Press, 2016).

74. Lawrence R. Jacobs and Desmond King, "The Political Crisis of the American State: The Unsustainable State in a Time of Unraveling," in *The Unsustainable American State*, ed. Lawrence Jacobs and Desmond King (New York: Oxford University Press, 2009), 3–33 (quotation on 26).

75. See, for example, Daniel P. Carpenter, *The Forging of Bureaucratic Autonomy: Reputations, Networks, and Policy Innovation in Executive Agencies, 1862–1928* (Princeton, NJ: Princeton University Press, 2001); Christopher Sneddon, *Concrete Revolution: Large Dams, Cold War Geopolitics, and the US Bureau of Reclamation* (Chicago: University of Chicago Press, 2015); Theodor Porter, *Trust in Numbers* (Princeton, NJ: Princeton University Press, 1995); Carpenter, *Reputation and Power*.

76. Peter Baldwin, "Beyond Weak and Strong: Rethinking the State in Comparative Policy History," *Journal of Policy History* 17, no. 1 (January 2005): 12–33; Desmond King and Robert C. Lieberman, "Ironies of State Building: A Comparative Perspective on the American State," *World Politics* 61, no. 3 (July 2009): 547–588.

77. AHR Editor, John Fabian Witt, Gary Gerstle, Julia Adams, William J. Novak, "AHR Exchange: On the 'Myth' of the 'Weak' American State," *American Historical Review* 115, no. 3 (June 2010): 766–800.

78. Gary Gerstle, *Liberty and Coercion: The Paradox of American Government from the Founding to the Present* (Princeton, NJ: Princeton University Press, 2017).

79. The Automotive Safety Foundation initiated and sponsored, for example, the National Committee on Urban Transportation in 1954. According to a member of the BPR, "the list of organizations which formally named representatives to the National Committee is impressive in its inclusion of virtually every association concerned with transportation in the urban area" (Holmes, "The State-of-the-Art in Urban Transportation Planning or How We Got Here," 383). On the importance of such private actors, see also the comments contained in Anonymous, *Traffic Engineering and Control in the USA* (Paris: European Productivity Agency/Organisation for European Economic Cooperation, 1955), especially 24.

80. Louis Galambos, "Recasting the Organizational Synthesis: Structure and Process in the Twentieth and Twenty-First Centuries," *Business History Review* 79, no. 1 (Spring 2005): 1–38.

81. It should be recalled here that several large transportation engineering firms, such as Parsons Brinckerhoff (now WSP) and AECOM, have been endowed, in fact, with substantial travel demand modeling units.

82. For an early collection of studies dealing with knowledge-intensive organizations, see Frank A. Dubinskas, ed., *Making Time: Ethnographies of High-Technology Organizations* (Philadelphia: Temple University Press, 1988). See also the relevant references in Kipping and Clark, *The Oxford Handbook of Management Consulting*.

83. Indeed, all major UTDM-related proprietary software companies have been patronizing user groups, the members of which meet on a regular basis. Concerning the role of the user in producing technical-scientific knowledge and know-how, see, for example, Nelly Oudshoorn and Trevor Pinch, "User-Technology Relationships: Some Recent Developments," in Hackett et al., *The Handbook of Science and Technology Studies*, 541–565; Sampsa Hyysalo, Neil Pollock, and Robin Williams, "Method Matters in the Social Study of Technology: Investigating the Biographies of Artifacts and Practices," *Science & Technology Studies* 32, no. 3 (2019): 2–25. Needless to say, the development of user groups has greatly benefited from the wide availability of cheap and convenient hardware (through personal computers) and software packages since the 1990s.

84. Business Software Alliance (BSA), *Open Source and Commercial Software: An In-depth Analysis of the Issues* (Washington, DC: Business Software Alliance, 2005).

85. Peter Gay, *Modernism: The Lure of Heresy: From Baudelaire to Beckett and Beyond* (New York: W. W. Norton & Company, 2008), 507.

# Index

*Note:* Page numbers in *italics* denote illustrations.